Dynamics of Urbanization and Ecosystem Services Provision

Dynamics of Urbanization and Ecosystem Services Provision

Editors

Luca Congedo
Francesca Assennato
Michele Munafò

MDPI • Basel • Beijing • Wuhan • Barcelona • Belgrade • Manchester • Tokyo • Cluj • Tianjin

Editors

Luca Congedo
Italian Institute for
Environmental Protection
and Research
Italy

Francesca Assennato
Italian Institute for
Environmental Protection
and Research
Italy

Michele Munafò
Italian Institute for
Environmental Protection
and Research
Italy

Editorial Office
MDPI
St. Alban-Anlage 66
4052 Basel, Switzerland

This is a reprint of articles from the Special Issue published online in the open access journal *Land* (ISSN 2073-445X) (available at: https://www.mdpi.com/journal/land/special_issues/urban_ecosystem_service).

For citation purposes, cite each article independently as indicated on the article page online and as indicated below:

LastName, A.A.; LastName, B.B.; LastName, C.C. Article Title. *Journal Name* **Year**, *Volume Number*, Page Range.

ISBN 978-3-0365-7584-1 (Hbk)
ISBN 978-3-0365-7585-8 (PDF)

Contents

land

Article

A Preliminary Study on the Impact of Landscape Pattern Changes Due to Urbanization: Case Study of Jakarta, Indonesia

Dikman Maheng [1,2,3,*], Assela Pathirana [4] and Chris Zevenbergen [1,2]

[1] Faculty of Civil Engineering and Geosciences, Delft University of Technology, Stevinweg 1, 2628CN Delft, The Netherlands; c.zevenbergen@un-ihe.org
[2] Department of Coastal and Urban Risk and Resilience, IHE Delft, Institute for Water Education, WestVest 7, 2611AX Delft, The Netherlands
[3] Department of Environmental Engineering, Universitas Muhammadiyah Kendari, Jl. Ahmad Dahlan 10, Kendari 93117, Indonesia
[4] UNDP Maldives Country Office, United Nations Development Programme (UNDP), 4th Floor, H. Aaage (Bank of Ceylon Building), Boduthakurufaanu Magu, Malé 20026, Maldives; assela@pathirana.net
* Correspondence: m.d.maheng@tudelft.nl

Abstract: Urbanization is changing land use–land cover (LULC) transforming green spaces (GS) and bodies of water into built-up areas. LULC change is affecting ecosystem services (ES) in urban areas, such as by decreasing of the water retention capacity, the urban temperature regulation capacity and the carbon sequestration. The relation between LULC change and ES is still poorly examined and quantified using actual field data. In most ES studies, GS is perceived as lumped areas instead of distributed areas, implicitly ignoring landscape patterns (LP), such as connectivity and aggregation. This preliminary study is one of the first to provide quantitative evidence of the influence of landscape pattern changes on a selection of urban ecosystem services in a megacity as Jakarta, Indonesia. The impact of urbanization on the spatiotemporal changes of ES has been identified by considering connectivity and aggregation of GS. It reveals that LP changes have significantly decreased carbon sequestration, temperature regulation, and runoff regulation by 10.4, 12.4, and 11.5%, respectively. This indicates that the impact of GS on ES is not only determined by its area, but also by its LP. Further detailed studies will be needed to validate these results.

Keywords: urbanization; land use–land cover change; landscape pattern; green space area; ecosystem services; Jakarta

Citation: Maheng, D.; Pathirana, A.; Zevenbergen, C. A Preliminary Study on the Impact of Landscape Pattern Changes Due to Urbanization: Case Study of Jakarta, Indonesia. *Land* **2021**, *10*, 218. https://doi.org/10.3390/land10020218

Academic Editors: Monika Kopecká and Luca Congedo

Received: 12 December 2020
Accepted: 19 February 2021
Published: 21 February 2021

1. Introduction

The urban population had increased to 3.9 billion (54% of the total population) by 2014 and will increase to around 6.3 billion (67% of the total population) by 2050 [1]. Population increase drives land use–land cover (LULC) change transforming green spaces as well as bodies of water into built-up areas resulting in decrease and loss of ecosystem services (ES) in urban areas [2–5]. Ecosystem services are the varied benefits to humans from the natural environment and healthy ecosystems. They can be categorized as provisioning services, regulating services, cultural services, and supporting services [6]. A city or an urban area can be seen as a single ecosystem where the urban population directly benefits from ecosystem services generated by urban ecosystems, such as green and blue infrastructure [7]. Ecosystem services in urban areas include air quality improvement, noise reduction, controlling air temperature, carbon storage and sequestration, water regulation, and recreation [7–13] and therefore enhance citizens well-being [14,15].

Many studies show that urbanization and LULC change are key anthropogenic drivers affecting urban ecosystem's functions and services [11,16–19]. Some examples of the impact of LULC change on ES due to urbanization are a decrease in the capacity to retain stormwater leading to urban flooding [20], and an increase in the land surface temperature

(LST) [21] in cities. Yuan et al. [22] showed that in Nanjing, China, there was a decrease in the provision and delivery of supply services, the supporting services, and the regulating services by 17, 0.9, and 4.3%, respectively, due to urbanization between 2000 and 2015. In Beijing, urban growth between 1985 and 2015 decreased carbon storage and water yield by 3.3 and 4%, respectively, while sediment export increased by 15.5% (Sun et al., 2018b). Another study from Atlanta Metropolitan area, US, showed that urban expansion between 1985 and 2012 reduced carbon storage by 23% and water purification capacity by 28 and 49%, for nitrogen and phosphorus, respectively, but increased sediment transport by 17% (Sun et al., 2018). Jaligot et al. [23] showed that the cultural services in Yaoundé, Cameroon, decreased by more than 90% due to urbanization between 2000 and 2018. LULC change from 1990 to 2017 in the West Bengal, India, resulted in some of the regulating services increasing while others decreased. Water regulation, water supply, and waste treatment increased by 57, 22, and 8%, respectively, while climate regulation and nutrient cycling decreased by 31.82 and 31.25%, respectively [24].

The relationship between urbanization, LULC change, and ES has been studied from various perspectives. Panagopoulos et al. [14] discussed the importance of urban ES and landscape planning to sustainable urban development. Das and Das [24] investigated the impact of LULC change on ES by quantifying the total ecosystem service value (ESV) based on the area of the ecosystem, and a value coefficient (VC) [25]. The extension of built-up areas in West Bengal decreased agricultural and vegetation covers. As a result, the total ESV of landscape decreased by 24.30% from 1990 to 2017. Wang et al. [26] investigated the impact of urbanization from 2000 to 2010 on multiple ES at hotspots and urban megaregion scale of Beijing–Tianjin–Hebei (BTH), the capital economic zone of China, by considering three components of urbanization: the built-up area proportion, the population density, and the gross domestic product (GDP) density. The ES in their study was calculated using the area of the ecosystem, for instance, the total food production was based on the area of cropland. It was revealed that providing services of food supply had an "inverse U" shape relation with the population and the GDP density, while urban expansion decreased regulating services. The impact of urbanization on ES was not only influenced by an increase in urban areas or a decrease in green spaces but also by landscape pattern changes, such as a number of patches (NP), aggregation index (AI), and "clumpiness" index, as discussed by Asadolahi et al. [27]. The ecosystem services in their study were based on the area of ecosystems, including water body, agriculture, rangeland, forest, and urban. Moreover, the influence of landscape patterns to ES provision was analyzed by comparing the landscape metric changes and ES. They saw that increasing food supply had a positive correlation with the good connectivity and connectedness of agricultural land. Haas et al. [28] discussed the relationship between a decrease in ES with an increase in built-up areas and landscape pattern changes due to urbanization in the period from 1989 to 2001 in Stockholm and Shanghai. They observed that an increase in built-up areas along with landscape pattern changes, such as a decrease in connectivity, an increase in fragmentation, and more complex shapes of the natural landscape, had a negative impact on ES. The ES in their study was based on the area of ecosystems, such as wetland, water, forest, agriculture, urban green space, high-density building, and low-density building. Moreover, the relationship between ES and landscape pattern changes was discussed by comparing the changes of the landscape metrics and ES. For urban flood mitigation, Bai et al. [29] conducted a field survey study in Luohe, China, which observed that inundated areas had a low proportion of green spaces and bare land. Moreover, landscape metrics analysis showed that the mean patch size of green spaces of flooded areas was smaller compared with that of roofs. Meanwhile, the connectivity of green spaces had no significant contribution to mitigating urban flooding. The importance of landscape patterns on surface runoff control was also observed by [30]. They directly integrated large patch index (LPI) and aggregation index (AI) of green spaces with runoff coefficients to estimate runoff volume controlled in Beijing, China. It was found that increasing runoff was related to a decrease in green spaces, as well as increasing the fragmentation and disaggregation of green spaces. The landscape patterns can also exhibit a relation with an increase in LST in the Olympic forest park of Beijing, China, as investigated by using the Pearson correlation analysis [31]. It was revealed that an increase in LST was related to increasing urbanization in the period 2000 to 2015. Moreover, there was a negative correlation between LST and the landscape metrics of large patch index (LPI) and aggregation index (AI) of green spaces.

A growing number of studies have investigated the impact of urbanization and LULC change on ES. These studies reveal that a decrease in green spaces affects ES, feeding the perception that there is a positive relationship between the surface area of green spaces and urban ecosystem functioning [4,32]. However, these studies generally ignore the impact of spatial heterogeneity of the land surfaces as they lump together the area of each land-use type. Only a few of them acknowledged the differences in landscape patterns but still did not consider landscape patterns in calculating the impact of green spaces. As a result, an area of green spaces, as shown in Figure 1, could still be perceived as a lumped area instead of a distributed area, where the green space units have different landscape patterns, such as connectivity, aggregation, mean patch size, and large patch index, which could affect ecosystem services [33,34].

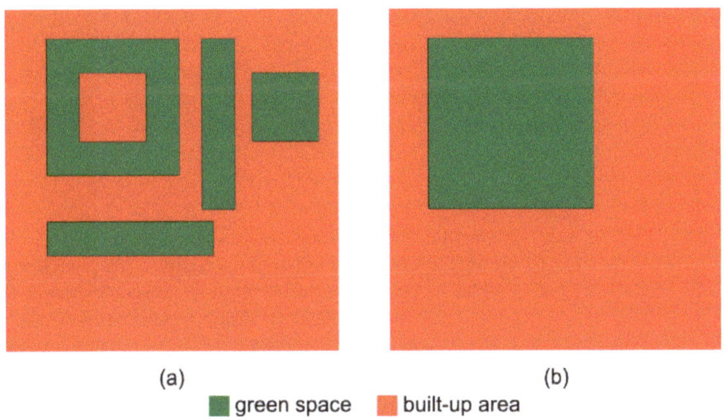

Figure 1. Schematic illustration of the loss of information due to lumping green areas. (**a**) Real landscape pattern; (**b**) the lumped area used in many previous ecosystem services (ES) studies.

In many existing studies, the impact of urbanization and LULC change on ES is generally discussed using the lumped area of green spaces. Some studies have discussed the impact by comparing the ES changes and the changes of landscape patterns, which show that landscape patterns can affect ES. However, there is a lack of studies that directly examine the impact of landscape pattern changes on the area of green spaces to ES calculation.

This preliminary study aims to identify the impact of landscape pattern changes on urban ecosystem services using two spatial configurations of green spaces, as shown in Figure 1. Firstly, the area of green spaces is assumed as a lumped area, as depicted in Figure 1b. Secondly, the area of green spaces is assumed as a distributed area, as shown in Figure 1a, which is represented by directly combining the area and landscape patterns defined by the aggregation and the connectivity of the green spaces and bodies of water. As a preliminary study, this study is designed to focus on the investigation of the landscape pattern changes and their influence on urban ES. It is expected that the results of this study can provide insight into the influence of landscape patterns on ES calculation. Moreover, the validation of the results is not a part of this study since there is a lack of field survey data for all municipalities in Jakarta. Hence, the validation would be considered as a part of further studies.

This study is divided into three main parts: (1) assessing the impact of urbanization on LULC change; (2) analyzing the impact of LULC change on the landscape patterns; and (3) analyzing the impact of the landscape pattern changes on ecosystem services.

2. Materials and Methods

This study was carried out in Jakarta, Indonesia, and it was conducted in six steps. First, LULC maps for 1995 and 2014 were generated using Landsat images. The LULC classification used six LULC classes, which were urban, suburban, grassland, cropland, trees, and water (sea and surface water). Second, the results of LULC classification were imported into the FRAGSTATS, the spatial pattern analysis program, for landscape metric analysis. The analysis considered four landscape metrics and the relationship between them: Class Area (CA), Land Proportion (PLAND), Aggregation Index (AI), and connectivity (COHESION). The landscape metrics analysis was carried out for five municipalities of Jakarta. Third, the impact of urbanization on LULC change was analyzed, which was based on the CA of LULC classes. Fourth, the impact of LULC change on the landscape patterns was analyzed, which was based on PLAND, AI, and COHESION. Fifth, ecosystem services were calculated using two methods: area-based estimation and landscape metric area (LMA). The area-based estimation followed commonly ecosystem services area calculation that uses only the area of green spaces, while the landscape metric area directly incorporated the landscape metric representing the spatial characteristic/configuration of green spaces to ecosystem services area calculation. Sixth, ecosystem services changes between 1995 and 2014 from two different methods were then analyzed using ecosystem services index (ESI).

2.1. Study Area

Jakarta is the capital of Indonesia, which has experienced rapid urbanization and the extensive transformation of vegetative green spaces into impermeable urban surfaces [35–38]. Urbanization in Jakarta is driven by several factors, including its strategic political and economic position as the capital city where important activities like the functions of the national government, education, manufacturing, and commerce are taking place [39]. Jakarta has six municipalities, which are Central Jakarta, North Jakarta, West Jakarta, South Jakarta, and East Jakarta, as shown in Figure 2, and one regency, which is the Thousand Islands (not shown in the figure). This study focuses on the municipalities that consist of 42 subdistricts, which are home of approximately 10 million people, with the population growth rate of 1.02% per year in 2015 [40]. East Jakarta is the most populated area, whereas West Jakarta is the most densely populated area with a population density of 19,018 per km^2. Located on the northwest coast of Java, Jakarta's climate is a tropical monsoon climate according to the Köppen climate classification system and it has two seasons, which are the rainy season from October through May and the dry season from June to September. During the rainy season, the average monthly rainfall is 300 mm, while it is 43 mm in August during the dry season [40].

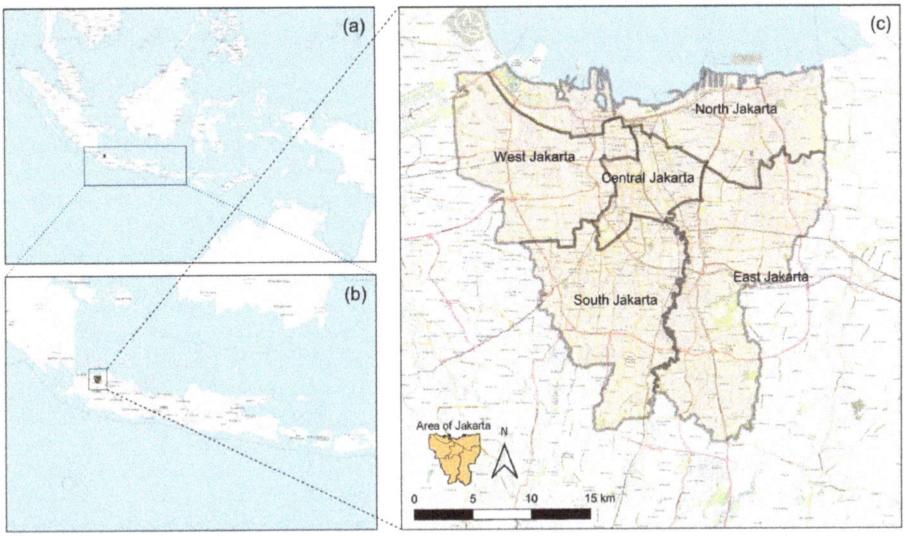

Figure 2. (**a**) Map of Indonesia; (**b**) Jakarta is located on Java island, which is the densest island in Indonesia; (**c**) Jakarta consists of five municipalities.

2.2. Data Sources

The impact of urbanization on LULC change was analyzed using land use–land cover maps of Jakarta for the year 1995 and 2014. The maps were prepared using Landsat images selected for the area of Jakarta and its surrounding. For the year 1995, the Landsat 5 TM L1TP from 24 August 1995, was used, while the Landsat 8 OLI TIRS L1TP obtained on 13 September 2014, was selected for the land use–land cover map of the year 2014. Both datasets had 30 m resolution. The land cover classification was based on the land cover classes used to investigate the relationship between green spaces and increasing urban temperature [41]. We used the semiautomatic classification in QGIS [42], a free and open-source geographic information system, to generate six land cover classes: urban, suburban, grassland, cropland, trees, and water. An urban area is an area where there is a dense mix of compact high-rise, compact midrise, or low-rise buildings, while a suburban area is an open arrangement of low-rise buildings with pervious land covers, such as grass, low plants, and scattered trees. A grassland area is defined as an area where the vegetation is dominated by grasses, and there are few or no other plants. A cropland area is an area mainly used for agricultural and farming purposes, where rice, corn, vegetables, and fruits are grown. An area of trees is an area where trees are dominant. Water represents both sea and surface water, such as rivers and lakes. Moreover, the LULC change analysis was based on the CA from five land cover classes: urban, trees, grassland, cropland, and bodies of water.

The boundary area of Jakarta and its municipalities was obtained online from the Geospatial Information Agency of Indonesia (BIG) database, www.tanahair.indonesia.go.id (accessed on 20 September 2019).

2.3. Landscape Metrics

Landscape metrics are quantified characteristics of the spatial pattern at different levels (i.e., patch level, class level, and landscape-level) used to describe landscape structures and LULC change [43–45]. Information about landscape structures is essential to analyze ecosystem structures, functions, and services [46] and to evaluate their spatial effects at the patch–class–landscape level [47,48]. Patch level metrics are important for defining individual patches and for characterizing the spatial context of the patches, which are used as the computational basis for other landscape metrics. At the class level, the metrics are calculated by integrating all patches from a given class. The class level calculation can be done by a simple aggregation of the area of patches to reflect the total area of a given class. Moreover, landscape-level metrics are the integration of all classes over the landscape [49]. This study uses four landscape metrics to capture the spatial features which affect urban ecosystem processes: class area (CA), connectivity (COHESION), land proportion (PLAND), and aggregation index (AI). CA is a measure of landscape composition; specifically, to indicate how much of the landscape is comprised of a patch type. The class area is the primary component used in the ecosystem service quantification [45]. The connectivity of a class is measured by the patch cohesion index (COHESION) that measures the physical connectedness of the corresponding patch type. Patch cohesion increases as the patch type become more clumped or aggregated in its distribution; hence, more physically connected [50]. The decrease in patch connectivity in a landscape could be a negative sign for ecosystem services [28]. The land proportion (PLAND) is the percentage of a landscape, which quantifies the proportional abundance of each patch type in the landscape, while the AI is used to understand the spatial aggregation of a patch in a landscape [31,45,51]. The landscape metrics in this study were calculated using FRAGSTATS 4.2, the spatial pattern analysis program, which is developed by McGarigal and Cushman of University of Massachusetts in the United States of America [48].

The landscape metrics used in this study were selected based on the premise that there is a correlation between ES and the area, the aggregation, and the connectivity of ecosystem [31,45,52]. In general, many ES studies focus on the relationship between ecosystem services and the area of green spaces revealing a positive correlation between the size of the green spaces and the weight of the ecosystem services provided [53–55]. However, some studies showed that ecosystem services were also influenced by landscape patterns [29,30,55,56]. In this study, ES was analyzed by considering the "effective area" of green spaces and bodies of water by taking into account the influence of aggregation and connectivity on the area of green spaces and bodies of water. The landscape metric combinations for this study are selected based on the FRAGSTATS 4.2 guidelines and the existing research on landscape metric and ES. There are some indices that can be used to analyze the relationship between landscape metrics and ES, such as Number of Patch (NP), Patch Density (PD), Large Patch Index (LPI), patch cohesion index (COHESION), land proportion (PLAND), CONTAGION, and Aggregation Index (AI). Among them, the AI was selected because it can give more accurate results compared to that of other aggregation indices, such as CONTAGION [51], while the COHESION was selected for connectivity analysis because it can represent the connectivity of a corresponding patch type of a particular class [28,50]. This resulted in the "Landscape Metric Area" (LMA) calculated by multiplying the metrics class area (CA), aggregation index (AI), and cohesion index (COHESION) for each land cover class, as follows:

$$LMA = CA \times AI \times COHESION \quad (1)$$

LMA was calculated at the municipality level following the Indonesian regulation for spatial planning. The standard focuses on the city level and defines the spatial configuration of different land use–land cover types, including the minimum proportion of open green spaces [57].

2.4. Urban Ecosystem Services

Many studies discussed ES in urban areas [7,9,28,45,58,59]. Back in the 1990s, one of the pioneering urban ecosystem services studies considered six ecosystem services: air filtering, microclimate regulation, noise reduction, rainwater drainage, sewage treatment, and recreation [7]. Derkzen et al. [9] studied ecosystem services bundles in Rotterdam, the Netherlands, considering six ecosystem services: air purification, carbon storage, noise reduction, runoff regulation, cooling, and recreation. The importance of urban ecosystem services inspires local and national governments to incorporate ES in their urban planning process and documents [58,60–62]. For example, the government of Indonesia requires a minimum of 30% open green spaces in each city and urban area [57,63].

In this study, ES changes analysis was limited to carbon sequestration, temperature regulation, and runoff regulation, due to the limitation of the medium resolution satellite images to capture several urban details, such as street corridors and small city parks. Furthermore, in the ES selection increasing carbon emissions [64,65], and recent climate-related hazard events in Jakarta, such as urban flooding in 2013, 2014, 2015, and 2020, have also been considered [66,67], as well as the potential increase in the urban temperature [37,68]. The ES estimation adopted a weighting factor used in [9], where the green space with high-temperature regulation capacity was given a weight of 1.0, whereas 0.5 was given for low capacity. In this study, the weighting factor was implemented for all ES estimations for which the factor was given on the basis of the importance of a particular green space to a particular ES. For example, grassland was given the weight of 0.5 for temperature regulation, but it was given 1.0 of the weight factors for runoff regulation.

2.4.1. Carbon Sequestration

Green spaces have an important role in the carbon cycle, capturing CO_2 from the atmosphere during photosynthesis, and releasing it during respiration. In an urban area, carbon sequestration mainly takes place in green spaces [69–71]. Each green space has a different carbon sequestration rate, and urban trees generally have a higher rate compared to that of other green spaces because of their higher leaf area index (LAI), which is the ratio of the area of trees' leaves per area of the land surface [72]. For example, the urban forest of Punggol forest in Singapore had the annual carbon sequestration rate of 1.6 ton/year/ha, where trees with high LAI such as *Delonix regia* and *Artocarpus heterophyllus* gave a significant contribution [53]. Studies from Indonesian cities showed that trees had a carbon sequestration rate higher than that of other vegetation, including grass and herbaceous. For instance, the carbon sequestration of tree-covered areas in a part of Jakarta was 129.92 kg/ha/hour, much higher than 2.74 kg/ha/hour for grass areas [73]. Moreover, trees in the city of Singaraja on Bali island had 112.751 tons/ha of carbon sequestration higher than 4.845 tons/ha of carbon sequestration of herbaceous [74].

In this study, carbon sequestration was based on the LMA of trees. The LMA of trees was given a weight of 1.0 considering its carbon sequestration, which is higher than that of other vegetation. Other types of green spaces were given a weight of 0, assuming no significant contribution to capturing carbon in the atmosphere.

2.4.2. Temperature Regulation

Increasing LST and air temperature in urban areas can be influenced by several factors, including a decrease in green spaces and bodies of water. The contribution of green spaces to the control of urban temperatures can be achieved through ecosystem functions such as evapotranspiration, trees shading, and modifying air movement, while the greatest contributing factor to the cooling effect is shading and evapotranspiration [75]. The effectivity of green spaces in providing the ecosystem functions is influenced by various factors, including the area and the spatial distribution of green spaces [76], the configuration and the shape of green spaces [52,77], and the connectivity of green spaces [31]. Among green spaces, trees have high cooling potential through shading and evapotranspiration, while other short vegetative covers such as grass and crops might have a lower capacity [9,78]. The temperature in urban areas can also be regulated through evaporation and heat absorption from bodies of water [79]. The cooling effect of bodies of water is influenced by the area of the bodies of water [80], and the effect decreases with the increase in distance from the water body, which is similar to the cooling effect from green spaces [81,82].

Temperature regulation in this study was estimated using the combination of the LMA from trees, grassland, cropland, and bodies of water. The LMA of trees was given a weight of 1.0, while a weight of 0.5 was given for the LMA of grassland, cropland, and bodies of water.

2.4.3. Runoff Regulation

Due to urbanization, cities and urban areas are more vulnerable to urban flooding. The increasing vulnerability is influenced by some factors such as a decrease in green spaces and bodies of water. As a result, many urban areas have an insufficient capacity for infiltration and interception, as well as an increase in the runoff coefficient. Urban flooding-related studies showed that the area of green spaces could be considered as one of the important factors that determine runoff regulation [30,83,84]. For instance, increasing tree-cover by nearly 11% in Beijing could increase the total runoff reduction volume by more than 30% [85]. Moreover, the capacity of green spaces in controlling surface runoff is also influenced by several factors, including the interception of rainfall. Among many types of green spaces, trees could provide the high interception capacity, particularly in the short–moderate rainfall or in the earlier periods of precipitation, and it decreases in the longer duration or during high rainfall intensity [86]. A field experiment from the city of Uruaçu, Goiás, Brazil, showed that interception rate during short, low-intensity precipitation was about 40%, while the rate was about 3.6% in the high intensity and long duration rainfall [87]. Besides interception, green spaces also provide runoff regulation through infiltration, which is influenced by numerous factors including rainfall intensity, green space types, soil characteristics, and green space coverage [19,88–90]. In addition, green spaces contribute to runoff coefficients by increasing flow resistance, which can be associated with its density [91]. Furthermore, bodies of water like rivers, lakes, and wetlands also have the potential to control surface runoff through storage functions [83].

The estimation of runoff regulation service in this study considered the combination of the LMA from trees, cropland, grass, and bodies of water. All green spaces were given a weight of 1.0 based on the assumption that green spaces and bodies of water had a similar contribution to runoff regulation from one or the combination of infiltration, interception, flow resistance, and storage functions.

2.5. Ecosystem Services Index (ESI)

This study used ecosystem services index (ESI) to assess the ES changes. The ESI was based on the ES calculated using the combination of the LMA from green spaces and bodies of water. The ES calculation was divided into two steps. First, the ES was only based on the area of green spaces and bodies of water called area-based estimation. Second, the ES was based on the LMA calculated using the following equation.

$$ES = aLMA_{tree} + bLMA_{grassland} + cLMA_{cropland} + dLMA_{water} \quad (2)$$

where *a*, *b*, *c*, and *d* were the weighting factors for each green space and bodies of water. The weighting factor was different for each green space in each ES estimation, as given in Table 1.

Table 1. The weighting factors (modified from Derkzen et al. [9]).

Weighting Factors	Carbon Sequestration	Temperature Regulation	Runoff Regulation
a (Trees)	1	1	1
b (Grassland)	0	0.5	1
c (Cropland)	0	0.5	1
d (Water)	0	1	1

The ESI was then calculated by normalizing the ES values using the following formula:

$$ESI_{ij} = \frac{ES_{ij} - ES_{min}}{ES_{max} - ES_{min}} \quad (3)$$

where ESI_{ij} is the index of ES, i for the municipality j, ES_{ij} is the value of ES, i for the municipality j, and ES_{min} and ES_{max} are the minimum and the maximum value of ES i, respectively.

3. Results

3.1. Land Use–Land Cover (LULC) Change Analysis

The high rate of land transformation marked urbanization in Jakarta from 1995 to 2014, as shown in Figure 3. It shows that there has been a substantial change in the spatial distribution of green spaces due to increasing urban areas. The growing urban areas, about 44.3%, lost about 58.1% of the areas with trees, 36.1% of grassland, 32.3% of cropland, 15.9% of suburban areas, and 6.1% of bodies of water (lakes and rivers).

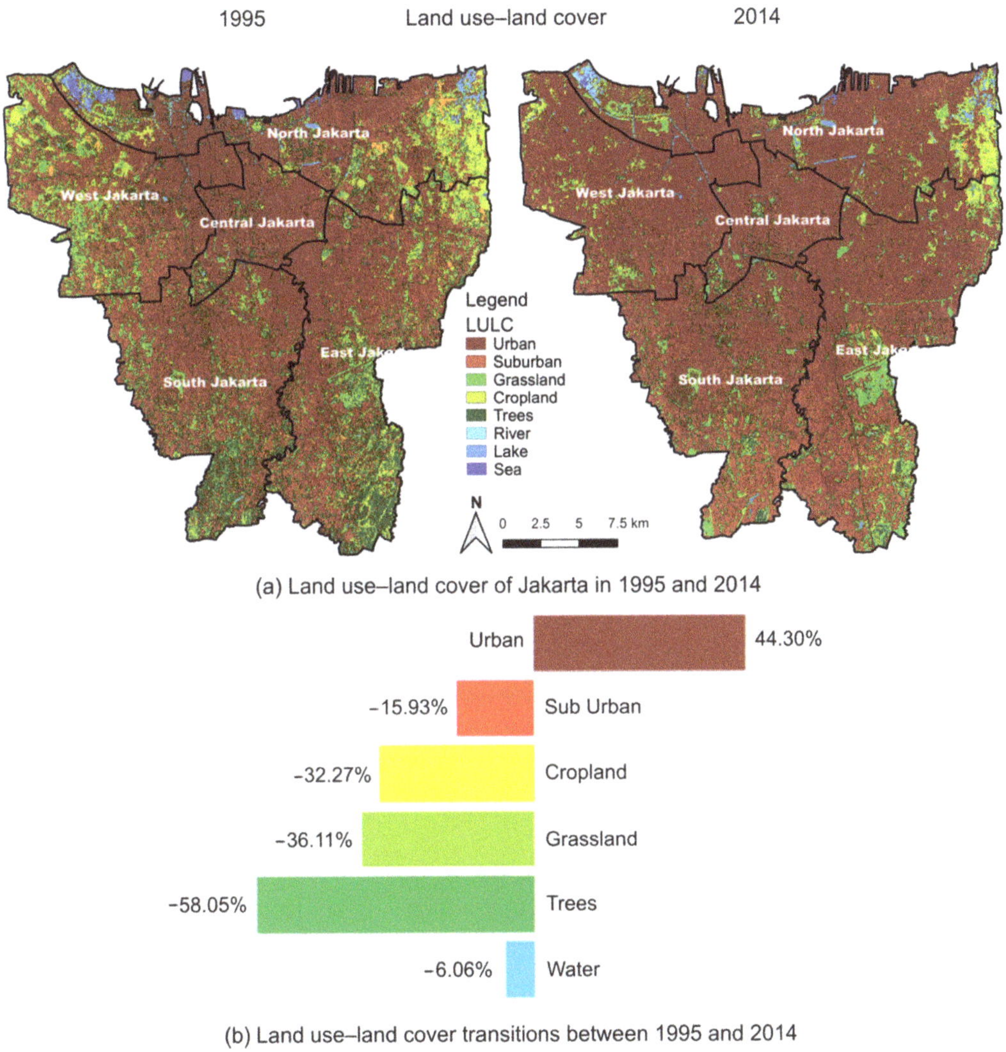

Figure 3. Land use–land cover change from 1995 to 2014.

The LULC change in the study was analyzed by focusing on the spatial extension of the urban areas and the decrease in vegetative cover and bodies of water from 1995 to 2014, as shown in Figure 4. In 1995, the spatial distribution of urban areas was concentrated in three municipalities, which were East Jakarta (65.3 km^2), followed by North Jakarta (63.9 km^2) and South Jakarta (58.5 km^2). On the other hand, the smallest urban area was about 35.4 km^2, which was observed in Central Jakarta. After 20 years, urban areas increased in all municipalities. The largest urban area in 2014 was about 97.8 km^2 in West Jakarta, which increased by about 93.7% from that in 1995. Meanwhile, urban areas in East Jakarta and North Jakarta increased by almost a half from that in 1995. On the other hand, the smallest urban area, by approximately 38.4 km^2, was identified in Central Jakarta, which increased by 8.2%.

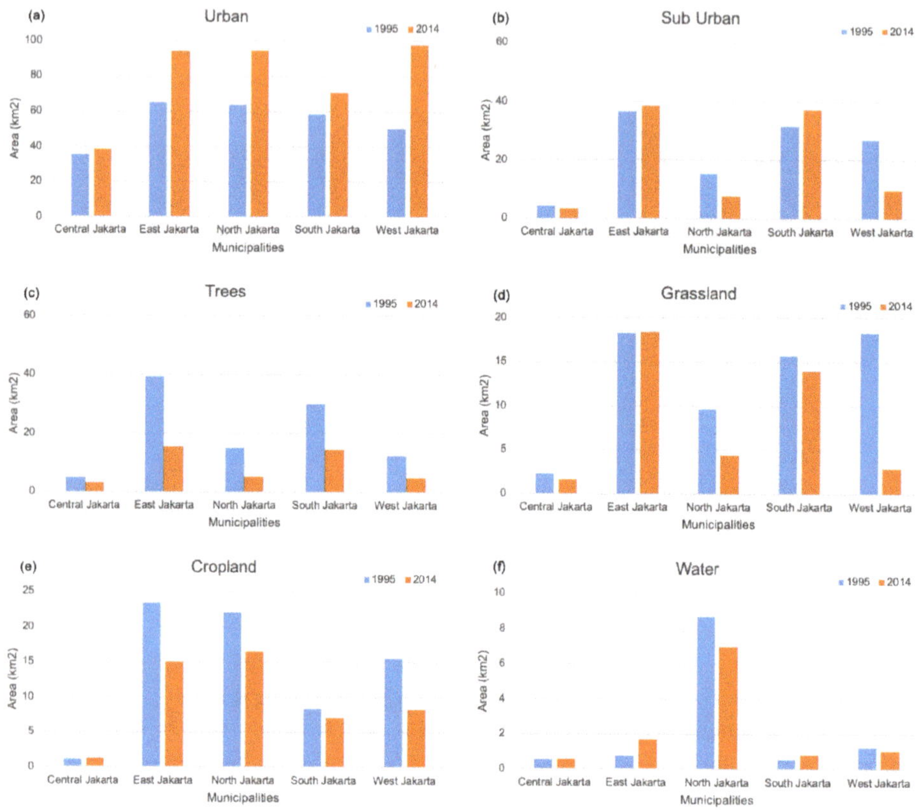

Figure 4. The area of land use–land cover of Jakarta in 1995 and 2014. (**a**) The area of urban class; (**b**) the area of suburban class; (**c**) the area of trees; (**d**) the area of grassland; (**e**) the area of cropland; (**f**) the area of bodies of water.

The increase in built-up areas was also observed in suburban areas from 1995 to 2014, as shown in Figure 4. In 1995, the largest suburban area was detected in East Jakarta (36.4 km^2), followed by South Jakarta (31.6 km^2) and West Jakarta (26.9 km^2). Central Jakarta had the smallest suburban area by about 4.30 km^2. The spatial distribution of suburban areas in 2014 was a bit different from urban areas, which increased in all municipalities. The changes of suburban areas were marked by a decrease in West Jakarta (−64.0%), followed by North Jakarta (−51.0%) and Central Jakarta (−19.6%), but they increased in East Jakarta and South Jakarta. The largest suburban area of 38.37 km^2 was identified in East Jakarta, which increased by 5.3% from that in 1995. This increase, however, was lower than 17.7% in South Jakarta. On the other hand, the smallest suburban area of 3.4 km^2 was detected in Central Jakarta.

The dynamic changes of the built-up areas affected the area of green spaces, where trees were by far the most affected, followed by grassland and cropland, as shown in Figures 4 and 5. In 1995, trees were mainly distributed in East Jakarta and South Jakarta, where the area of trees was more than two times of that in the other municipalities, such as in North Jakarta and West Jakarta. The largest area of trees (39.4 km^2) was observed in East Jakarta, while the smallest area of 5.0 km^2 was in Central Jakarta. In 2014, the area of trees decreased above the average in all municipalities except in South Jakarta and Central Jakarta. A significant decrease of 65.8% was observed in North Jakarta, followed by West Jakarta (−62.2%) and East Jakarta (−61.1%). However, the area of trees in that area was still higher than that in other municipalities. For instance, East Jakarta had the largest area of trees (15.4 km^2), which was higher than 14.5 km^2 in South Jakarta. On the other hand, the smallest area of trees (3.2 km^2) was detected in Central Jakarta.

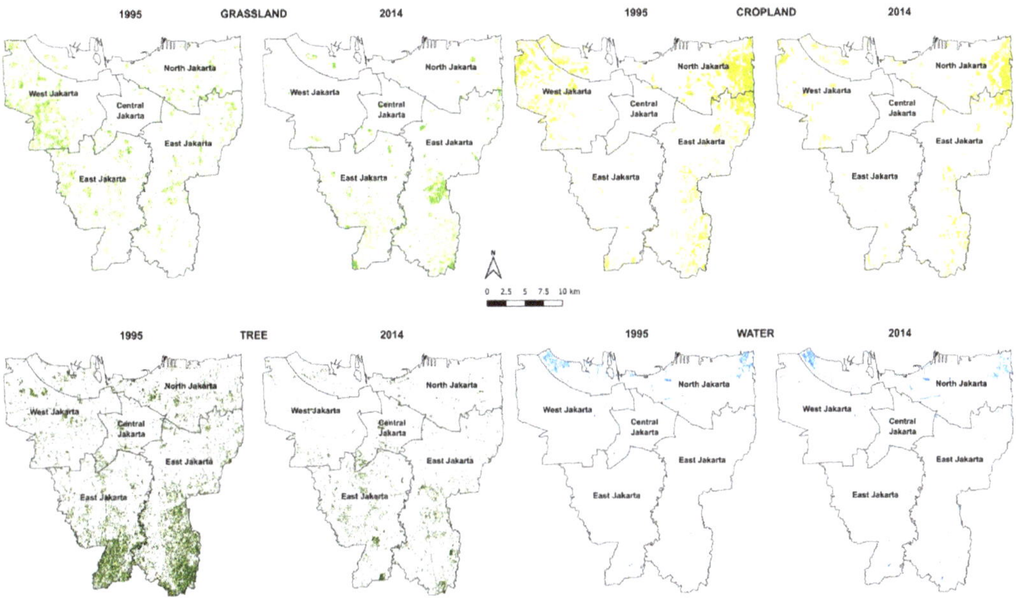

Figure 5. The spatial and temporal distribution of green spaces and bodies of water in 1995 and 2014.

Besides trees, areas of grassland were also impacted by an increase in the built-up areas, as depicted in Figures 4 and 5. In 1995, the widespread occurrence of areas of grassland was mainly detected in East Jakarta, West Jakarta, and South Jakarta. The largest grassland area (18.3 km^2) was observed in East Jakarta, which was similar to that of West Jakarta. On the other hand, Central Jakarta had the smallest area of about 2.3 km^2. In 2014, the grassland area remained almost the same in East Jakarta, while a significant decrease of 84.5% was observed in West Jakarta, followed by North Jakarta (−54.5%). The smallest area of grassland was observed in Central Jakarta (1.6 km^2), followed by West Jakarta (2.8 km^2) and North Jakarta (4.4 km^2).

A decrease in the area of green spaces was also found in the area of cropland during urbanization from the year 1995 to 2014, as shown in Figures 4 and 5. The spatial distribution of cropland in 1995 was dominant in East Jakarta and North Jakarta, where the areas of cropland were 23.4 km^2 and 22.1 km^2, respectively. In 2014, there was a decrease in the area of cropland. The highest decrease was identified in West Jakarta, where the area of cropland decreased by 47.1%, followed by East Jakarta (−35.9%) and North Jakarta (−25.2%). Moreover, the spatial distribution of cropland in 2014 was similar to that in 1995, where the dominant distribution of cropland was observed in North Jakarta (16.5 km^2), followed by East Jakarta (14.9 km^2).

The existence of bodies of water between 1995 and 2014 was identified by a small decrease by −6.1%, as shown in Figures 4 and 5. The spatial distribution of the bodies of water was mainly observed in North Jakarta, representing lakes, fishponds, and wetlands. As a result, North Jakarta had the largest area of bodies of water between 1995 and 2014, where small patches of bodies of water, including rivers, creeks and small wetlands, were evenly distributed in all municipalities.

3.2. Landscape Metrics Analysis

The rapid urbanization in Jakarta not only changed the spatial distribution of green spaces but also altered the landscape patterns indicated by changes in the proportion, the aggregation, and the connectivity of green spaces, as well as of bodies of water. In 1995, landscape metrics analysis showed the highest proportion of trees at the municipality level of about 20% in East Jakarta and South Jakarta, while other municipalities had about 10% of trees in their area. Meanwhile, trees showed a dispersed distribution in Central Jakarta, North Jakarta, and West Jakarta, where the AI was lower than 50%. However, trees had good connectivity, as shown in Figure 6. It shows that trees had the COHESION higher than 60% in all municipalities, with the highest values being observed in South Jakarta.

In 2014, the proportion of trees decreased in all municipalities, where the highest decrease of 65.8% was observed in North Jakarta, followed by West Jakarta and East Jakarta, where the proportion decreased by 62.2 and 61.1%, respectively. Meanwhile, the highest trees proportion of 9.9% was identified in South Jakarta, followed by East Jakarta (8.3%) and Central Jakarta (6.6%), while a similar proportion was observed in West Jakarta and North Jakarta. The decrease in the proportion of trees was followed by a decrease in the aggregation and the connectivity of trees except in Central Jakarta. It was found that trees in Central Jakarta were more aggregated and connected in a small proportion. The aggregation and the connectivity of trees in Central Jakarta increased by 12.1 and 6.7%, respectively. In other municipalities, trees were more disaggregated, as indicated by a decrease in the AI values. For instance, the AI in East Jakarta and South Jakarta decreased by 27.4 and 26.3%, respectively. Furthermore, the connectivity of trees also decreased in East Jakarta and South Jakarta, where the connectivity decreased by 19.7 and 23.8%, respectively. However, the connectivity of trees was still high in all municipalities since the COHESION was higher than 50%.

Meanwhile, the dominant proportion of the urban areas also influenced the landscape patterns of the grassland, as shown in Figure 7. In 1995, the highest proportion of grassland was about 14.5% in West Jakarta, followed by South Jakarta (10.9%) and East Jakarta (9.9%). The proportion of grassland in West Jakarta was found higher than the proportion of trees in the same year, which showed a different green space pattern in a municipality. Meanwhile, the low AI percentage indicated that grassland had dispersed distribution in all municipalities. The connectivity of grassland in 1995 was almost similar in four municipalities where the connectivity varied from 71.4% to 72.4%. The highest connectivity of grassland was observed in West Jakarta, where COHESION was 84.9%.

In 2014, all municipalities saw a decrease in the proportion of grassland. The highest decrease was observed in West Jakarta, where the proportion decreased from 14.5% in 1995 to 2.2% in 2014. In addition, the location of the highest proportion shifted from West Jakarta to East Jakarta. The highest grassland proportion of 9.9% was identified in East Jakarta, which was slightly higher than 9.6% in South Jakarta. On the other hand, a similarly low proportion was observed in Central Jakarta, North Jakarta, and West Jakarta, where the proportion of grassland was 3.3, 3.2, and 2.2%, respectively. Furthermore, the area of grassland underwent a small amount of disaggregation in all municipalities except in East Jakarta, where the area of grassland was aggregated by 12.7%. The connectivity of grassland in 2014 was also affected by urbanization as it decreased in four municipalities, but it increased in East Jakarta. The highest decrease was 29.3%: this was observed in West Jakarta, where the connectivity reduced from 84.9% in 1995 to 60.0% in 2014. Moreover, a small decrease was found in Central Jakarta, South Jakarta, and North Jakarta, where the connectivity decreased by 5.9, 5.3, and 3.5%, respectively.

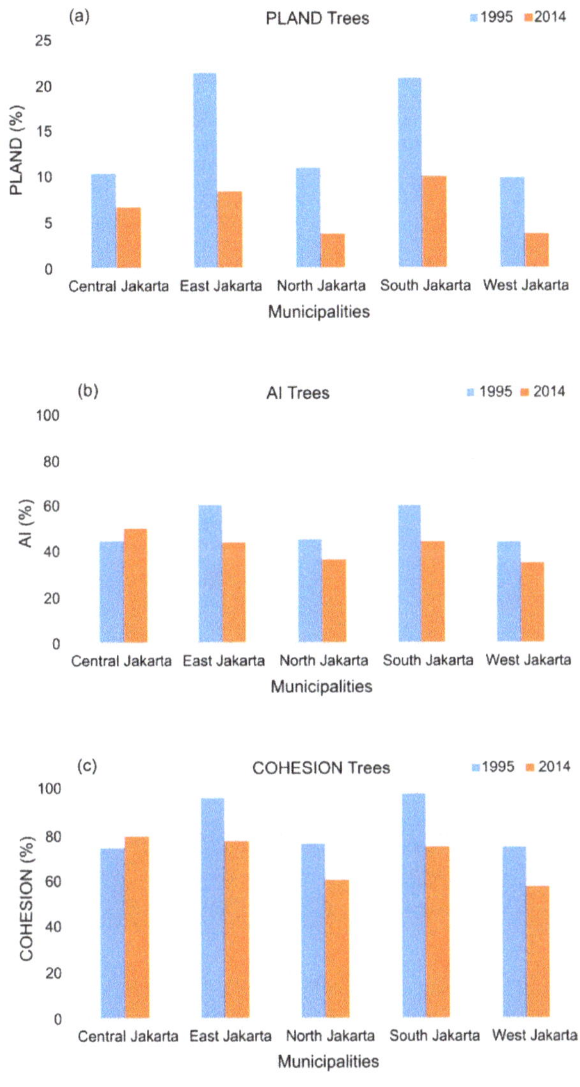

Figure 6. Landscape metrics of trees. (**a**) Land proportion; (**b**) aggregation index; (**c**) cohesion.

Moreover, the landscape patterns of cropland were also subject to dynamic changes due to urbanization, as shown in Figure 8. In 1995, the highest proportion of cropland was mainly observed in North Jakarta, East Jakarta, and West Jakarta. The highest cropland proportion of 15.9% was observed in North Jakarta, while the lowest cropland proportion of 2.3% was observed in Central Jakarta. Meanwhile, the cropland distribution in the four municipalities exhibited a scattered pattern, as indicated by the AI percentage. For instance, cropland in Central Jakarta, East Jakarta, and South Jakarta was below 50%. Moreover, the connectivity of crops was high in North Jakarta, West Jakarta, and East Jakarta, where the COHESION was 95.3, 87.9, and 84.6%, respectively.

In 2014, the proportion of cropland decreased sharply. The highest decrease of 47.1% was identified in West Jakarta, followed by East Jakarta and North Jakarta, where cropland decreased by 35.9 and 25.2%, respectively. However, North Jakarta still had the highest cropland proportion of 11.9%, while the smallest cropland proportion of 2.6% was observed in Central Jakarta. The decrease in the proportion of cropland affected its aggregation in some municipalities in 2014. For instance, cropland in West Jakarta was disaggregated by 13.9%, followed by North Jakarta and Central Jakarta, where cropland was disaggregated by 11.3 and 13.8%, respectively. Furthermore, cropland had a small decrease in the connectivity, but the percentage was still high. For instance, the highest decrease of 8.4% was observed in West Jakarta, where the connectivity of crops was 87.9%.

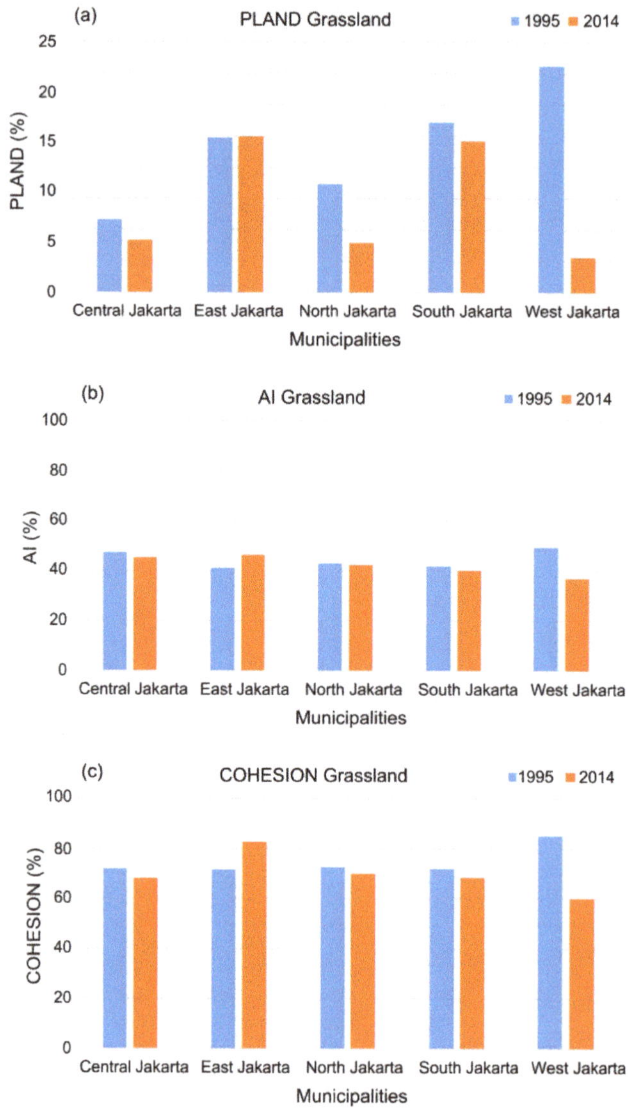

Figure 7. Landscape metrics of grassland. (**a**) Land proportion; (**b**) aggregation index; (**c**) cohesion.

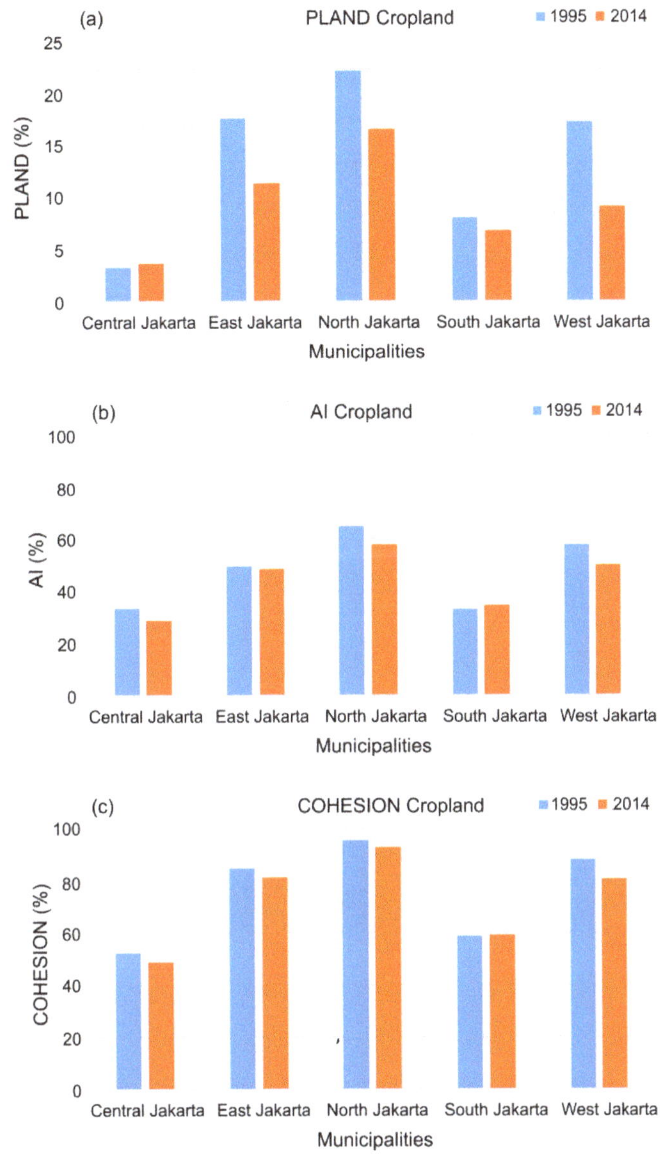

Figure 8. Landscape metrics of cropland. (**a**) Land proportion; (**b**) aggregation index; (**c**) cohesion.

In contrast with the landscape patterns of green spaces, urbanization did not have a significant impact on the landscape patterns of bodies of water, as depicted in Figure 9. The proportion of bodies of water in 2014 was similar to that in 1995. The dynamic changes of the landscape patterns of bodies of water were marked by a small increase in the proportion of bodies of water in East Jakarta and South Jakarta, while this proportion decreased in North Jakarta and West Jakarta. However, the aggregation and the connectivity of bodies of water increased in all municipalities in 2014, as shown in Figure 9.

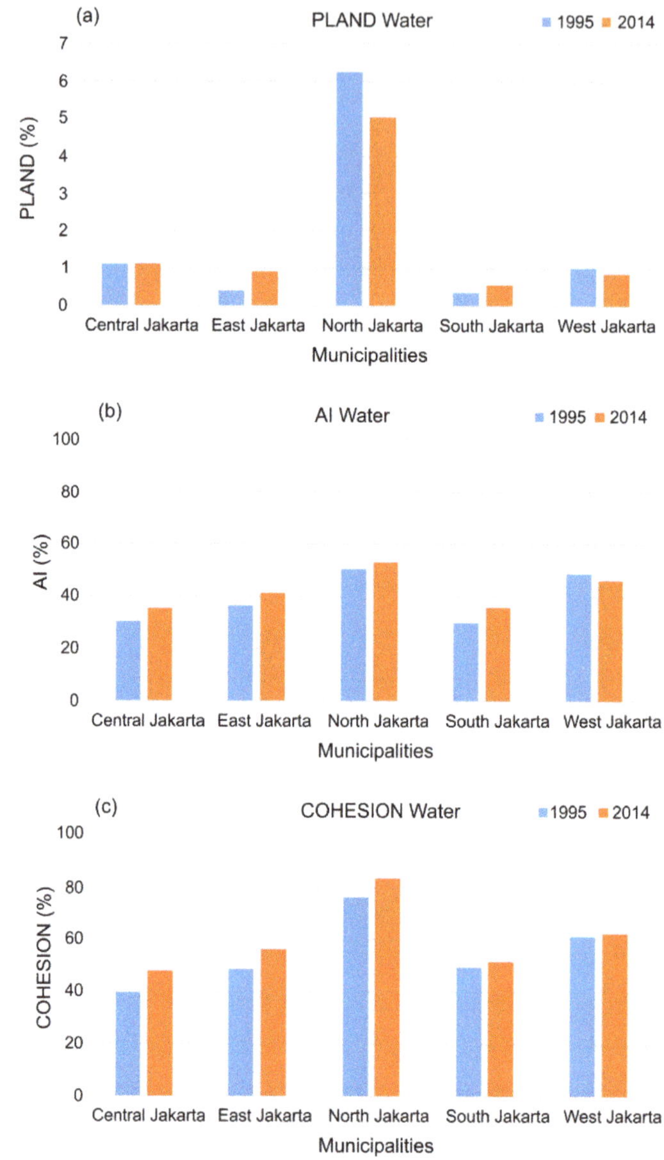

Figure 9. Landscape metrics of water. (**a**) Land proportion; (**b**) aggregation index; (**c**) cohesion.

3.3. Spatial and Temporal Distribution of ES Changes

The spatial and temporal distribution of ES changes were analyzed using Equations (2) and (3). The results showed that the landscape pattern changes had a significant impact on the decrease in the carbon sequestration, temperature regulation, and runoff regulation.

The spatiotemporal changes of the carbon sequestration are given in Figure 10. The picture shows that the area-based estimation and the LMA-based estimation give an almost similar pattern of the carbon sequestration, but the decrease rates are different. The area-based estimation that used a lumped area of trees gave a higher carbon sequestration capacity compared to that from the LMA-based estimation, which used an effective area. The area-based estimation showed that East Jakarta had the highest carbon sequestration in 2014, followed by South Jakarta, North Jakarta, West Jakarta, and Central Jakarta. The highest decrease in the carbon sequestration was identified in North Jakarta (−65.8%), followed by West Jakarta (−62.2%), East Jakarta (−61.1%), South Jakarta (−51.8%), and Central Jakarta (−35.5%). On the other hand, the landscape pattern changes in the LMA-based estimation gave a significant change in the carbon sequestration capacity. For instance, the LMA-based estimation suggested that North Jakarta had the highest decrease in the carbon sequestration (−78.4%), followed by East Jakarta (−77.3%), West Jakarta (−77.1%), South Jakarta (72.9%), and Central Jakarta (−22.8%). Furthermore, East Jakarta had the highest carbon sequestration in 2014, followed by South Jakarta, Central Jakarta, North Jakarta, and West Jakarta.

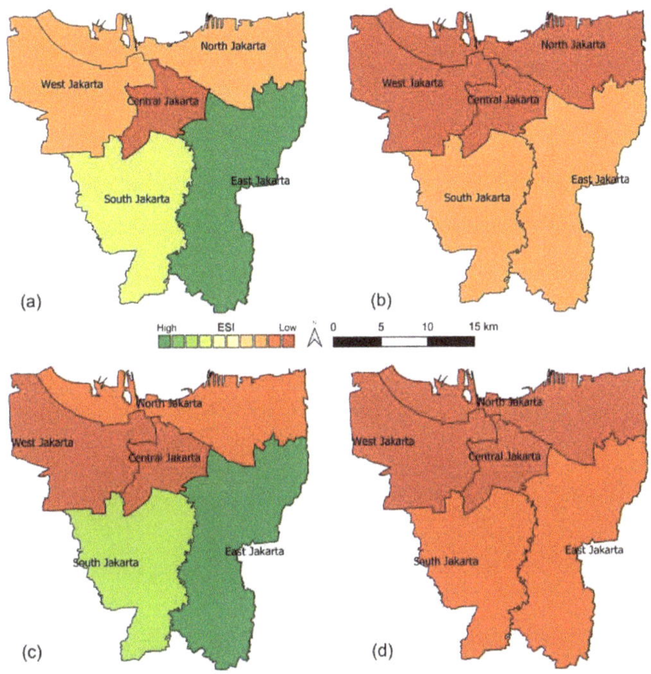

Figure 10. ESI for carbon sequestration in the year 1995 and 2014. (**a**) ESI in 1995 without LMA; (**b**) ESI in 2014 without LMA; (**c**) ESI in 1995 with LMA; and (**d**) ESI in 2014 with LMA.

The different spatial patterns of the temperature regulation indicate the influence of the landscape patterns of green spaces and bodies of water, as shown in Figure 11. It shows that the temperature regulation from the LMA-based estimation had a higher decrease rate compared to that from the area-based estimation. The impact of the landscape pattern changes on the temperature regulation was reflected in the LMA-based estimation that suggested the highest decrease rate in West Jakarta (−74.6%), followed by South Jakarta (−64.1%), East Jakarta (−61.1%), North Jakarta (−50.1%), and Central Jakarta (−23.7%). Moreover, the highest temperature regulation in 2014 was in East Jakarta, followed by North Jakarta, South Jakarta, West Jakarta, and Central Jakarta. On the other hand, the area-based estimation showed a similar spatial distribution of the temperature regulation, but the decrease rate was lower than that in the LMA-based estimation. For instance, the decrease rate of the temperature regulation with the area-based estimation in East Jakarta was −44.7% lower than that from the LMA-based estimation. This decreased pattern was also observed in all municipalities.

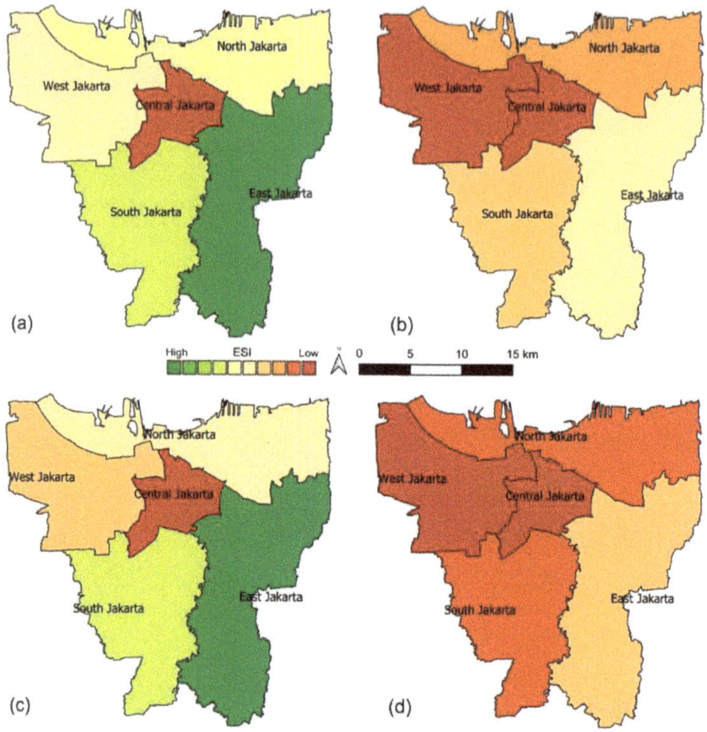

Figure 11. Ecosystem services index (ESI) for temperature regulation in the year 1995 and 2014. (**a**) ESI in 1995 without landscape metric area (LMA); (**b**) ESI in 2014 without LMA; (**c**) ESI in 1995 with LMA; and (**d**) ESI in 2014 with LMA.

Furthermore, the dynamic changes of the landscape patterns affect the spatial distribution of the runoff regulation, as shown in Figure 12. It shows that the area-based estimation indicated a decrease of the runoff regulation between 1995 and 2014 in all municipalities where the highest decrease was found in West Jakarta (−64.6%), followed by North Jakarta (−40.5%), East Jakarta (−38.4%), South Jakarta (−33.6%), and Central Jakarta (−25.8%). Meanwhile, East Jakarta had the highest runoff regulation in 2014, followed by South Jakarta and North Jakarta, while the lowest runoff regulation was in Central Jakarta. The LMA-based estimation, on the other hand, showed a higher decrease rate of the runoff regulation. By considering landscape metrics in the effective area of green spaces, the LMA-based estimation suggested that the highest decrease rate was in West Jakarta (−74.6%), followed by South Jakarta (−57.9%), East Jakarta (−51.8%), North Jakarta (−46.3%), and Central Jakarta (−25.0%). Furthermore, the highest runoff regulation was in East Jakarta, followed by North, South Jakarta, West Jakarta, and Central Jakarta.

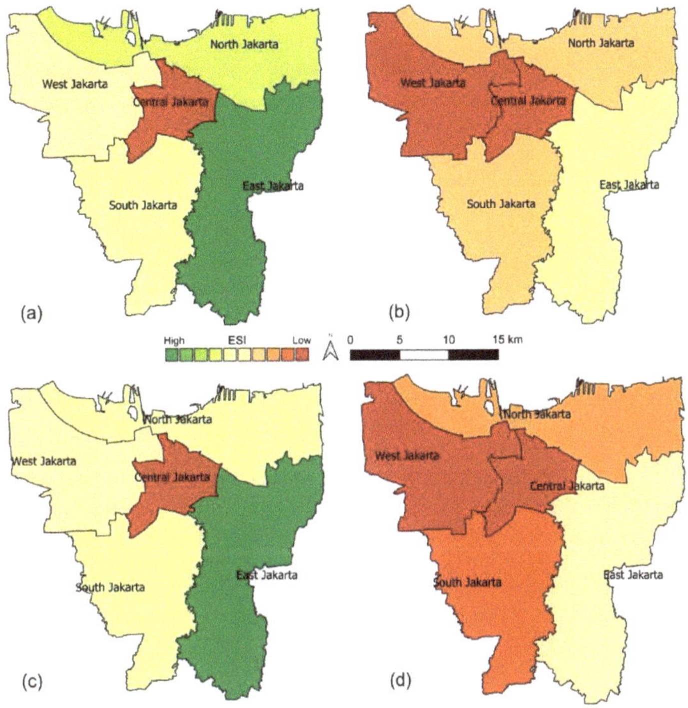

Figure 12. ESI for runoff regulation in the year 1995 and 2014. (**a**) ESI in 1995 without LMA; (**b**) ESI in 2014 without LMA; (**c**) ESI in 1995 with LMA; and (**d**) ESI in 2014 with LMA.

4. Discussion

4.1. Landscape Pattern Changes and Ecosystem Services

The LULC change analysis shows a considerable change in the spatial distribution and the landscape patterns of green spaces between 1995 and 2014, which in turn affect ecosystem services, as given by the LMA-based estimation. An increase in built-up areas results in a decrease in the size and the proportion of green spaces, as well as in the disaggregation and disconnection of green spaces. A change in the landscape patterns affects the capacity of green spaces to provide ecosystem services [8,45].

This study indicates that the decrease in the carbon sequestration between 1995 and 2014 was influenced by a decrease in the area of green spaces, as well as a decrease in the aggregation and the connectivity of trees. For instance, Central Jakarta was covered with about 3.2 km^2 of trees in 2014, which is smaller than the area of trees in North Jakarta (namely 5.2 km^2). However, the carbon sequestration capacity in Central Jakarta is higher than that in North Jakarta, which can be explained by differences in the landscape patterns. In Central Jakarta, the trees are more aggregated, as indicated by 49.6% of the AI index, while the AI of the trees in North Jakarta is 36.1%. Moreover, the trees in Central Jakarta are more connected, as seen from 78.9% of the COHESION, which is higher than 59.9% in North Jakarta. However, the influence of the connectivity and the aggregation of the trees is not always significant compared to its area. For instance, Central Jakarta has a smaller area and higher connectivity of trees, but carbon sequestration in that area is lower than that in East Jakarta, which has a higher area of trees. Furthermore, the aggregation of trees in East Jakarta is slightly lower than that in Central Jakarta and South Jakarta. The relationship between the carbon sequestration with the area and the landscape patterns of the green spaces is consistent with a previous study in Jakarta, which showed that a higher carbon sequestration rate was observed in the location of a large area of trees [73].

The essential influence of the area of trees in the carbon sequestration was also observed in tropical cities such as Singapore [53] and in cities at high latitudes such as in Espoo, Southern Finland [70]. Moreover, it is important to underline that the carbon sequestration discussed in this study and the examples from other cities are limited to a city or an urban scale. The urban carbon sequestration may not give a significant contribution to the global carbon sequestration process because the total carbon uptake from urban vegetation is smaller than carbon emissions in urban areas, as reported in some cities, including Mexico City and Singapore [92], Beijing [93], and Cebu City, Philippines [94].

The significant influence of the landscape patterns on ecosystem services is also observed when comparing the temperature regulation estimates based on the area and the landscape patterns of green spaces, and bodies of water. Between 1995 and 2014, temperature regulation was found to decrease in all municipalities where the contribution of green spaces is more significant than that of bodies of water, which have small areas and proportions as a result of the land cover classification. As a result, the bodies of water in this study only provide a small contribution to temperature regulation as well as to other ES. Moreover, temperature regulation is more influenced by the area of green spaces than by their connectivity. For instance, the lowest temperature regulation is identified in Central Jakarta and West Jakarta, where the area of green spaces was small in 2014, while the connectivity is high. On the other hand, the highest temperature regulation in 2014 was observed in East Jakarta, where the large area and the high connectivity of green spaces are observed. In addition, the proportion of green spaces in this municipality was relatively higher than that in other municipalities except for West Jakarta. A positive relationship between the increasing urban temperature and the dynamic changes of the landscape patterns has been found in this study which is consistent with the spatial distribution of LST and air temperature in Jakarta discussed in Ramdhoni and Rushayati [37]. In their study, the high LST (higher than 30 °C) was distributed in all municipalities. At the same time, the high air temperature was identified in Central Jakarta, East Jakarta, and North Jakarta, mainly where the small and the fragmented green spaces are observed in our study. The relationship between LST and the landscape patterns of green spaces was also observed in three megacities in Southeast Asia: Jakarta, Bangkok, and Manila [54]. It revealed that Jakarta had the highest mean LST of the three cities considered. It was concluded that this high LST was caused by the relatively large, complex, and more aggregated impervious built-up areas of Jakarta. Furthermore, the impact of the decreasing aggregation and connectivity of green spaces was also identified in the urban center of the Olympic Forest Park in Beijing, China. The observed increase in LST correlated with an increase in the green space fragmentation and a decrease in the green space connectivity [31].

Urbanization in Jakarta between 1995 and 2014 decreased runoff regulation in all municipalities, which could be related to a decrease in green spaces and bodies of water, and the landscape pattern changes. The changes in the runoff regulation are mainly influenced by the dynamic changes of the area and the landscape patterns of green spaces, while the influence of bodies of water is relatively small. The most significant decrease in runoff regulation is observed in West Jakarta, where there has been a high decrease in the area and the proportion of green spaces, which mean West Jakarta has the lowest runoff regulation. Furthermore, in all municipalities the connectivity of green spaces is relatively high while the proportion of green spaces is relatively low and more disaggregated, as shown by a decrease in the aggregation index. The low surface runoff regulation in North Jakarta and West Jakarta is consistent with the area inundated by the 2013 flood [95]. Moreover, the capacity of green spaces to control surface runoff can also be influenced by local physical characteristics such as the slope of the land. However, these characteristics have not been addressed in this study. Increasing the area of green spaces for surface runoff control can be considered in the gentlest sloping area with the high topographic wetness index (TWI), while this is not an appropriate solution in an area with the steep slope and the low TWI [96]. The capacity of green spaces to minimize surface runoff is also associated with LPI and AI. In Beijing, for instance, a decrease of LPI and AI reduced the runoff reduction capacity from 23% in 2000 to 17% in 2010 [30]. The low connectivity and aggregation could be associated with a decrease in the capacity of green spaces to minimize surface runoff. A study from Texas, US, showed that large area, less fragmentation, and high connectivity of green spaces could play an important role in anticipating the mean annual peak runoff. On the other hand, urban development could increase the built-up land cover class, which has more edges, more complex shapes, and is more connected, which facilitates high runoff [97].

4.2. Limitations of This Study and Future Perspectives

The spatial analysis in this study used a medium satellite image resolution of 30 m from Landsat 5 and 8 to analyze three ecosystem services: carbon sequestration, temperature regulation, and runoff regulation. The Landsat products provide a cost-effective option for spatial analysis purposes, such as the LULC change analysis and the landscape metrics analysis. However, some specific urban details such as city parks, small wetland areas, and lakes, as well as street corridors, could not be captured properly. Meanwhile, this study's results have not yet been validated, which could limit the applicability of the proposed method. Even though it indicates a more realistic ES calculation, the proposed method could potentially bias the results due to different spatial scales used in this study and field survey data.

In future studies, the use of high spatial resolution satellite images for land use–land cover classification is recommended. This increases the spatial analysis quality, and it allows researchers to include more aspects of ecosystem services analysis, such as air purification from trees on street corridors or tourism and educational ecosystem services from the city and the neighborhood parks. The proposed ES estimation could be further improved in the future by incorporating empirical values based on runoff coefficients derived from land cover classes. For validation purposes, future studies should consider methods which apply remote sensing techniques using a range of spatial resolutions combined with actual land management information to identify carbon sequestration rates, land surface temperatures, and flood-prone areas.

5. Conclusions

The present study shows that the landscape pattern changes in the period 1995–2014 have a significant impact on urban ES in Jakarta. The observed landscape pattern changes reduce the carbon sequestration, the temperature regulation, and the runoff regulation with an average value of 66, 55, and 51%, respectively, during that period. The impact of the landscape pattern changes on a decrease of ES is higher compared to the area-based calculation in which the average decrease of the carbon sequestration is 55%, followed by the temperature regulation (44%) and the runoff regulation (41%). This indicates that the ability of green spaces to provide ES is strongly influenced by the spatial characteristics or the landscape patterns of green spaces.

From 1995 to 2014, it is observed that urbanization has not only reduced the number and total surface area of green spaces but also has changed its spatial characteristics. After 20 years, an increase in built-up areas has affected the spatial characteristics of green spaces marked by a scattered distribution, and a decrease in the aggregation and the connectivity of green spaces, making green spaces more fragmented. The average decrease in aggregation is about 8.7 and 9.6% for its connectivity, which can be associated with a decrease in urban ES. The reduced number and area of green spaces along with the landscape pattern changes and a decrease in urban ES indicate that there is a relationship between those factors, which need to be considered as part of the ES calculation.

This study proposes a new method for ES calculation that explicitly acknowledges the landscape patterns of green spaces resulting in the landscape metric area (LMA) of an urban area. The LMA reflects the area which is defined as green space that provides an effective contribution to ES. The spatial characteristics of green spaces are not static, but change over time. This study has shown that urbanization is a factor that impacts the separation and disaggregation as well as the disconnection of green spaces—and thus to a large extent the spatial distribution of green spaces. Ignoring the influence of the landscape patterns may result in an overestimation of the ES of an urban area, as shown in this study.

The importance of the area of green spaces to ES provision has been discussed in many previous studies, which feed the general perception that there is a positive relationship between the area of green spaces and ES. The findings of our study contribute to the existing body of knowledge in this field and suggest that ES calculations should consider not only the area of green spaces but also the landscape patterns, which represent the spatial configuration of green spaces. By considering the landscape patterns in this study, the proposed method indicates a more realistic ES calculation compared to commonly area-based approaches in which green spaces can be perceived as a lumped area. Our study is one of the first to provide quantified and convincing evidence of the influence of the landscape pattern changes on urban ecosystem services in a megacity such as Jakarta. We acknowledge that further detailed studies will be needed and that the present study will provide incentives and directions to initiate further research. It is recommended to further improve this new method by using land cover classification from a high satellite image resolution, considering more landscape metric combinations, incorporating empirical values such as runoff coefficients or albedo values based on LULC, using field survey data or using model simulations to verify the ES calculation.

Author Contributions: Conceptualization, D.M.; methodology, D.M.; formal analysis, D.M.; investigation, D.M.; writing—original draft preparation, D.M., A.P. and C.Z.; writing—review and editing, D.M., A.P. and C.Z.; visualization, D.M. All authors have read and agreed to the published version of the manuscript.

Funding: This research was financially supported by the Indonesia Endowment Fund for Education (BUDI-LPDP) for the doctoral scholarship for Dikman Maheng. The Laminga fund of the Delft University of Technology for providing the travel grant to present a part of this research at the IWA Water and Development Congress and Exhibition 2019.

Institutional Review Board Statement: Not applicable.

Informed Consent Statement: Not Applicable.

Data Availability Statement: The data presented in this study are available on request from the corresponding author.

Acknowledgments: The authors acknowledge the use of data from Geospatial Information Agency of Indonesia (BIG), and Landsat 5 and Landsat 8 images from U.S. Geological Survey. The authors appreciate all anonymous reviewers who provide valuable and constructive comments which have improved the quality of the paper. The authors also thank Mohanasundar Radakrishnan for the English correction of the first draft of the paper.

Conflicts of Interest: The authors declare no conflict of interest.

References

1. United Nations Trends in Urbanization. In *World Urbanization Prospects*; United Nations: New York, NY, USA, 2014; Volume 12, pp. 7–12. ISBN 9789211515176.
2. Dou, Y.; Kuang, W. A comparative analysis of urban impervious surface and green space and their dynamics among 318 different size cities in China in the past 25 years. *Sci. Total Environ.* **2020**, *706*, 135828. [CrossRef]
3. Wu, Y.; Tao, Y.; Yang, G.; Ou, W.; Pueppke, S.; Sun, X.; Chen, G.; Tao, Q. Impact of land use change on multiple ecosystem services in the rapidly urbanizing Kunshan City of China: Past trajectories and future projections. *Land Use Policy* **2019**, *85*, 419–427. [CrossRef]
4. Xu, Q.; Dong, Y.-X.; Yang, R. Influence of land urbanization on carbon sequestration of urban vegetation: A temporal cooperativity analysis in Guangzhou as an example. *Sci. Total Environ.* **2018**, *635*, 26–34. [CrossRef] [PubMed]
5. Ye, Y.; Bryan, B.A.; Zhang, J.; Connor, J.D.; Chen, L.; Qin, Z.; He, M. Changes in land-use and ecosystem services in the Guangzhou-Foshan Metropolitan Area, China from 1990 to 2010: Implications for sustainability under rapid urbanization. *Ecol. Indic.* **2018**, *93*, 930–941. [CrossRef]

6. Alcamo, J.; Ash, N.J.; Butler, C.D.; Callicot, J.B.; Capistrano, D.; Carpenter, S.R.; Castilla, J.C.; Chambers, R.; Chopta, K.; Cropper, A.; et al. Ecosystem and Their Services. In *Ecosystems and Human Well-Being: A Framework for Assessment*; Reid, W.V., Ed.; Island Press: Washington, DC, USA, 2003; pp. 49–70. ISBN 1559634022.
7. Bolund, P.; Hunhammar, S. Ecosystem services in urban areas. *Ecol. Econ.* **1999**, *29*, 293–301. [CrossRef]
8. Burkhard, B.; Kroll, F.; Nedkov, S.; Müller, F. Mapping ecosystem service supply, demand and budgets. *Ecol. Indic.* **2012**, *21* (Suppl. C), 17–29. [CrossRef]
9. Derkzen, M.L.; Van Teeffelen, A.J.A.; Verburg, P.H. REVIEW: Quantifying urban ecosystem services based on high-resolution data of urban green space: An assessment for Rotterdam, the Netherlands. *J. Appl. Ecol.* **2015**, *52*, 1020–1032. [CrossRef]
10. Gkatsopoulos, P. A Methodology for Calculating Cooling from Vegetation Evapotranspiration for Use in Urban Space Microclimate Simulations. *Procedia Environ. Sci.* **2017**, *38*, 477–484. [CrossRef]
11. Gómez-Baggethun, E.; Barton, D.N. Classifying and valuing ecosystem services for urban planning. *Ecol. Econ.* **2013**, *86*, 235–245. [CrossRef]
12. Tratalos, J.; Fuller, R.A.; Warren, P.H.; Davies, R.G.; Gaston, K.J. Urban form, biodiversity potential and ecosystem services. *Landsc. Urban Plan.* **2007**, *83*, 308–317. [CrossRef]
13. Van Oudenhoven, A.P.E.; Petz, K.; Alkemade, R.; Hein, L.; De Groot, R.S. Framework for systematic indicator selection to assess effects of land management on ecosystem services. *Ecol. Indic.* **2012**, *21*, 110–122. [CrossRef]
14. Panagopoulos, T.; Duque, J.A.G.; Dan, M.B. Urban planning with respect to environmental quality and human well-being. *Environ. Pollut.* **2016**, *208*, 137–144. [CrossRef] [PubMed]
15. Loures, L.; Santos, R.; Panagopoulos, T. Urban parks and sustainable city planning—The case of Portimão, Portugal. *Population* **2007**, *15*, 171–180.
16. Clerici, N.; Cote-Navarro, F.; Escobedo, F.J.; Rubiano, K.; Villegas, J.C. Spatio-temporal and cumulative effects of land use-land cover and climate change on two ecosystem services in the Colombian Andes. *Sci. Total Environ.* **2019**, *685*, 1181–1192. [CrossRef] [PubMed]
17. Depietri, Y.; Renaud, F.G.; Kallis, G. Heat waves and floods in urban areas: A policy-oriented review of ecosystem services. *Sustain. Sci.* **2011**, *7*, 95–107. [CrossRef]
18. Dupras, J.; Marull, J.; Parcerisas, L.; Coll, F.; Gonzalez, A.; Girard, M.; Tello, E. The impacts of urban sprawl on ecological connectivity in the Montreal Metropolitan Region. *Environ. Sci. Policy* **2016**, *58*, 61–73. [CrossRef]
19. Fox, D.M.; Witz, E.; Blanc, V.; Soulié, C.; Penalver-Navarro, M.; Dervieux, A. A case study of land cover change (1950–2003) and runoff in a Mediterranean catchment. *Appl. Geogr.* **2012**, *32*, 810–821. [CrossRef]
20. Farrugia, S.; Hudson, M.D.; McCulloch, L. An evaluation of flood control and urban cooling ecosystem services delivered by urban green infrastructure. *Int. J. Biodivers. Sci. Ecosyst. Serv. Manag.* **2012**, *9*, 136–145. [CrossRef]
21. Marando, F.; Salvatori, E.; Sebastiani, A.; Fusaro, L.; Manes, F. Regulating Ecosystem Services and Green Infrastructure: Assessment of Urban Heat Island effect mitigation in the municipality of Rome, Italy. *Ecol. Model.* **2019**, *392*, 92–102. [CrossRef]
22. Yuan, Y.; Wu, S.; Yu, Y.; Tong, G.; Mo, L.; Yan, D.; Li, F. Spatiotemporal interaction between ecosystem services and urbanization: Case study of Nanjing City, China. *Ecol. Indic.* **2018**, *95*, 917–929. [CrossRef]
23. Jaligot, R.; Kemajou, A.; Chenal, J. Cultural ecosystem services provision in response to urbanization in Cameroon. *Land Use Policy* **2018**, *79*, 641–649. [CrossRef]
24. Das, M.; Das, A. Dynamics of Urbanization and its impact on Urban Ecosystem Services (UESs): A study of a medium size town of West Bengal, Eastern India. *J. Urban Manag.* **2019**, *8*, 420–434. [CrossRef]
25. Costanza, R.; d'Arge, R.; de Groot, R.; Farber, S.; Grasso, M.; Hannon, B.; Limburg, K.; Naeem, S.; O'Neill, R.V.; Paruelo, J.; et al. The value of the world's ecosystem services and natural capital. *Nature* **1997**, *387*, 253–260. [CrossRef]
26. Wang, J.; Zhou, W.; Pickett, S.T.; Yu, W.; Li, W. A multiscale analysis of urbanization effects on ecosystem services supply in an urban megaregion. *Sci. Total Environ.* **2019**, *662*, 824–833. [CrossRef] [PubMed]
27. Asadolahi, Z.; Salmanmahiny, A.; Sakieh, Y.; Mirkarimi, S.H.; Baral, H.; Azimi, M. Dynamic trade-off analysis of multiple ecosystem services under land use change scenarios: Towards putting ecosystem services into planning in Iran. *Ecol. Complex.* **2018**, *36*, 250–260. [CrossRef]
28. Haas, J.; Furberg, D.; Ban, Y. Satellite monitoring of urbanization and environmental impacts—A comparison of Stockholm and Shanghai. *Int. J. Appl. Earth Obs. Geoinf.* **2015**, *38*, 138–149. [CrossRef]
29. Bai, T.; Mayer, A.L.; Shuster, W.D.; Tian, G. The Hydrologic Role of Urban Green Space in Mitigating Flooding (Luohe, China). *Sustainability* **2018**, *10*, 3584. [CrossRef]
30. Zhang, B.; Xie, G.-D.; Li, N.; Wang, S. Effect of urban green space changes on the role of rainwater runoff reduction in Beijing, China. *Landsc. Urban Plan.* **2015**, *140*, 8–16. [CrossRef]
31. Amani-Beni, M.; Zhang, B.; Xie, G.-D.; Shi, Y. Impacts of Urban Green Landscape Patterns on Land Surface Temperature: Evidence from the Adjacent Area of Olympic Forest Park of Beijing, China. *Sustainability* **2019**, *11*, 513. [CrossRef]
32. Jaganmohan, M.; Knapp, S.; Buchmann, C.M.; Schwarz, N. The Bigger, the Better? The Influence of Urban Green Space Design on Cooling Effects for Residential Areas. *J. Environ. Qual.* **2016**, *45*, 134–145. [CrossRef] [PubMed]
33. Mitchell, M.G.; Suarez-Castro, A.F.; Martinez-Harms, M.; Maron, M.; McAlpine, C.; Gaston, K.J.; Johansen, K.; Rhodes, J.R. Reframing landscape fragmentation's effects on ecosystem services. *Trends Ecol. Evol.* **2015**, *30*, 190–198. [CrossRef] [PubMed]

34. Mitchell, M.G.E.; Bennett, E.M.; Gonzalez, A. Linking Landscape Connectivity and Ecosystem Service Provision: Current Knowledge and Research Gaps. *Ecosystems* **2013**, *16*, 894–908. [CrossRef]
35. Nagasawa, R.; Fukushima, A.; Yayusman, L.F.; Novresiandi, D.A. Urban Expansion and Its Influences on The Suburban Land Use Change in Jakarta Metropolitan Region (JABODETABEK). *Urban Plan. Des. Res.* **2015**, *3*, 7. [CrossRef]
36. Pravitasari, A.E. *Study on Impact of Urbanization and Rapid Urban Expansion in Java and Jabodetabek, Megacity in Indonesia*; Kyoto University: Kyoto, Japan, 2015.
37. Ramdhoni, S.; Rushayati, S.B.; Prasetyo, L.B. Open Green Space Development Priority Based on Distribution of air Temperature Change in Capital City of Indonesia, Jakarta. *Procedia Environ. Sci.* **2016**, *33*, 204–213. [CrossRef]
38. Rustiadi, E.; Zain, A.M.; Trisasongko, B.H.; Carolita, I. *Land Cover Change in Jabotabek Region*; Himiyama, Y., Mather, A., Bicik, I., Milanova, E.V., Eds.; International Geographical Union Commission on Land Use/Cover Change: Hokkaido, Japan, 2002.
39. Firman, T. The continuity and change in mega-urbanization in Indonesia: A survey of Jakarta–Bandung Region (JBR) development. *Habitat Int.* **2009**, *33*, 327–339. [CrossRef]
40. *BPS DKI Jakarta Jakarta in Figures 2016*; BPS Provinsi DKI Jakarta/Jakarta Statistics Bureau: Jakarta, Indonesia, 2016.
41. De Ridder, K.; Adamec, V.; Bañuelos, A.; Bruse, M.; Bürger, M.; Damsgaard, O.; Dufek, J.; Hirsch, J.; Lefebre, F.; Pérez-Lacorzana, J.; et al. An integrated methodology to assess the benefits of urban green space. *Sci. Total Environ.* **2004**, *334–335*, 489–497. [CrossRef]
42. Leroux, L.; Congedo, L.; Bellón, B.; Gaetano, R.; Bégué, A. Land Cover Mapping Using Sentinel-2 Images and the Semi-Automatic Classification Plugin: A Northern Burkina Faso Case Study. In *QGIS and Applications in Agriculture and Forest*; Wiley: Hoboken, NJ, USA, 2018; pp. 119–151.
43. Arowolo, A.O.; Deng, X.; Olatunji, O.A.; Obayelu, A.E. Assessing changes in the value of ecosystem services in response to land-use/land-cover dynamics in Nigeria. *Sci. Total Environ.* **2018**, *636*, 597–609. [CrossRef] [PubMed]
44. Gökyer, E. Understanding landscape structure using landscape metrics. In *Advances in Landscape Architecture*; Özyavuz, M., Ed.; IntechOpen: London, UK, 2013.
45. Mugiraneza, T.; Ban, Y.; Haas, J. Urban land cover dynamics and their impact on ecosystem services in Kigali, Rwanda using multi-temporal Landsat data. *Remote. Sens. Appl. Soc. Environ.* **2019**, *13*, 234–246. [CrossRef]
46. Daniels, B.; Zaunbrecher, B.S.; Paas, B.; Ottermanns, R.; Ziefle, M.; Roß-Nickoll, M. Assessment of urban green space structures and their quality from a multidimensional perspective. *Sci. Total Environ.* **2018**, *615*, 1364–1378. [CrossRef] [PubMed]
47. Estoque, R.C.; Murayama, Y. Landscape pattern and ecosystem service value changes: Implications for environmental sustainability planning for the rapidly urbanizing summer capital of the Philippines. *Landsc. Urban Plan.* **2013**, *116*, 60–72. [CrossRef]
48. McGarigal, K.; Cushman, S.; Ene, E. *FRAGSTATS v4: Spatial Pattern Analysis Program for Categorical and Continuous Maps. Computer Software Program*; University of Massachusetts: Amherst, MA, USA; Available online: http://www.umass.edu/landeco/research/fragstats/fragstats.html (accessed on 30 September 2019).
49. De Smith, M.J.; Goodchild, M.F.; Longley, P. *Geospatial Analysis: A Comprehensive Guide to Principles, Techniques and Software Tools*; Troubador Publishing Ltd.: Leicester, UK, 2007; ISBN 1-905886-60-8.
50. Kong, F.; Yin, H.; Nakagoshi, N.; Zong, Y. Urban green space network development for biodiversity conservation: Identification based on graph theory and gravity modeling. *Landsc. Urban Plan.* **2010**, *95*, 16–27. [CrossRef]
51. He, H.S.; Dezonia, B.E.; Mladenoff, D.J. An aggregation index (AI) to quantify spatial patterns of landscapes. *Landsc. Ecol.* **2000**, *15*, 591–601. [CrossRef]
52. Bao, T.; Li, X.; Zhang, J.; Zhang, Y.; Tian, S. Assessing the Distribution of Urban Green Spaces and its Anisotropic Cooling Distance on Urban Heat Island Pattern in Baotou, China. *ISPRS Int. J. Geo-Inf.* **2016**, *5*, 12. [CrossRef]
53. Chen, J.; Goh, J. Carbon Accounting in Local-Scale Land Use and Land Cover Change. *Cons. J. Sustaianble Dev.* **2017**, *17*, 46–74.
54. Estoque, R.C.; Murayama, Y.; Myint, S.W. Effects of landscape composition and pattern on land surface temperature: An urban heat island study in the megacities of Southeast Asia. *Sci. Total Environ.* **2017**, *577*, 349–359. [CrossRef] [PubMed]
55. Li, X.; Zhou, W.; Ouyang, Z.; Xu, W.; Zheng, H. Spatial pattern of greenspace affects land surface temperature: Evidence from the heavily urbanized Beijing metropolitan area, China. *Landsc. Ecol.* **2012**, *27*, 887–898. [CrossRef]
56. Ramesh, R.; Chen, Z.; Cummins, V.; Day, J.; D'Elia, C.; Dennison, B.; Forbes, D.; Glaeser, B.; Glaser, M.; Glavovic, B.; et al. Land–Ocean Interactions in the Coastal Zone: Past, present & future. *Anthropocene* **2015**, *12*, 85–98. [CrossRef]
57. Werner, C. Green open spaces in Indonesian cities: Schisms between law and practice. *Pacific Geogr.* **2014**, *41*, 26–31.
58. Cortinovis, C.; Geneletti, D. Ecosystem services in urban plans: What is there, and what is still needed for better decisions. *Land Use Policy* **2018**, *70*, 298–312. [CrossRef]
59. Vincent, S.U.; Radhakrishnan, M.; Hayde, L.; Pathirana, A. Enhancing the Economic Value of Large Investments in Sustainable Drainage Systems (SuDS) through Inclusion of Ecosystems Services Benefits. *Water* **2017**, *9*, 841. [CrossRef]
60. Rotterdam, C. *Rotterdam Programme on Sustainability and Climate Change 2015–2018*; City of Rotterdam: Rotterdam, The Netherlands, 2015.
61. *Urban Ecology Programme 2011–2026*; Department of Environmental Affairs and Transport: Oslo, Norway, 2011.
62. *Environmental Action 2016–2021: Strategy and Action Plan*; City of Sydney Council: Sydney, Australia, 2017; Available online: https://apo.org.au/sites/default/files/resource-files/2017-03/apo-nid174586.pdf (accessed on 20 February 2021).
63. Arifin, H.S.; Nakagoshi, N. Landscape ecology and urban biodiversity in tropical Indonesian cities. *Landsc. Ecol. Eng.* **2011**, *7*, 33–43. [CrossRef]

64. Rusiawan, W.; Tjiptoherijanto, P.; Suganda, E.; Darmajanti, L. System Dynamics Modeling for Urban Economic Growth and CO_2 Emissions: A Case Study of Jakarta, Indonesia. *Procedia Environ. Sci.* **2015**, *28*, 330–340. [CrossRef]
65. Surahman, U.; Kubota, T.; Wijaya, A. Life cycle assessment of energy and CO2emissions for residential buildings in Jakarta, Indonesia. *IOP Conf. Ser. Mater. Sci. Eng.* **2016**, *128*, 012002. [CrossRef]
66. Siswanto, G.J.; Van Der Schrier, G.; Lenderink, G.; Hurk, B.V.D. Trends in High-Daily Precipitation Events in Jakarta and the Flooding of January. *Bull. Am. Meteorol. Soc.* **2015**, *96*, 131–135. [CrossRef]
67. Wijayanti, P.; Zhu, X.; Hellegers, P.; Budiyono, Y.; Van Ierland, E.C. Estimation of river flood damages in Jakarta, Indonesia. *Nat. Hazards* **2017**, *86*, 1059–1079. [CrossRef]
68. Darmanto, N.S.; Varquez, A.C.G.; Kawano, N.; Kanda, M. Future urban climate projection in a tropical megacity based on global climate change and local urbanization scenarios. *Urban Clim.* **2019**, *29*, 100482. [CrossRef]
69. Churkina, G. Carbon cycle of urban ecosystems. In *Carbon Sequestration in Urban Ecosystems*; Lal, R., Augustin, B., Eds.; Springer: Berlin/Heidelberg, Germany, 2011; p. 315. ISBN 978-94-007-2365-8.
70. Kuittinen, M.; Moinel, C.; Adalgeirsdottir, K. Carbon sequestration through urban ecosystem services. *Sci. Total Environ.* **2016**, *563–564*, 623–632. [CrossRef]
71. Nero, B.F.; Callo-Concha, D.; Anning, A.; Denich, M. Urban Green Spaces Enhance Climate Change Mitigation in Cities of the Global South: The Case of Kumasi, Ghana. *Procedia Eng.* **2017**, *198*, 69–83. [CrossRef]
72. Forman, R.T.T. *Urban Ecology: Science of Cities*; Cambridge University Press: Cambridge, UK, 2014; ISBN 9781139030472.
73. Azaria, L.; Wibowo, A.; Shidiq, I.P.A. Rokhmatuloh Carbon Sequestration Capability Analysis of Urban Green Space Using Geospatial Data. *E3S Web Conf.* **2018**, *73*, 03009. [CrossRef]
74. Oviantari, M.V.; Gunamantha, I.M.; Ristiati, N.P.; Santiasa, I.M.P.A.; Astariani, P.P.Y. Carbon sequestration by above-ground biomass in urban green spaces in Singaraja city. *IOP Conf. Ser. Earth Environ. Sci.* **2018**, *200*, 012030. [CrossRef]
75. Oke, T.R.; Mills, G.; Christen, A.; Voogt, J.A. *Urban Climates*; Cambridge University Press: Cambridge, UK, 2017; ISBN 9781139016476.
76. Maheng, D.; Ducton, I.; Lauwaet, D.; Zevenbergen, C.; Pathirana, A. The Sensitivity of Urban Heat Island to Urban Green Space—A Model-Based Study of City of Colombo, Sri Lanka. *Atmosphere* **2019**, *10*, 151. [CrossRef]
77. Sodoudi, S.; Zhang, H.; Chi, X.; Müller, F.; Li, H. The influence of spatial configuration of green areas on microclimate and thermal comfort. *Urban For. Urban Green.* **2018**, *34*, 85–96. [CrossRef]
78. Amani-Beni, M.; Zhang, B.; Xie, G.-D.; Xu, J. Impact of urban park's tree, grass and waterbody on microclimate in hot summer days: A case study of Olympic Park in Beijing, China. *Urban For. Urban Green.* **2018**, *32*, 1–6. [CrossRef]
79. Steeneveld, G.; Koopmans, S.A.; Heusinkveld, B.G.; Theeuwes, N. Refreshing the role of open water surfaces on mitigating the maximum urban heat island effect. *Landsc. Urban Plan.* **2014**, *121*, 92–96. [CrossRef]
80. Theeuwes, N.E.; Solcerová, A.; Steeneveld, G.J. Modeling the influence of open water surfaces on the summertime temperature and thermal comfort in the city. *J. Geophys. Res. Atmos.* **2013**, *118*, 8881–8896. [CrossRef]
81. Chen, Y.-C.; Tan, C.-H.; Wei, C.; Su, Z.-W. Cooling Effect of Rivers on Metropolitan Taipei Using Remote Sensing. *Int. J. Environ. Res. Public Health* **2014**, *11*, 1195–1210. [CrossRef] [PubMed]
82. Gunawardena, K.; Wells, M.; Kershaw, T. Utilising green and bluespace to mitigate urban heat island intensity. *Sci. Total Environ.* **2017**, *584*, 1040–1055. [CrossRef]
83. Alves, A.; Vojinovic, Z.; Kapelan, Z.; Sanchez, A.; Gersonius, B. Exploring trade-offs among the multiple benefits of green-blue-grey infrastructure for urban flood mitigation. *Sci. Total Environ.* **2020**, *703*, 134980. [CrossRef] [PubMed]
84. Yang, B.; Lee, D.K.; Heo, H.K.; Biging, G. The effects of tree characteristics on rainfall interception in urban areas. *Landsc. Ecol. Eng.* **2019**, *15*, 289–296. [CrossRef]
85. Yao, L.; Chen, L.; Wei, W.; Sun, R. Potential reduction in urban runoff by green spaces in Beijing: A scenario analysis. *Urban For. Urban Green.* **2015**, *14*, 300–308. [CrossRef]
86. *ASCE Hydrology Handbook (Manual N0. 28)*, 2nd ed.; American Society of Civil Engineers: New York, NY, USA, 1996; ISBN 978-0-7844-0138-5.
87. Alves, P.L.; Formiga, K.T.M.; Traldi, M.A.B. Rainfall interception capacity of tree species used in urban afforestation. *Urban Ecosyst.* **2018**, *21*, 697–706. [CrossRef]
88. Highfield, W.E. Section 404 Permitting in Coastal Texas: A Longitudinal Analysis of the Relationship Between Peak Streamflow and Wetland Alteration. *Environ. Manag.* **2012**, *49*, 892–901. [CrossRef]
89. Te Chow, V.; Maidment, D.R.; Mays, L.W. *Applied Hydrology*; McGraw-Hill Series in Water Resources and Environmental Engineering; McGraw-Hill: Singapore, 1988; ISBN 0070108102.
90. Yang, X.; You, X.-Y.; Ji, M.; Nima, C. Influence factors and prediction of stormwater runoff of urban green space in Tianjin, China: Laboratory experiment and quantitative theory model. *Water Sci. Technol.* **2013**, *67*, 869–876. [CrossRef] [PubMed]
91. Shang, H.; Zhang, K.; Wang, Z.; Yang, J.; He, M.; Pan, X.; Fang, C. Effect of varying wheatgrass density on resistance to overland flow. *J. Hydrol.* **2020**, *591*, 125594. [CrossRef]
92. Velasco, E.; Roth, M.; Norford, L.; Molina, L.T. Does urban vegetation enhance carbon sequestration? *Landsc. Urban Plan.* **2016**, *148*, 99–107. [CrossRef]
93. Zhao, S.; Tang, Y.; Chen, A. Carbon Storage and Sequestration of Urban Street Trees in Beijing, China. *Front. Ecol. Evol.* **2016**, *4*, 1–8. [CrossRef]

94. Pansit, N.R. Carbon storage and sequestration potential of urban trees in Cebu City, Philippines. *Mindanao J. Sci. Technol.* **2019**, *17*, 98–111.
95. Padawangi, R.; Douglass, M. Water, Water Everywhere: Toward Participatory Solutions to Chronic Urban Flooding in Jakarta. *Pac. Aff.* **2015**, *88*, 517–550. [CrossRef]
96. Kim, H.; Lee, D.-K.; Sung, S. Effect of Urban Green Spaces and Flooded Area Type on Flooding Probability. *Sustainability* **2016**, *8*, 134. [CrossRef]
97. Kim, H.W.; Park, Y. Urban green infrastructure and local flooding: The impact of landscape patterns on peak runoff in four Texas MSAs. *Appl. Geogr.* **2016**, *77*, 72–81. [CrossRef]

land

Article

Spatial–Temporal Impacts of Urban Sprawl on Ecosystem Services: Implications for Urban Planning in the Process of Rapid Urbanization

Xiaoyan Li [1,*], Gulinaer Suoerdahan [1], Zhenyu Shi [1], Zihan Xing [1], Yongxing Ren [1] and Ran Yang [1,2]

1 College of Earth Sciences, Jilin University, Changchun 130061, China; glnes19@mails.jlu.edu.cn (G.S.); shizy20@mails.jlu.edu.cn (Z.S.); xingzh20@mails.jlu.edu.cn (Z.X.); ryx20@mails.jlu.edu.cn (Y.R.); ranyang19@mails.jlu.edu.cn (R.Y.)
2 Key Laboratory of Wetland Ecology and Environment, Northeast Institute of Geography and Agroecology, Chinese Academy of Sciences, Changchun 130102, China
* Correspondence: lxyan@jlu.edu.cn; Tel.: +86-0431-8850-2065

Abstract: Rapid urbanization drives land cover change, affecting urban ecosystems and inducing serious environmental issues. The study region of Changchun, China was divided into three urbanization categories according to different urbanization levels and the characteristics of urban sprawl and changes and relationships between typical ecosystem services (ESs) under rapid urbanization were analysed. The results showed that Changchun has undergone considerable urban expansion since 2000, which has significantly impacted all ESs in terms of spatial and temporal heterogeneity. Habitat suitability and crop yield have relatively stronger service capacity in the study area. Since the expansion of large-scale infrastructures, the mean ES values of developed urban areas are the lowest among the three zones, except for water retention and sandstorm prevention in 2015, when the balance between all services decreased. Over the past 16 years, habitat suitability in developing urban areas has decreased to a large extent due to urban sprawl. Because of the improvement in agricultural science and technology, crop yield in three regions increased, while the area of cropland reduced from 1720 km^2 to 1560 km^2 (9.3%). Synergies between habitat suitability and carbon storage and habitat suitability and soil retention were detected in three areas. A trade-off between habitat suitability and water retention was detected in three areas. The interactions between crop yield and carbon storage, habitat suitability, and soil retention were more complex in this study region. In addition to water retention, urbanization index has a negative correlation with ESs. According to the results, some suggestions to alleviate ES loss during the process of rapid urbanization were proposed, which may guide scientific urban planning for sustainable urban development.

Keywords: urbanization; ecosystem services; Changchun; China

Citation: Li, X.; Suoerdahan, G.; Shi, Z.; Xing, Z.; Ren, Y.; Yang, R. Spatial–Temporal Impacts of Urban Sprawl on Ecosystem Services: Implications for Urban Planning in the Process of Rapid Urbanization. *Land* **2021**, *10*, 1210. https://doi.org/10.3390/land10111210

Academic Editors: Luca Congedo, Francesca Assennato and Michele Munafò

Received: 19 September 2021
Accepted: 4 November 2021
Published: 8 November 2021

1. Introduction

Urbanization is a global universal process and an inevitable phase towards modernization for the development of most countries [1]. According to a report released by the United Nations, the rate of global urbanization increased from 30% to 55% between 1950 and 2018 and will continue to increase to 68% by 2050 [2]. In China, the urbanization rate, which was only 10.64% in 1949, exceeded 50% for the first time in 2011 and reached 59.58% in 2018 [3]. There are many factors influencing urban expansion, including rural land reform, large-scale industrialization, incentive policy, migration, economic reform, and market-led urbanization, etc [3]. With the rapid development of urbanization, rural populations moved into cities and secondary and tertiary industries continued to gather in cities. China's seventh national census shows that 902 million people live in the city in 2021, reaching 63.89% of the total population [4]. The population agglomeration in urban areas in particular has led to a surge in demand for urban land resources. Statistics about urban development released by the World Bank showed that urban land consumption outpaces population growth by as much as 50%, and is expected to add 1.2 million km^2 of new urban built up areas to the world in the next three decades [5]. Such a sprawl puts pressure on land and natural resources and a large number of natural resources are consumed.

Statistics from the United Nations showed that cities accounting for less than 2% of the earth's surface consume 78% of the energy generated and produce 60% of greenhouse gas emissions [6]. According to the Land Rent Theory, the price and demand for land change with the increase in distance from the city centre. As a result, commercial, office, and residential activities, which are more profitable and have a higher rent, outbid and replace agriculture and ecosystem services to maximize access to the dense city centre [7]. ESs became the first casualty of urban expansion and economic development. ESs are a range of important goods and benefits that humans obtain from ecosystems, which are a key bridge linking human welfare with the environment, therefore providing an important framework within which to understand and manage urban sustainability [8,9]. ESs can be classified into four categories: provisioning services, regulation services, supporting services, and cultural services [6]. Assessment and evaluation of ESs provide decision-making information for the effective management of the earth's ecosystem and a vital framework to understand and manage urban sustainability [10,11].

Since urbanization was one of the principle driving factors of ES change, the way that urbanization influences ecosystems is a worldwide concern [12]. Recent studies have revealed that, driven by numerous factors, including physical environment, economics, socio-political issues, science, and technology, as well as scale and degree of urbanization, the impact mechanisms of urban processes on ESs are quite complex, showing negative or positive correlations in different urban areas [3,13–16]. A study by Wang et al. in Guangdong-Hong Kong showed that urbanization is negatively correlated with natural habitats, water retention, soil conservation, and carbon sequestration, both in urban and rural areas [9]. Sun et al. found that in the process of urbanization, carbon storage, habitat quality index, water yield, and soil conservation decreased but food supply grew steadily [17]. Furthermore, studies have also shown that ESs increased in the early stage of urbanization and declined in the later stage, forming an inverted "U"-shaped relationship [18]. Therefore, there is still no universally consistent conclusion, and it is particularly necessary to research the impact of urbanization on ESs at local and regional levels. Some studies have also explored the impact of the urban–rural spatial gradient on ESs, especially for developed coastal cities [18–20]. However, few studies have been conducted on the influences of urbanization on ESs in the Black Soil Belt of northeast China.

As the provincial city of Jilin Province and the core city of the Urban Agglomeration of Harbin–Changchun, Changchun has seen an unexpectedly rapid urban expansion in the past two decades, especially after the implementation of the Strategy of Revitalizing Northeast China, which provides a good opportunity for regional development, as well as posing a major threat to the fragile ecosystems of the region [21]. At the same time, located in the typical Black Soil Belt and the important granary of China, Changchun plays a vital role in national food security. Rapid sprawling urbanization transformed a large number of high-quality croplands into artificial surfaces [21], which caused the challenges for cropland protection and food safety. Maintaining a balance between environmental sustainability and rapid urbanization becomes the principal issue that Changchun faces for future development. Many initiatives have been developed to protect cropland and natural landscapes to optimize land use structure. However, in order to minimize the loss of ESs during urbanization, a full understanding and quantitative analysis of impact of the urbanization processes on ESs are necessary.

For these reasons, the aims of this paper are: (1) to analyse dynamic characteristics of urbanization of Changchun since 2000; (2) to evaluate six ESs changes and bundle relationships during urbanization; and (3) to explore impacts of urbanization process on ES changes. The study aims to provide quantitative information for the balance and sustainability of ESs under rapid urbanization, which provides a basis for the development planning of Changchun.

2. Materials and Methods

2.1. Study Area

Changchun (43°26′31″–44°36′29″ N, 125°1′44″–125°48′58″ E), the capital city of Jilin Province, is located in the middle part of the Songnen Plain, covering a total area of 3.1×10^3 km^2 within an urban built-up region (Figure 1). In this area, altitude varies from 67 to 368 m with small undulation in topography. The area has a semi-humid continental monsoon climate, characterized by a cold winter and cool summer with a mean annual precipitation of 570–700 mm and a mean annual temperature of 4.5–4.8 °C. In this region, rainfall occurs primarily in June–August. At the same time, Changchun belongs to the famous Black Soil Belt of northeast China and zonal soil types include mollisol, chernozem, and dark brown soil, in which corn is the dominant crop.

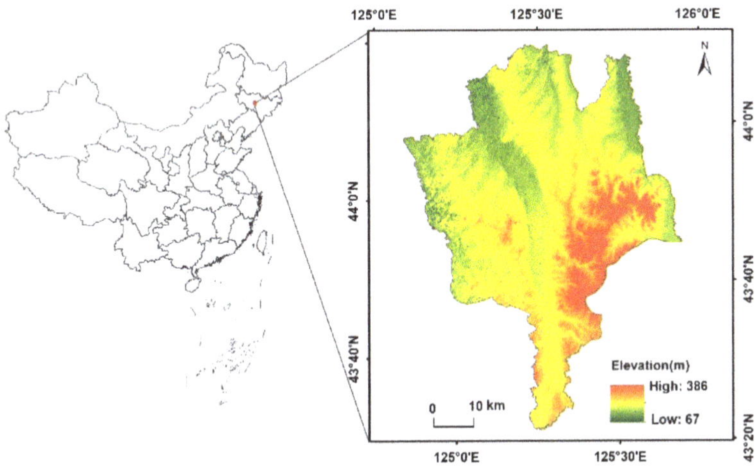

Figure 1. Location of the Changchun, Jilin Province, China.

As a famous old industrial base in China, urbanization of Changchun was once higher than the national average level. With the emergence of abuses in the development of resource-based cities, the economic development of Changchun has stagnated. The proposal and implementation of the Strategy of Revitalizing Northeast China in 2003 propelled Changchun into an economic boom and rapid urbanization, which has inevitably accelerated land cover changes, soil resource loss, the heat island effect, and other environmental issues [21].

2.2. Data Source and Processing

In this study, we collected and used multi-source geographic data to quantitatively assess ES changes during urban sprawl. The main data sources and their descriptions are as the follows (Table 1):

Digital elevation model (DEM) data with a 30 m resolution were derived from the ASTER GDEM product (http://www.jspaceystems.or.jp/ersdac/GDEM/E, accessed on 15 April 2021). Meteorological data in 2000 and 2015 were acquired from the website of the China Meteorological Data Sharing Service System (CMDSSS, http://cdc.cma.gov.cn/, accessed on 15 April 2021). Climate datasets were spatially interpolated to grid data with a 30 m spatial resolution by the Kringing method in ArcGIS 10.3 software before being used in the ES models. Soil type data in vector format with a scale of 500,000 were provided by the Soil and Fertilizer Station of Jilin Province (SFSJP). Night-time light imagery data were downloaded from the website of NOAA/NGDC (https://ngdc.noaa.gov, accessed on 15 April 2021). Night-time light imagery data for different periods arrived from different sensors: data of the Operational Linescan System (OLS) arrived from the Defence Meteorological Satellite Program (DMSP) in 2000 and data from the Visible Infrared Imaging Radiometer (VIIRS) arrived from the NPOESS Preparatory Project (NPP) in 2015; therefore, resampling, mutual correction, and interannual fusion of the two periods of data were carried out according to the methods described in the literature to ensure data consistency [26,27]. Evapotranspiration products from a moderate resolution imaging spectroradiometer (MODIS) with a 500 m resolution were accessed from the NASA website and were used to calculate the water retention services. Land cover datasets in 2000 and 2015, interpreted from Landsat Thematic Mapper (TM) and Operational Land Imager (OLI) data were provided by the Northeast Institute of Geography and Agroecology, Chinese Academy of Sciences (IGA, CAS). Land cover types were divided into seven categories according to Wu [28]: cropland, grassland, forestland, wetland, built-up land, water body, and others. Based on the ground investigation data, the overall accuracy of the land cover classification data was higher than 90% for the two periods.

Table 1. Main data sources used in the study.

Datasets	Description	Resolution	Time	Data Sources
DEM	ASTER GDEM product	Raster, 30 m	2009	http://www.jspacesystems.or.jp/ersdac/GDEM/E, accessed on 15 April 2021 [22]
Climatic data	Daily temperature, Daily rainfall, Daily wind speed, Daily sunshine duration	Point	2000–2015	CMDSSS (http://data.cma.cn), accessed on 15 April 2021 [23]
Soil data	Soil map	Vector, 1:500,000	2016	SFSJP
Remote sensing image	Night data, DMSP/OLS Night data, NPP/VIIRS	Raster, 1000 m Raster, 750 m	2000 2015	https://ngdc.noaa.gov accessed on 15 April 2021 [24]
	Evapotranspiration data, MODIS 16A2	Raster, 500 m	2000, 2015	https://ladsweb.modaps.eosdis.nasa.gov/ accessed on 15 April 2021 [25]
Land cover data	Interpreted data of Landsat TM/OLI Image	Vector	2000, 2015	IGA, CAS

2.3. *Methods*

2.3.1. Urban Expansion Zoning

DMSP/OLS or NPP/VIIRS sensors can detect the low-intensity light emitted by urban lights and even small-scale residential areas and traffic flow at night, which make urban areas different from dark rural areas. Therefore, night-time light imagery data, as an intuitive representation of human activities, have been widely used in modelling urban sprawl [26]. Generally, higher Digital Number (DN) values indicate higher levels of urbanization. According to previous studies, a DN value of 50 was used as a threshold for distinguishing urbanization areas from rural areas. Accordingly, the study area was divided into three areas according to different levels of urbanization [1]: developed urban, developing urban, and rural area. For developed urban area, the DN values of the whole study region were both greater than 50 in 2000 and 2015; for developing urban area, the DN value in 2000 was less than 50, but greater than 50 in 2015; for rural area, the DN values were both less than 50 in 2000 and 2015.

2.3.2. Estimation of ESs

In this study, six vital ESs, including one provision service (crop yield), four regulation services (carbon storage, sandstorm prevention, soil retention, and water retention), and one supporting service (habitat suitability) were selected because they are closely related to the dwellers' health and security, and are sensitive to land cover changes.

The approaches used to quantify each service are briefly described in the following subsections:

Carbon Storage Service

Carbon storage is an important indicator of climate regulation [29]. The InVEST model was used to evaluate carbon stored in carbon pools, including above ground biomass, below ground biomass, soil, and dead organic matter [30]. The total carbon storage is the sum of those four carbon sources:

$$C_i = C_{i(above)} + C_{i(below)} + C_{i(dead)} + C_{i(soil)} \tag{1}$$

where C_i represents the total carbon storage of grid cell i.

Carbon storage was calculated based on the specific ecosystem classification [28]. Because the InVEST model was designed for worldwide application, some functions are general without considering regional heterogeneity [30]. For better results, the biomass carbon stocks per unit area of each land cover type were derived from the local research results and long-term observation data from field stations in northeast China [31].

Soil Retention Service

Soil retention is an important regulating ES, especially in the case of rapid urban sprawl and significant increases in impervious surfaces in the Black Soil Belt. With frequent human interference, regional soil erosion is becoming more and more serious, and the soil retention capacity of the study area has changed greatly. The Revised Universal Soil Loss Equation [32] was used to estimate the soil retention service of different ecosystems. Areas of land with high soil retention present less risk of erosion loss [33]. Soil retention is the difference between potential soil erosion and actual soil erosion; the formula is as follows [34]:

$$A = A_p - A_a = R \times K \times LS \times (1 - C \times P) \tag{2}$$

where A is the annual soil loss (t km^{-2} a^{-1}); A_p is the potential annual soil loss (t km^{-2} a^{-1}); A_a is the actual annual soil loss (t km^{-2} a^{-1}); R indicates the erosion factor of rainfall runoff (MJ mm km^{-2} ha^{-1} a^{-1}), which is calculated according to the daily rainfall; K refers to the soil erosion factor (t h JM^{-1} mm^{-1}); LS denotes the slope length and gradient, representing the terrain factor; C indicates the dimensionless vegetation cover factor, and P represents the soil retention measure.

All factors were spatialized into a grid with a 30 m resolution in ArcGIS 10.3 software and grid calculations were performed to obtain the annual soil loss for each unit.

Water Retention Service

Water retention is one of the most critical services of the ecosystem, which closely connects the growing population with the ecosystem [34]. The amount of water retention for each pixel on the landscape was calculated using the water balance equation [35]. The model is expressed as follows [35]:

$$WR = \sum_{i=1}^{j} (P_i - R_i - AET_i) \times A_i \tag{3}$$

$$R_i = P_i \times \alpha \tag{4}$$

where WR is the total amount of water retention, P_i means annual precipitation (mm), R_i indicates the surface runoff (mm), AET_i represents annual actual evapotranspiration (mm), A_i is area (km^2) of the ecosystems defined by land cover types, α is average surface runoff coefficient collected from the published research results in this area [36], i is the index of ecosystem types, and j denotes the total number of ecosystem types. Precipitation data were calculated using meteorological data from CMDSSS (http://data.cma.cn, accessed on 15 April 2021). The evapotranspiration data were calculated using MODIS products and meteorological records [35].

Habitat Suitability

Habitat suitability plays a crucial role in biological diffusion and biodiversity conservation, representing the ecosystems' ability to provide suitable conditions for the sustainable survival of individuals and populations [37]. The habitat suitability index was used to estimate the habitat and vegetation supporting ecosystems across the whole landscape, which considered the relative quality of the ecosystem type, the availability of the water source, the sensitivity to disturbing factors, and the abundance of food. The equation was as follows [38]:

$$HSI = \sum_{1}^{i} w_i f_i \tag{5}$$

where *HSI* represents the total value of the comprehensive index of habitat suitability in the study year, w_i is the weight for the *i*th factor assigned by experts, f_i is the standardized value of the *i*th factor involved in the assessment of habitat suitability, and *i* is the number of evaluated factors.

Spatialization and weighted summation for all factors were performed utilizing ArcGIS 10.3 software.

Sandstorm Prevention

Sandstorm prevention of the ecosystem was calculated using the revised wind erosion equation (RWEQ) [39]. The RWEQ is an empirical model for estimating long-term soil loss caused by wind erosion [34]. In this model, climate, topography, vegetation, and soil properties were considered to quantify wind erosion [40]. Potential and actual wind erosion intensity were estimated, and the difference between them was used as the sand fixation capacity of the ecosystem to assess the function of wind and sand fixation in the ecosystem. The formulas involved are as follows [39]:

$$\triangle Q = Q_0 - Q_v \tag{6}$$

$$Q(x) = Q_{max}\left\{1 - exp\left[-(x/s)^2\right]\right\} \tag{7}$$

$$S = 150.71 \cdot (WF \times EF \times SCF \times K' \times C)^{-0.3711} \tag{8}$$

$$Q_{max} = 109.8 \cdot (WF \times EF \times SCF \times K' \times C) \tag{9}$$

$$S = 150.71 \cdot [WF \times EF \times SCF \times K' \times (1 - C)]^{-0.3711} \tag{10}$$

where ΔQ means the amount of sand fixation (t km^{-2} a^{-1}), Q_0 refers to the potential soil erosion without vegetation cover (t km^{-2} a^{-1}), and Q_v indicates the actual soil erosion considering the current land cover and management conditions (t km^{-2} a^{-1}). $Q(x)$ represents the amount of soil transported by wind at point x, Q_{max} indicates the maximum sand transport capacity by wind, and *S* describes critical field length. *WF*, *EF*, *SCF*, K', and *C* represent weather factor, soil erodible fraction, soil crust factor, surface roughness factor, and vegetation factor, respectively.

Crop Yield

Located in the typical Black Soil Belt of Northeast China, the food supply of the study area is a crucial service for regional food security. Crop yield was selected to estimate food provisioning services of Changchun. Crop yields data were obtained from the statistical yearbook of Changchun in 2001 and 2016.

2.3.3. Relationship between ESs

According to previous studies on the relationship between ESs [1,37,41], the changes of ES in divided urban area were analysed and compared to reveal the influences of urbanization on the relationships between ES changes.

There are three relationship types between ESs: synergy, trade-off, and neutral. Synergy represents a win–win or lose–lose situation of two ESs with a consistent trend of increases or losses; trade-off indicates a win-lose or lose-win situation in which one ES rises at the expense of the other; and neutral means that a change in one ES will not cause variation in another [1]. In this study, 754 random points (432 in rural area, 209 in developing area, and 113 in developed areas) for each ES layer, both in 2000 and 2015, were extracted using ArcGIS 10.3 software. Before comparing the 2000 and 2015 ecosystem services, all data were standardized using the following equations:

$$S_{v2000} = \frac{v_{2000} - \min(v_{2000}, v_{2015})}{\max(v_{2000}, v_{2015}) - \min(v_{2000}, v_{2015})} \quad (11)$$

$$S_{v2015} = \frac{v_{2015} - \min(v_{2000}, v_{2015})}{\max(v_{2000}, v_{2015}) - \min(v_{2000}, v_{2015})} \quad (12)$$

where S_{v2000} and S_{v2015} are the standardized values for each ES in 2000 and 2015, respectively, and v_{2000} and v_{2015} are the ES values in 2000 and 2015, respectively.

3. Results

3.1. Urban Sprawl of Changchun during 2000–2015

Based on night light data, the scope of urban and industrial land in the study area in 2000 and 2015 was extracted and the dynamic of urban expansion was obtained via the spatial analysis module of ArcGIS 10.3. The results showed that Changchun has undergone considerable urban expansion since 2000 (Figure 2). The area of urban and industrial land was 25, 2067 ha in 2000, and 487,064 ha in 2015, respectively, an increase of 234,997 ha in the past 16 years with a 14,687 ha yearly increase (8.45%). The average DN of night light data for the study area increased from 12.21 in 2000 to 22.55 in 2015. The mean DN was 4.21, 48.76, and 60.31 in rural area, developing urban area, and developed urban area in 2000, respectively. The average DN increased by 2.52 in developed urban area, and by 10.74 in developing urban area in the last 16 years. The average DN in rural area changed the most, increasing by 13.64 from 2000 to 2015 (Table 2).

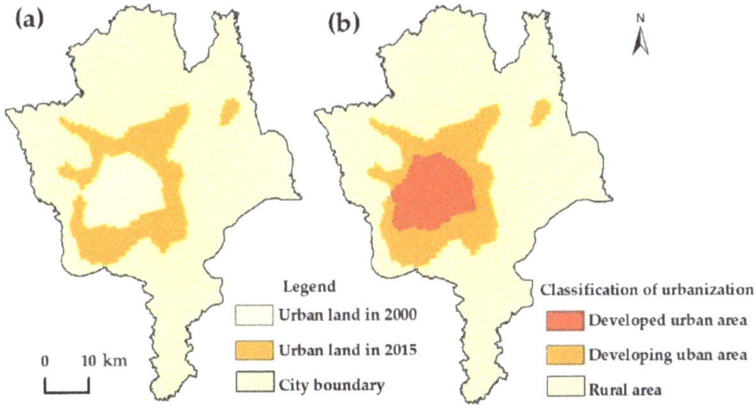

Figure 2. Urbanization (**a**) in 2000–2015 and (**b**) classification of urbanization in study area.

Table 2. The average DN of night light data in different areas in 2000 and 2015.

Time	Mean	Rural Area	Developing Urban Area	Developed Urban Area
2000	12.21	4.21	48.76	60.31
2015	22.55	17.85	59.50	62.83

3.2. Changes of ESs in Changchun

3.2.1. Spatial–Temporal Changes for Overall ESs

The spatial distribution of carbon storage, soil retention, and habitat suitability services presented similar patterns with high values in eastern areas, where the slope gradient was relatively high, and the dominant land cover type was forestland (Figure 3). The values range from 0 to 37.1 t, 0 to 4.2 $\times$ 10^3 t $km^{-2}a^{-1}$, and 0 to 60.2 for these three ecosystem services, respectively. The sandstorm prevention deceased from west to east, and the areas with high values of water retention were concentrated in the central parts (Figure 3). From 2000 to 2015, the three ESs-carbon storage, soil retention, and water retention-increased by 2.01%, 124%, and 28.03% in the study area, while habitat suitability and sandstorm prevention services decreased by 3.55%, and 1.64%, respectively (Figure 4). In the study area, the area of cropland was reduced from 1720.4 km^2 to 1560.4 km^2 by 9.3%, but the crop yield service increased from 668.5 $\times$ 10^6 t to 753.4 $\times$ 10^6 t by 12.87% (Figure 4).

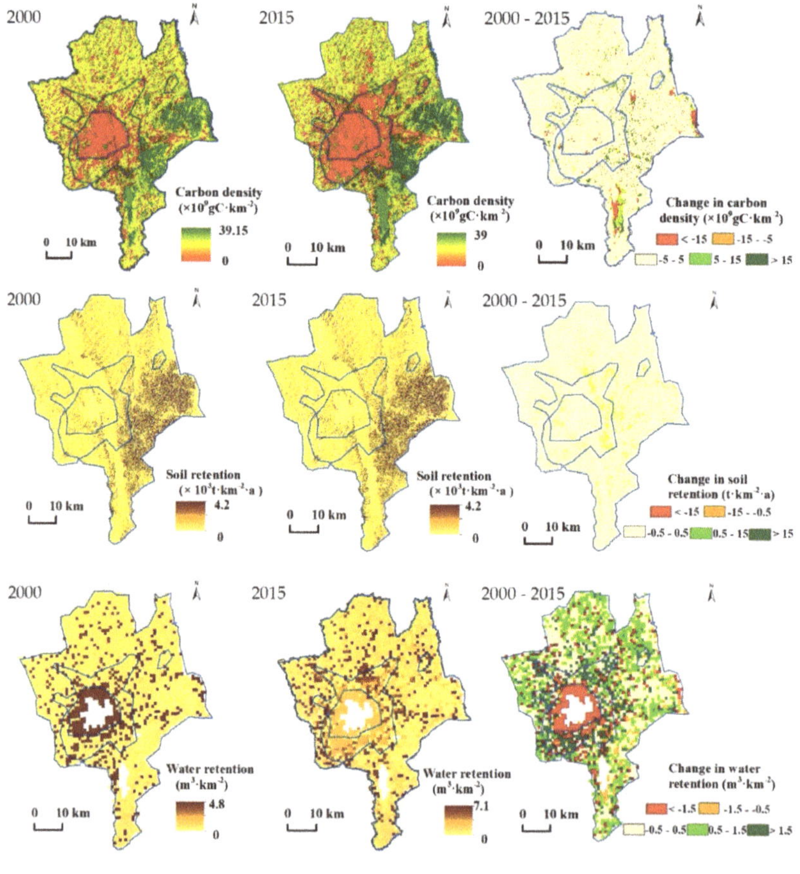

Figure 3. *Cont.*

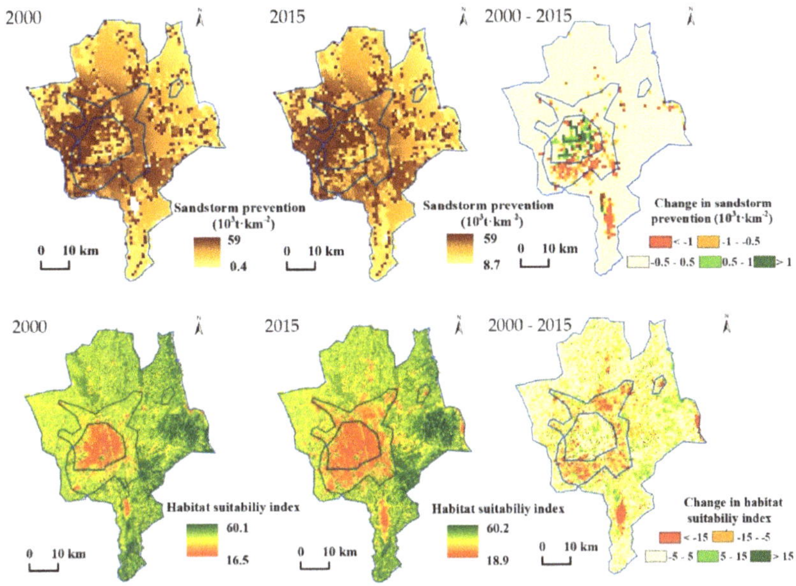

Figure 3. Spatial Distribution of ESs in 2000, 2015, and changes of Changchun.

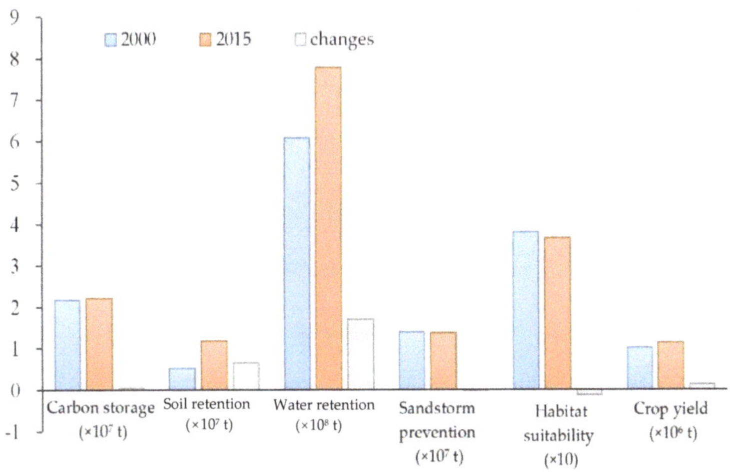

Figure 4. ESs in 2000 and 2015.

The relative service capacities are shown using radar plots based on standardized values (Figure 5). Each axis of the plot indicates a different ES; the outermost point on the axes means the highest level of ES, and the provisioning service decreases towards the centre. The symmetry of the plot represents relative balance of the ESs; consequently, the larger and more symmetrical they are, the higher the overall potential benefits of ESs [1]. It can be seen that six service capacities in the study area are extremely unbalanced. Habitat suitability and crop yield are relatively stronger than the others, while the service capacity of soil retention is the weakest.

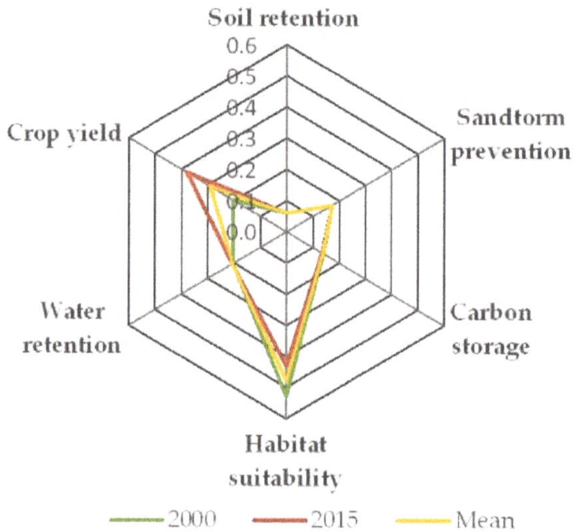

Figure 5. Radar maps of the standardized ESs.

3.2.2. Relationships among ES Changes

Interactions between ESs were analysed to evaluate the differential changes in ESs in the process of urbanization and to detect the urbanization problems presented in Changchun (Table 3). Habitat suitability significantly corresponded with carbon storage, soil retention, and water retention. Synergies between habitat suitability and carbon storage, and habitat suitability and soil retention were detected in the three areas, that is, habitat suitability deteriorated with the decrease in carbon storage and soil retention. However, the correlation coefficient decreased from developed urban area to rural area in terms of carbon storage and increased from a developed urban area to a rural area for soil retention. A trade-off between habitat suitability and water retention was detected in the three zones, and the correlation coefficient of the rural area was the highest.

Table 3. Relationships between pairs of ESs.

		Carbon Storage	Habitat Suitability	Soil Retention	Sandstorm Prevention	Water Retention	Crop Yield
Carbon storage	developed	1					
	developing	1					
	rural	1					
Habitat suitability	developed	0.528 **	1				
	developing	0.495 **	1				
	rural	0.430 **	1				
Soil retention	developed	−0.023	0.188 *	1			
	developing	0.191 **	0.198 **	1			
	rural	0.306 **	0.247 **	1			
Sandstorm prevention	developed	−0.009	−0.010	−0.070	1		
	developing	−0.002	−0.015	−0.102	1		
	rural	0.006	−0.036	−0.028	1		
Water retention	developed	−0.058	−0.063	0.013	−0.015	1	
	developing	−0.088	−0.181 **	−0.061	−0.043	1	
	rural	−0.121 **	−0.260 **	−0.043	0.055	1	
Crop yield	developed	−0.114	−0.339 **	−0.211 *	−0.025	0.105	1
	developing	0.024	0.051	0.055	−0.036	−0.108	1
	rural	−0.172 **	0.176 **	−0.104 **	−0.015	−0.051	1

Note: * indicates significance at the 5% level, ** indicates significance at the 1% level.

The interactions between crop yield and carbon storage, habitat suitability, and soil retention were more complex. There was only a weak trade-off between crop yield and carbon storage in rural areas, while the correlation between the other two areas was not significant. For crop yield and habitat suitability, a higher trade-off in developed urban areas but a weaker trade-off in rural areas can be seen.

The correlation coefficients of sandstorm prevention with all of the other five ESs were insignificant in three areas. Insignificant results were also detected between water retention and soil retention, and water retention and crop yield.

3.3. Impact of Urbanization on ESs

3.3.1. ES Changes in Different Urbanization Areas

Average values of ESs in the three urban zones were calculated to compare the impact of rapid urban sprawl on ESs (Table 4). It can be seen that the mean values of soil retention, carbon density, habitat suitability, and crop yield in rural areas are significantly higher than those in the other two areas, both in 2000 and 2015. All of the mean ES values for urban areas are the lowest among the three areas, except for water retention and sandstorm prevention in 2015. For developed and developing urban areas, carbon density and habitat suitability decreased, while the other four ecological services improved slightly from 2000 to 2015. For rural area, only sandstorm prevention decreased from 6.0×10^3 t·km^{-2} to 4.44×10^3 t·km^{-2} during the study period.

We can also learn from radar maps that the relative balance of all of the services in three areas is quite different. For developed urban area (Figure 6a), habitat suitability and sandstorm prevention are stronger than other services and the balance between all services decreased from 2000 to 2015. For developing urban area (Figure 6b), habitat suitability is significantly higher than other ESs in 2000 but decreased to a large extent in 2015. For rural area (Figure 6c), crop yield increased significantly from 2000 to 2015, while other ESs hardly changed. The relative balance of all of the services in developing urban area is the most symmetrical among the three areas, but the overall potential ES benefit is not higher (Figure 6d).

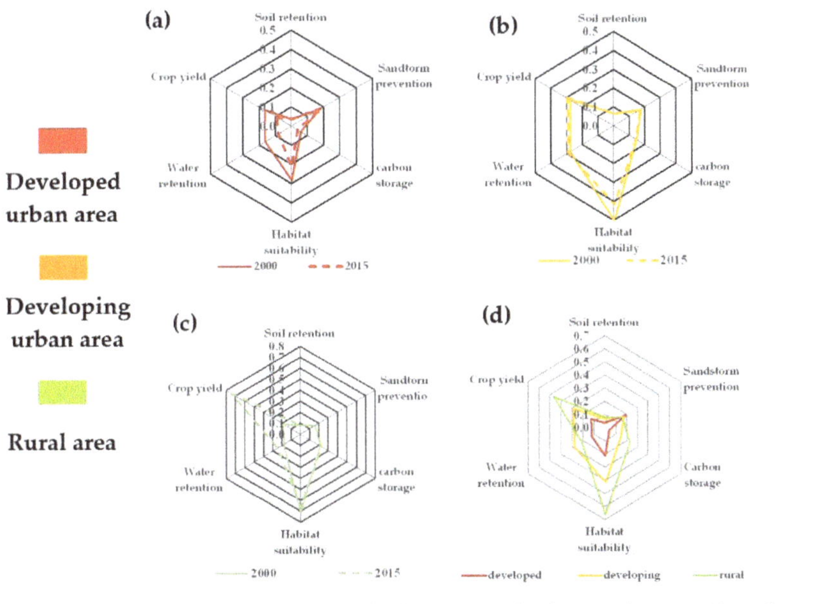

Figure 6. Radar maps of ESs in (**a**) developed urban area, (**b**) developing urban area, (**c**) rural area, and (**d**) the mean value of ESs in three areas.

Table 4. Mean values of ESs in the three urban areas.

	Soil Retention (10^3 t·km^{-2})		Sandstorm Prevention (10^3 t·km^{-2})		Carbon Density (10^9 gC·km^{-2})		HabitatSuitability		Water Retention (10^5 t·km^{-2})		Crop Yield (10^3 kg·ha^{-1})	
	2000	2015	2000	2015	2000	2015	2000	2015	2000	2015	2000	2015
Developed	0.68	1.75	4.10	4.94	1.83	1.52	28.64	28.30	3.99	4.58	1.51	2.28
Developing	1.33	3.28	4.08	4.50	4.30	4.19	40.38	36.17	2.19	2.98	3.79	4.16
Rural	1.88	4.14	6.0	4.44	7.29	7.53	45.21	45.71	1.66	2.15	5.97	6.79

3.3.2. Correlation Analysis between Urbanization and ESs

A total of 1500 sample points were randomly extracted, including 120 in developed urban areas, 221 in developing urban areas, and 1159 in rural areas. Taking nightlight satellite measurements as a proxy for urbanization, correlation analysis were analysed between urbanization and ecosystem services in 2000 and 2015, respectively. Habitat suitability were negatively correlated with urbanization in all three areas (Table 5). The relationship between water retention and urbanization was negative in developed urban areas, while it was positive in developing urban areas and rural areas. The relationship between crop yield and urbanization was negative in developed and developing urban areas, while it was insignificant in rural areas. Due to the extensiveness and continued growth of water-impervious surfaces at the loss of forest and crop land in developing urban areas, urbanization is negatively correlated with carbon density in 2015. There was no significant correlation between soil conservation, sandstorm prevention and urbanization in all areas.

Table 5. The correlation coefficients of ESs with urbanization in different urban areas in 2000 and 2015.

	Developed Area		Developing Area		Rural Area	
	2000	2015	2000	2015	2000	2015
Soil retention	−0.067	−0.081 *	−0.036	−0.117	−0.064 *	−0.077 *
Sandstorm prevention	−0.021	0.071	0.113	0.112	0.04	0.067 *
Carbon density	−0.081	−0.014	−0.104	−0.207 **	−0.071 *	−0.076 *
Habitat suitability	−0.373 **	−0.309 **	−0.397 **	−0.467 **	−0.17 **	−0.233 **
Wate retention	−0.252 **	−0.428 **	0.218 **	0.08	0.126 **	0.248 **
Crop yield	−0.359 **	−0.314 **	−0.172 *	−0.147 *	0.064 *	−0.054

Note: * indicates significance at the 5% level, ** indicates significance at the 1% level.

4. Discussion

4.1. *Typical Impact of Urban Expansion on ESs in Changchun*

Many studies have shown that changes in land cover driven by urbanization contribute greatly to spatial and temporal changes in ESs [12,33,41,42]. The influences brought about by urbanization not only reflected the urban–rural gradient changes, but also presented a spatial spill over effect [36]. In this study, we also found that rapid and extensive changes in land cover from rural to urban area in Changchun led to a remarkable influence on the spatial distribution of ESs. Some ESs, including carbon storage, soil retention, habitat suitability, and crop yield, increased with the distance from developed urban areas, which was consistent with the previous research results with regard to gradient change along the urban–rural gradient [1]. We found that water retention represented the opposite trend: the mean value in developed urban area was much higher than that in rural area. This is mainly because the evapotranspiration of impervious surface is small, which leads to larger water retention values.

Due to different environmental and land cover patterns in different areas, the dynamic trends of ESs are quite different. According to previous studies, an increased rate of water-impervious surfaces in developing urban areas is the highest compared to the other two types of areas; consequently, the greatest reduction in regulation services as well as food production services occurs in developing urban areas [1,38]. In our study, the mean value of carbon storage and habitat suitability declined, while other services increased in developing urban areas.

Crop yield increased in all three areas, which is consistent with the results obtained by Sun et al. in the Atlanta metropolitan area [16]. Although urbanization took up a large amount of cropland, especially in developing urban areas, the progress in agricultural science and technology and the improvement of farming management greatly increased crop yield per unit area, which made the overall level and average level of the crop yield in the three areas improve. In addition, the conversion of some dry farmland into paddy fields in rural areas also promoted the largest increase in crop yield among three areas. However, in the study of Qi et al., it was found that the food supply in this area increased first and then decreased [43]. Cropland protection cannot be ignored.

Urban sprawl declined the area of cropland, as well as grassland. Accordingly, the average amount of carbon storage in developing urban areas and developed urban areas reduced, while that in rural areas increased due to the implementation of the Grain to Green Program (the area of forest land increased by 13.90%, and water body increased by 128.95%), which offset the decrease in carbon storage due to the increase in impervious surface (Figure 7). It is worth mentioning that because of burgeoning spiritual demands, the three regions pay more attention to the environment and aesthetics. As a result, there was an increase in green infrastructure from 2000 to 2015, and the forestland area in the three regions increased by 13.90%, 88.9%, and 29.1%, compared with 2000, respectively. However, due to the rapid growth of artificial surface (the rate of increase was 26.55%, 91.9%, and 10.43% for rural area, developing urban area, and developed urban area, respectively, from 2000–2015), developing urban area showed the largest decline in their habitat suitability index from 40.38 to 36.17 (Figure 7). In rural area, though rural residential land expanded significantly due to urban sprawl, habitat suitability still improved a small amount because forestland and water body provide better habitat suitability services than cropland. Both water retention and soil retention services presented an inclining trend in all three areas and the largest increase in water retention occurred in developed urban area because of these areas saw the largest increase in forestland and bodies of water. Furthermore, the reduced evapotranspiration due to the large increase in impervious areas also resulted in the largest increase in water retention occurring in developed urban area. Since the increasing forestland as well as the high- density buildings reduced the wind speed and surface sand, the largest increase in sandstorm prevention services occurred in developed urban areas.

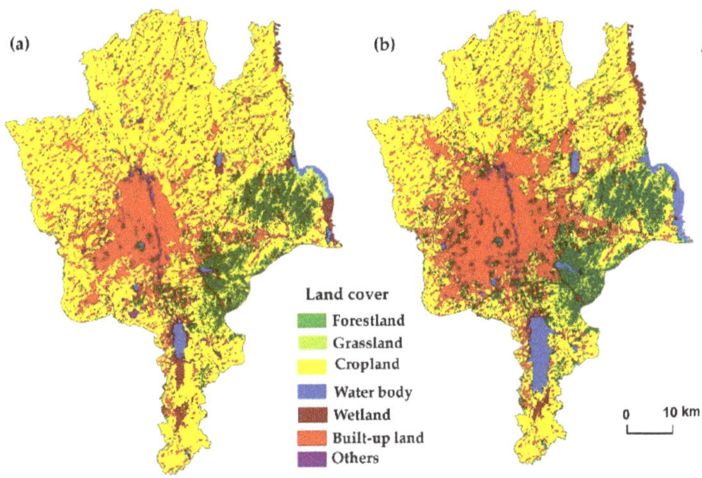

Figure 7. Land cover of Changchun in (**a**) 2005, and in (**b**) 2015.

Furthermore, correlation analyses showed that the urbanization index and habitant suitability, and crop yield had a negative correlation in Changchun, which are consistent with previous studies [9,41] and indicates that urbanization exerts significant influence on ecosystem services. In our study, water retention is significantly negatively correlated in urbanized areas, but positively correlated in developing urban areas and rural areas. However, the study of sun et al. in Guangdong-Hong Kong-Macao "Megacity" found that water retention was negatively correlated only in urban areas, but insignificant in rural areas [9]. This is because in addition to land use, other factors such as topography and meteorology, likely contributed to the relationship between urbanization and service capacity. In developed urban area, the terrain is so flat and high-density artificial surface caused large runoff and consequent reduction of water retention. While in developing urban area and rural area, water retention is more impacted by slope gradient and heterogeneity of evapotranspiration and precipitation. Carbon density only significantly negatively correlated with urbanization in developing areas, but not significantly in developed area and rural area. It shows that the large increase of urban land has a significant impact on carbon storage in the developing area than others.

4.2. *Effects of Urbanization on the Relationships among ESs*

Correlation analyses of multiple ESs could provide instructive information for the balanced development of ESs overall. The factors leading to changes in ES relationships varied spatiotemporally due to the different characteristics of the natural environment and the degree of human disturbance. This study mainly focused on the changes in ES relationships caused by land cover transformation.

Studies have shown that there are general trade-off relationships between provision services and regulation services, such as crop yield and carbon storage, water retention, and soil retention [44–46]. However, in our study, crop yield presented a more complex relationship with other ESs in different urbanization areas. Due to the large proportion of cropland (the proportion was 67.01% in 2000 and 53.08% in 2015), cropland plays a vital role as a carbon sink in rural area. Although there was an increase in forestland areas, carbon storage had barely changed due to the decrease in large area of cropland. However, impacted by advances in agricultural science and technology, and agricultural management factors, crop yield showed rapidly increasing trend, and became the most prominent ES in rural area (Figure 6), which caused a weak trade-off between crop yield and carbon storage in rural area. Large-scale expansion of urban areas leads to loss of large areas of suitable habitats, resulting in the largest reduction in habitat suitability in developing urban areas in this study. These two diverse trends resulted in a high trade-off between crop yield and habitat suitability in this region.

In addition, we found that habitat suitability significantly corresponded with carbon storage, soil retention, and water retention. In contrast, land cover changes driven by urbanization transformed most land surfaces into water-impervious surfaces, which raised water retention but reduced the soil retention and carbon storage. Therefore, habitat suitability has a trade-off relationship with water retention, but a synergistic relationship with soil retention and carbon storage.

4.3. *Implications for Urban Planning*

Due to rapid urban sprawl, coordinating development activities without destroying ecological sustainability has become a preliminary and vital issue in relation to urban management. Many studies have tried to solve this situation through delimiting ecological protection areas with high ESs [47]. However, the benefit of urbanization often promotes development activities in ecological protection zones, which lead to a failure to protect of these areas. The "Guiding Opinions on the Overall Delimitation and Implementation of Three Control Lines in National Territory Spatial Planning" issued by the Office of the State Council of China in 2019 emphasized the importance of the demarcation of the ecological protection red line, permanent basic cropland, and the urban development boundary [48]. Therefore, the contradiction between urban development and ecological protection can be mitigated by strict constraints on development activities in delineated areas. Taking ESs and urbanization as a whole system and analysing their interactions at local scale can provide more constructive suggestions for ecological protection in the process of urbanization. According to urbanization and its effect on ESs, our detailed recommendations for local urban development planning are as follows:

First, Changchun, as the central city of the Harbin–Changchun Urban Agglomeration, is also the region where black soil is concentrated. The trade-off between food security and economic development is always the core problem of regional development. From 2000 to 2015, due to urbanization, 156.91 km^2 of cropland was transformed into urban built-up land, accounting for 8.2% of the total amount of cropland in 2000. Despite the improvement in agricultural science, technology, and management, crop yield showed an increasing trend in each of the three areas (Table 3). However, this benefit can only be maintained in a limited range. In the long run, the protection of cropland area is necessary. Therefore, urban sprawl should give full consideration to the red line of basic agriculture protection, as well as the sustainable use of black soil, so as to achieve a win–win situation of food security and economic development.

Second, we found that the increase in soil conservation services in the three areas and carbon storage in rural area due to Changchun's rapid urbanization benefited from the increase in forestland and water body with high ESs. Therefore, in urban landscape plan, we should pay more attention to the construction of green infrastructure, especially in developing urban areas where the area of water- impervious surfaces increases extensively, so as to reverse the phenomenon of the continuous loss of ESs in the process of urban sprawl.

Finally, it has been proven that the allocations and patterns of land cover have a significant impact on ESs. However, commonly, human activities changed land use types and tend to generate the most profit [41]. Therefore, it is critical that policy makers and planners should understand how to design the land use structure in order to promote sustainable urban development in a more reasonable and beneficial way.

5. Conclusions

Taking Changchun as study area, we studied the impact of rapid urbanization on changes in ESs in inland agricultural areas. We assessed six ESs and their interrelation in different urbanization areas. The urbanization changes and their impact on these ESs were analysed by integrating remote sensing and geospatial analysis methods.

Changchun experienced rapid urbanization between 2000 and 2015. At the same time, the ESs of Changchun changed significantly over the past 16 years, with increases in carbon storage, soil retention, water retention, and crop yield, and decreases in habitat suitability and sandstorm prevention service. Urbanization has a significant impact on ESs, and ESs change regularly in different urbanization areas. The relationships between ESs were not only simple trade-offs or synergies, but presented more complex issues, and the correlation characteristics were different in different urbanization areas. Habitant suitability and crop yield have a significantly negative relationship with urbanization. The increase in forestland and water with high ESs slowed down the decrease in ESs due to urbanization. In view of the impact of urbanization on ecosystem services in Changchun, we put forward three suggestions to avoid or minimize the ecological problems caused by urbanization. In this study, only the impact of land cover changes on ecosystem services was considered. The comprehensive impact of climate, terrain, and other factors on ESs should be considered in future work.

Author Contributions: Conceptualization, X.L.; methodology, X.L.; formal analysis, Y.R.; investigation, R.Y.; data curation, Z.S. and Z.X.; writing—original draft preparation, X.L.; writing—review and editing, X.L.; visualization, G.S. All authors have read and agreed to the published version of the manuscript.

Funding: This work was supported by the National Natural Science Foundation of China (No. 42171328) and the Science and Technology Development Program of Jilin Province (Nos. 20200201048JC and 20200301014RQ).

Data Availability Statement: The data are not publicly available for privacy reasons.

Conflicts of Interest: The authors declare no conflict of interest.

References

1. Li, B.J.; Chen, D.X.; Wu, S.H.; Zhou, S.L.; Wang, T.; Chen, H. Spatio-temporal assessment of urbanization impact on ecosystem services: Case study of Nanjing City, China. *Ecol. Indic.* **2016**, *71*, 416–427. [CrossRef]
2. Kuang, B.; Lu, X.H.; Han, J.; Fan, X.Y.; Zuo, Z. How urbanization influence urban land consumption intensity: Evidence from China. *Habitat Int.* **2020**, *100*, 102103. [CrossRef]
3. Gu, C.L.; Hu, L.Q.; Cook, I.G. China's urbanization in 1949–2015: Processes and driving forces. *Chin. Geogr. Sci.* **2017**, *27*, 847–859. [CrossRef]
4. The Central People's Government of the People's Republic of China. Available online: http://www.gov.cn/shuju//hgjjyxqk/detail.html?q=3 (accessed on 25 October 2021).
5. The World Bank. Available online: https://www.worldbank.org/en/topic/urbandevelopment/overview (accessed on 25 October 2021).
6. The United Nations. Available online: https://www.un.org/en/climatechange/climate-solutions/cities-pollution (accessed on 25 October 2021).
7. Ding, C.R.; Zhao, X.S. Land market, land development and urban spatial structure in Beijing. *Land Use Policy* **2014**, *40*, 83–90. [CrossRef]
8. Costanza, R.; d'Arge, R.; de Groot, R.; Farber, S.; Grasso, M.; Hannon, B.; Limburg, K.; Naeem, S.; O'Neill, R.V.; Paruelo, J.; et al. The value of the world's ecosystem service and natural capital. *Nature* **1997**, *387*, 253–260. [CrossRef]
9. Wang, W.J.; Wu, T.; Li, Y.Z.; Xie, S.L.; Han, B.L.; Zheng, H.; Ouyang, Z.Y. Urbanization impacts on natural habitat and ecosystem services in the Guangdong-Hong Kong-Macao "Megacity". *Sustainability* **2020**, *12*, 6675. [CrossRef]
10. Wu, J.G. Landscape sustainability science: Ecosystem services and human well-being in changing landscapes. *Landsc. Ecol.* **2013**, *28*, 999–1023. [CrossRef]
11. Athukorala, D.; Estoque, R.C.; Murayama, Y.J.; Matsushita, B. Impacts of urbanization on the Muthurajawela Marsh and Negombo Lagoon, Sri Lanka: Implications for landscape planning towards a sustainable urban wetland ecosystem. *Remote Sens.* **2021**, *13*, 316. [CrossRef]
12. Deng, C.X.; Liu, J.Y.; Nie, X.D.; Li, Z.W.; Liu, Y.J.; Xiao, H.B.; Hu, X.Q.; Wang, L.X.; Zhang, Y.T.; Zhang, G.Y.; et al. How trade-offs between ecological construction and urbanization expansion affect ecosystem services. *Ecol. Indic.* **2021**, *122*, 107253. [CrossRef]
13. Larondelle, N.; Haase, D. Urban ecosystem services assessment along a rural–urban gradient: Across-analysis of European cities. *Ecol. Indic.* **2013**, *29*, 179–190. [CrossRef]
14. Mach, M.E.; Martone, R.G.; Chan, K.M.A. Human impacts and ecosystem services: Insufficient research for trade-off evaluation. *Ecosyst. Serv.* **2015**, *16*, 112–120. [CrossRef]
15. Amundson, R.; Berhe, A.A.; Hopmans, J.W.; Olson, C.; Sztein, A.E.; Sparks, D.L. Soil and human security in the 21st century. *Science* **2015**, *348*, 1261071. [CrossRef]
16. Wan, L.L.; Ye, X.Y.; Lee, J.; Lu, X.Q.; Zheng, L.; Wu, K.Y. Effects of urbanization on ecosystem service values in a mineral resource-based city. *Habitat Int.* **2015**, *46*, 54–63. [CrossRef]
17. Sun, X.S.; Crittenden, J.C.; Li, F.; Lu, Z.M.; Dou, X.L. Urban expansion simulation and the spatio-temporal changes of ecosystem services, a case study in Atlanta Metropolitan area, USA. *Sci. Total Environ.* **2018**, *622*, 974–987. [CrossRef] [PubMed]
18. Aguilera, M.A.; Tapia, J.; Gallardo, C.; Nunez, P.; Varas-Belemmi, K. Loss of coastal ecosystem spatial connectivity and services by urbanization: Natural-to-urban integration for bay management. *J. Environ. Manag.* **2020**, *276*, 111297. [CrossRef]
19. María, D.A.; Juan, M.B.; Javier, G.S. Relationships between coastal urbanization and ecosystems in Spain. *Cities* **2017**, *68*, 8–17.
20. García-Nieto, A.P.; Geijzendorffer, I.R.; Baró, F.; Roche, P.K.; Bondeau, A.; Cramer, W. Impacts of urbanization around Mediterranean cities: Changes in ecosystem service supply. *Ecol. Indic.* **2018**, *91*, 589–606. [CrossRef]
21. Li, X.Y.; Yang, L.M.; Ren, Y.X.; Li, H.Y.; Wang, Z.M. Impacts of urban sprawl on soil resources in the Changchun–Jilin Economic Zone, China, 2000–2015. *Int. J. Environ. Res. Public Health* **2018**, *15*, 1186. [CrossRef] [PubMed]
22. ASTER GDEM Project. Available online: http://www.jspacesystems.or.jp/ersdac/GDEM/E (accessed on 15 April 2021).
23. National Meteorological Science Data Center. Available online: http://data.cma.cn (accessed on 15 April 2021).
24. National Oceanic and Atmospheric Administration. Available online: https://ngdc.noaa.gov (accessed on 15 April 2021).
25. National Aeronautics and Space Administration. Available online: https://ladsweb.modaps.eosdis.nasa.gov (accessed on 15 April 2021).
26. Li, X.P.; Gong, L. Calibration and fitting of DMSP/OLS and VIIRS/DNB night light images. *Bull. Surv. Mapp.* **2019**, *7*, 138–146. (In Chinese)
27. Zhang, B.F.; Miao, C.H.; Song, Y.N.; Wang, J.J. Correction of DMSP/OLS Stable Night Light Images in China. *J. Geo-Inf. Sci.* **2020**, *8*, 1679–1691. (In Chinese)

28. Wu, B.F. *Land Cover of China*; Science Press: Beijing, China, 2017.
29. Xu, Q.; Yang, R.; Dong, Y.X.; Liu, Y.X.; Qiu, L.R. The influence of rapid urbanization and land use changes on terrestrial carbon sources/sinks in Guangzhou, China. *Ecol. Indic.* **2016**, *70*, 304–316. [CrossRef]
30. Kumari, G.; Jian, Y.; Lei, F. Assessing ecosystem services from the forestry-based reclamation of surface mined areas in the North fork of the Kentucky river watershed. *Forests* **2018**, *9*, 652–674.
31. Wang, Z.M.; Mao, D.H.; Li, L.; Jia, M.M.; Dong, Z.Y.; Miao, Z.H.; Ren, C.Y.; Song, C.C. Quantifying changes in multiple ecosystem services during 1992–2012 in the Sanjiang Plain of China. *Sci. Total Environ.* **2015**, *514*, 119–130. [CrossRef]
32. Renard, K.G.; Foster, G.R.; Weesies, G.A.; Porter, J.P. RUSLE: Revised universal soil loss equation. *J. Soil Water Conserv.* **1991**, m.
33. Sajedipour, S.; Zarei, H.; Oryan, S. Estimation of environmental water requirements via an ecological approach: A case study of Bakhtegan Lake, Iran. *Ecol. Eng.* **2017**, *100*, 246–255. [CrossRef]
34. Jiang, C.; Nath, R.; Labzovskii, L.; Wang, D.W. Integrating ecosystem services into effectiveness assessment of ecological restoration program in northern China's arid areas: Insights from the Beijing-Tianjin Sandstorm Source Region. *Land Use Policy* **2018**, *75*, 201–214. [CrossRef]
35. Yang, Y.Y.; Zheng, H.; Kong, L.Q.; Huang, B.B.; Xu, W.H.; Ouyang, Z.Y. Mapping ecosystem services bundles to detect high-and low-value ecosystem services areas for land use management. *J. Clean. Prod.* **2019**, *225*, 11–17. [CrossRef]
36. Xiang, H.X.; Wang, Z.M.; Mao, D.H.; Zhang, J.; Zhao, D.; Zeng, Y.; Wu, B.F. Surface mining caused multiple ecosystem services losses in China. *J. Environ. Manag.* **2021**, *290*, 112618. [CrossRef] [PubMed]
37. Jackson, B.; Pagella, T.; Sinclair, F.; Orellana, B.; Henshaw, A.; Reynolds, B.; Mcintyre, N.; Wheater, H.; Eycott, A. Polyscape: A GIS mapping framework providing efficient and spatially explicit landscape-scale valuation of multiple ecosystem services. *Landsc. Urban Plann.* **2013**, *112*, 74–88. [CrossRef]
38. Sun, X.; Lu, Z.M.; Li, F.; Crittenden, J.C. Analyzing spatio-temporal changes and trade-offs to support the supply of multiple ecosystem services in Beijing, China. *Ecol. Indic.* **2018**, *94*, 117–129. [CrossRef]
39. Jiang, C.; Wang, F.; Zhang, H.Y.; Dong, X.L. Quantifying changes in multiple ecosystem services during 2000–2012 on the Loess Plateau, China, as a result of climate variability and ecological restoration. *Ecol. Eng.* **2016**, *97*, 258–271. [CrossRef]
40. Du, H.Q.; Xue, X.; Wang, T.; Deng, X.H. Assessment of wind-erosion risk in the watershed of the Ningxia-Inner Mongolia Reach of the Yellow River, northern China. *Aeolian Res.* **2015**, *17*, 193–204. [CrossRef]
41. Lyu, R.F.; Zhang, J.M.; Xu, M.Q.; Li, J.J. Impacts of urbanization on ecosystem services and their temporal relations: A case study in Northern Ningxia, China. *Land Use Policy* **2018**, *77*, 163–173. [CrossRef]
42. Sun, X.; Li, F. Spatiotemporal assessment and trade-offs of multiple ecosystem services based on land use changes in Zengcheng, China. *Sci. Total Environ.* **2017**, *609*, 1569–1581. [CrossRef] [PubMed]
43. Qi, Z.F.; Ye, X.Y.; Zhang, H.; Yu, Z.L. Land fragmentation and variation of ecosystem services in the context of rapid urbanization: The case of Taizhou city, China. *Stoch. Environ. Res. Risk. Assess.* **2014**, *28*, 843–855. [CrossRef]
44. Jopke, C.; Kreyling, J.; Maes, J.; Koellner, T. Interactions among ecosystem services across Europe: Bagplots and cumulative correlation coefficients reveal synergies, trade-offs, and regional patterns. *Ecol. Indic.* **2015**, *49*, 36–52. [CrossRef]
45. Zhou, Z.X.; Li, J.; Guo, Z.Z.; Li, T. Trade-offs between carbon, water, soil and food in Guanzhong-Tianshui economic region from remotely sensed data. *Int. J. Appl. Earth Obs. Geoinf.* **2017**, *58*, 145–156. [CrossRef]
46. Wu, J.J.; Jin, X.R.; Feng, Z.; Chen, T.Q.; Wang, C.X.; Feng, D.R.; Lv, J.Q. Relationship of ecosystem services in the Beijing-Tianjin-Hebei region based on the production possibility frontier. *Land* **2021**, *10*, 881. [CrossRef]
47. Zhang, Y.; Liu, Y.F.; Zhang, Y.; Liu, Y.; Zhang, G.X.; Chen, Y.Y. On the spatial relationship between ecosystem services and urbanization: A case study in Wuhan, China. *Sci. Total Environ.* **2018**, *637*, 780–790. [CrossRef]
48. Li, J.S.; Sun, W.; Li, M.Y.; Meng, L.L. Coupling coordination degree of production, living and ecological spaces and its influencing factors in the Yellow River Basin. *J. Clean. Prod.* **2021**, *298*, 126803. [CrossRef]

land

MDPI

Article

Green Infrastructure, Urbanization, and Ecosystem Services: The Main Challenges for Russia's Largest Cities

Oxana Klimanova [1], Olga Illarionova [1], Karsten Grunewald [2,*] and Elena Bukvareva [3]

1 Department of World Physical Geography and Geoecology, Faculty of Geography, Moscow State University, 119991 Moscow, Russia; oxkl@yandex.ru (O.K.); heatherpaw95@gmail.com (O.I.)
2 Leibniz Institute of Ecological Urban and Regional Development (Dresden), 01217 Dresden, Germany
3 Biodiversity Conservation Center, 117312 Moscow, Russia; bukvareva@gmail.com
* Correspondence: k.grunewald@ioer.de

Abstract: Globally, the process of urbanization is transforming land use and, as a consequence, reducing the efficiency of ecosystem services (ESs), which ensure a healthy and comfortable urban environment. In cities, green infrastructure (GI) is a key source of urban ESs. Russia is a highly urbanized country: 70% of its population live in towns or cities. As cities continue to expand, they are swallowing unsealed lands that support ESs. In this paper, we present the results of an analysis of the current state and recent changes in urban GI in Russia's 16 largest cities, including a biophysical evaluation of key urban ESs, using remote sensing data and freely available statistics. Our analysis identifies a weak correlation between GI area, ES volume, and favorable climate conditions as well as a stronger correlation between ESs and the increasing rate of urbanization. Considering Russia's high level of urbanization, the key importance of ESs for the well-being of citizens, and ongoing reductions of urban GI, major revisions to urban planning policy are required. Indicators of urban biodiversity, GI, and ESs should be incorporated into decision-making processes. In particular, it is vital that the accounting and monitoring of GI and ESs are established in all of Russia's larger cities.

Keywords: air pollution removal; ecosystem assessment; food security; heat mitigation; urban green areas; urban recreation; urban ecosystem services

Citation: Klimanova, O.; Illarionova, O.; Grunewald, K.; Bukvareva, E. Green Infrastructure, Urbanization, and Ecosystem Services: The Main Challenges for Russia's Largest Cities. *Land* **2021**, *10*, 1292. https://doi.org/10.3390/land10121292

Academic Editors: Luca Congedo, Francesca Assennato and Michele Munafò

Received: 28 October 2021
Accepted: 20 November 2021
Published: 25 November 2021

1. Introduction

Urbanization is one of the main drivers behind land use transformation around the world, and thus also for the decline in ecosystem services (ESs) [1]. Russia has a high level of urbanization. More than 70% of the population live in urban settlements, namely 1117 towns and cities of various sizes [2]. More specifically, 23% (or about 34 million people) reside in cities with over one million inhabitants, where over 30% of the national GDP is generated. Moscow, which is now Europe's largest city and the only megacity in Russia, generates around 15% of the national GDP.

Urban expansion generally occurs when (suburban) natural vegetated lands are converted into impermeable surfaces, thereby disturbing natural processes and the functioning of ecosystems [3–5]. Natural habitats are destroyed and fragmented not only by the construction of residential or commercial sites but also by ancillary infrastructure and transport networks [6–8]. Russia's largest cities are no exception. Between 2005 and 2019, the area of urbanized land in Russia increased from 78,000 to 84,000 km^2 (National report on the use and state of land categories in the Russian Federation in 2005 and 2019). In the same period, many of the most populated cities also expanded spatially. For instance, Moscow grew by 1480 km^2 in 2012, Ekaterinburg by 650 km^2 in 2015, and the urban precincts of Volgograd expanded by 290 km^2 in 2015.

Some of the main drivers of urban expansion are political decision-making and special federal programs. In particular, the national project "Housing and Urban Environment" (2018–2024) foresees the expansion of residential areas in Moscow and Saint Petersburg as well as other large cities. In 2020, around 25.5% of the total volume of newly commissioned housing in Russian cities was provided by the 16 largest cities (with a population of over one million people) (Table 1). It is the aim of the "Housing and Urban Environment" project that both new and old urban areas should provide all the requirements of a comfortable urban environment, which should include large green areas in built-up zones and well-established recreational areas. Consideration of these two factors is promoting an urban pattern of multistory buildings and artificially created green areas in the largest cities.

Table 1. The area and population of Russia's largest cities, according to the Federal State Statistics Service, 2020.

City	Urban Area in km^2	Population (mill.), 2020	Population Growth 2000–2020 (%)	Population Density per km^2
Moscow	2432	12.6	24.7	5091
Saint Petersburg	1439	5.4	11.4	3670
Novosibirsk	481	1.6	14.6	3333
Ekaterinburg	401	1.5	15	3631
Nizhniy Novgorod	317	1.3	−7	3981
Chelyabinsk	504	1.2	8.8	2337
Kazan	635	1.2	11.9	1940
Krasnoyarsk	378	1.1	23.8	2865
Omsk	580	1.1	−11.6	1752
Perm	806	1.1	44.2	1806
Rostov-on-Don	355	1.1	11.1	3169
Samara	543	1.1	1.2	2155
Ufa	667	1.1	2.3	1673
Volgograd	861	1	2.2	1179
Voronezh	601	1	14.5	1730
Krasnodar [1]	295	0.9	29	3050

[1] After Krasnodar reached the milestone of one million people in 2019, the population subsequently dropped to 940,000 in 2020. Nonetheless, the city is still the 16th most populated in Russia and is predicted to reach the million mark again in a few years.

Clearly, any expansion in urban development will provoke contradictory opinions and social conflict between the local authorities, citizens, and ecological activists [8]. This is also true of Russia's cities, in particular Moscow [9]. It is important to note that due to favorable natural conditions and historical tradition, the country's largest cities are rather green. A network of green areas since the 1930s was incorporated into the first Soviet urban planning system, the so-called General Plans (or Genplan), an official project document that forms the basis for urban planning, reconstruction, and development in many Russian cities. While this extensive ecological framework is greatly appreciated by local residents, the quality of green spaces is often severely undermined by pollution. According to national standards, about nine million people (or 6% of the country's population) live in cities with a high level of air pollution. The use of chemical substances in wintertime to remove snow and ice from roads also damages urban vegetation [10].

For all these reasons, the preservation of green areas is a hot topic amongst decision-makers and citizens in the largest cities. Clearly, there is a need for new approaches to integrate the value of green areas into urban planning [11]. In recent years, the concept of urban ecosystem services has become an important instrument in decision-making processes of urban spatial development or socio-economic sustainability [12–14].

In post-Soviet Russia, the study of urban ESs and biodiversity is still at a relatively early stage despite previous advances in urban environmental science [9]. The very few recent publications on the topic have generally focused on various individual aspects. The first systematic attempt to assess urban ecosystem services in Russia was undertaken as a part of the TEEB-Russia Project (http://teeb.biodiversity.ru/en/; accessed on 20 October 2020) [9].

In this article, we present the results of a high-resolution survey of green infrastructure (GI) in Russia's 16 largest cities before discussing the connection between GI and the urban climate as well as changes in urban GI in recent years. Subsequently, we present and discuss the results of our physical evaluation of selected ESs as well as the extent to which city residents are provided with ESs. Finally, we discuss the main methodological problems of ES accounting in Russia's cities and propose some initial approaches on how to integrate urban ESs into the process of urban planning against a backdrop of urbanization.

2. Materials and Methods

2.1. General Approaches to Assessing Urban Green Infrastructure and Ecosystem Services

Ecosystem services are understood as all kinds of benefits, both tangible and intangible, that humans obtain from ecosystems and their resident flora and fauna. Many classifications of urban ESs found in the literature (for example [15–18]) are based on existing systems specified by the Millennium Ecosystem Assessment [11,19,20]. These classifications encompass a wide variety of ESs, which can be calculated by the direct quantitative evaluation of biophysical factors (assuming that data on supplied, consumed, and demanded ESs are detailed in statistical reports) or the indirect quantitative evaluation of biophysical factors. The latter is applied when direct statistical data is lacking. In this case, cartographic data and previous studies can be used to evaluate selected indicators based on simple calculations or GIS modelling. Any assessment of carbon storage, biological productivity and diversity, or the accurate evaluation of air and water purification requires continuous periodical statistical data and monitoring, which is generally unavailable in Russia at the urban level. For this reason, in our study, we only used indirect assessment methods to assess a limited number of ESs [9].

In the TEEB-Russia project, ESs are assessed as provided (potential), demanded, and consumed [21–23]. The provided ES is a volume produced by ecosystems regardless of the presence (or absence) of consumers. The demanded ES is the amount needed to meet the demands of the population and regional economy. The consumed ES is the ES yield, which is materially or immaterially used by the population or from which people derive current benefits. The ratio between the demanded/consumed volumes indicates the level of ES use in the city and the extent to which the needs of the local population are satisfied.

Urban ecosystems have been defined as the vegetated space (which may include water bodies) in areas where a significant proportion of land is built-up or sealed [10,12,24] and where population density is high [25]. The concept of green infrastructure is closely related to that of the "urban ecological network" [26] popular in national urban planning schools of the 1980s. Green infrastructure, however, encompasses a wider range and thus volume of elements. Despite its frequent usage by scholars and journalists, the term "ecological network" is not legally recognized in Russia. To characterize urban vegetated areas, national schools of urban planning traditionally use the term "green plantings", which are defined by GOST 28329-89 as "a complex of tree, shrub and herbaceous vegetation in a specific area". This definition does not consider the main emergent features of the ecological network, in particular its integrity, connectivity, and hierarchy of green elements, which ensure its function of stabilizing the local environment. Unlike green plantings, ecosystem services are not specified in Russian planning documents; the term is not even used in General Plans. However, the term "ecological functions", which sometimes coincides with ecosystem services, is often mentioned. In our research, urban ESs refer to services provided by ecosystems that are situated within the administrative boundaries of the city. For our analysis, we used the administrative boundaries as indicated by the OpenStreetMap portal.

Our case studies are of Russia's 16 largest cities with populations over one million (Figure 1), and which have very different climates and different rates of population growth and social and economic development (Table 1). According to [27], the majority of the largest cities (Saint Petersburg, Ekaterinburg, Nizhniy Novgorod, Samara, Omsk, Chelyabinsk, Rostov-on-Don, Krasnoyarsk) are referred to the most industrial group with the highest index of anthropogenic impact (IAI) and the severest water pollution. Moscow is not among the most industrial cities, but it also has the highest IAI and air pollution. Other cities (Kazan, Ufa, Volgograd, Perm, Voronezh) have medium values of IAI and are mostly affected by waste accumulation. Krasnodar is the only city with a population over one million people that has a low IAI. The GNP values and functional specialization of the studied cities vary greatly. These factors obviously influence the GI state and, consequently, ESs. These particular correlations ought to be considered in further studies.

Figure 1. The largest cities of Russia.

We conducted a correlation analysis between natural conditions, the urbanization rate, and GI quantity. We also analyzed the GI quantity and composition dynamics for the top five cities, since changes to GI are directly connected to the ES supply.

Rosstat (the Federal State Statistics Service) provides open data for ES evaluation on the urban okrugs, a particular type of municipality that includes two to three settlements within a contiguous territory. However, the boundaries of urban okrugs do not always match city boundaries. In some regions of Russia with a high level of urbanization and industrial development (for example, in the Urals), these differences can be drastic.

There are specific difficulties in assessing the GI and ESs at the city level. Firstly, this local level of research requires higher accuracy and detail than research conducted at the regional level. Secondly, it is almost impossible to make conclusions on urban ESs in the country as a whole based on the results of a few cities; instead, it is necessary to study a significant number of case studies. Unfortunately, research at the local level is quite complex and time-consuming. Moreover, the quality and quantity of the required base data for the calculations will vary depending on the city or may even be entirely missing. For these reasons, we decided to assess only those ESs that could be estimated from our own modeling or using open official data. GI parameters are a valuable asset for ES assessment, since in the urban environment, it is such green elements that provide all ecosystem services. The following assessments are mostly based on the GI parameters described in the previous section.

2.2. Methods of GI Composition Assessment

We examined three main types of urban GI: tree vegetation, non-tree vegetation, and agricultural land.

The general inventory of GI was realized in two steps. Firstly, we used images taken on a cloudless summer day of the Landsat 5, 7, and 8 series for the years 2000–2020 with a spatial resolution of 30 m. The selected images were from the period 30 May to 3 September. These were used to create a Normalized Difference Vegetation Index (NDVI) raster in the ArcMap 10.3 program. Subsequently, a supervised classification was conducted to distinguish two classes of land cover: (1) non-vegetated surface (NDVI = 0.18–0.30) and (2) vegetated surface (NDVI > 0.30). NDVI values for dense/healthy tree vegetation, which is the most valuable for ESs [23], are usually quite high. In our case, the diapason for this type of vegetation was 0.35–0.50.

In the second stage, we made use of spatial data on tree cover taken from the "Global Forest Change" website of Maryland University. This data enabled us to verify our tree vegetation raster for the various cities, which are situated in different climate zones. In some cases (particularly cities in the steppe zone), NDVI data was insufficient to create an inventory of all green infrastructure due to the diverse condition of agricultural lands, which are the main providers of the food provisioning ESs.

We extracted spatial data from OpenStreetMap (OSM) by selecting "farmland" categories from the OSM "landuse" key. According to the definitions of OSM key elements, "farmland" is an area used for tillage, including croplands, pastures, and orchards, etc. We extracted this selected data and added it as agricultural lands to the maps, defined as a unique value in the legend, and subtracted the area from the total non-tree vegetation area. The verification of these OSM shapes using a high-resolution "Base map" in ArcGIS 10.3 showed that the OSM "farmland" category did not capture all agricultural lands in Rostov-on-Don and Samara. For this reason, all the missing polygons were defined manually through visual inspection, supplemented by land use data from the open geoportal Wikimapia. The maps and materials created on this basis were used to establish our inventory of ecosystems that perform services.

The transformation in GI (share of green area within the city borders) and composition (the percentage of tree vegetation, non-tree vegetation, and agricultural lands) during the period 2000–2020 was assessed by comparing the inventory results for these years using the Overlay function.

Within the TEEB-Russia project, we have already assessed the transformation of GI quantity and composition in all 16 cities for the period 2000–2016; however, only a few cities displayed any significant changes in this period. We updated our results from TEEB-Russia [28] for the sake of this study. The update concerned the investigated period and the studied area. For this publication, we analyzed the GI transformation between 2000 and 2020 for only the five largest cities, namely Moscow, Saint Petersburg, Novosibirsk, Ekaterinburg, and Nizhniy Novgorod, because our previous studies revealed some significant changes in green area (in general, most cities gained or lost less than 1% of green area over 16 years).

2.3. Methods of ES Assessment

Ecosystem assets supply ESs and are defined as "contiguous spaces of a specific ecosystem type characterized by a distinct set of biotic and abiotic components and their interactions" [29]. Due to the lack of data on the quality of urban ecosystems, in our assessment of the 16 largest cities, we only used indicators of ecosystem assets area described in the previous section:

- Area of urban tree cover derived from the map of [30];
- Area of non-tree cover derived from NDVI and classified Landsat Images;
- Area of agricultural land, derived from OpenStreetMap data.

In this way we could calculate the area of these GI types and, according to the valuation method presented in Table 2, derive the values of the provided ES volumes. In our work, we assume that the volume of ESs is determined by the efficient area of GI. For each ES, a different green area is considered efficient, i.e., capable of efficiently performing the specific service. For example, we assumed that food provisioning is mostly performed by agricultural lands, heat mitigation by large forest massifs, etc. As there are no freely available statistics for a full assessment of these ecosystem services in Russia, we used cartographic modeling. The only exception was food provisioning, for which we partly used statistical data on the annual yield and urban demand. Heat mitigation is usually assessed by monitoring near-surface temperatures or by modeling surface temperatures by the means of thermal images. Due to the lack of meteorological data for most areas of the cities, we assessed this service by calculating the area of GI that can be potentially used for heat mitigation, i.e., green massifs of tree vegetation of size 500 ha or more. Finally, the ES of daily and weekend recreation was estimated according to the recreational capacity of GI and its availability to local residents.

Table 2. Valuation methods and parameters to assess ESs by GI area.

Ecosystem Services	Parameters for Biophysical Assessment	Data Source	Valuation Methods
Food provisioning	Food provision, t/year	Database of Federal State Statistics Service	Determined by the area of agricultural land and its productivity
Temperature regulation in the city	Area of cooling effect, ha	Remote Sensing Data Coefficients from literature [14,31]	Determined by the area of urban green infrastructure and mean area with cooling effect
Removal of air pollutants	Amount of O_3, SO_2, NO_2, CO and PM10 µm (t/year), absorbed by tree vegetation × tree cover area (m^2)	Remote Sensing Data Coefficients from literature [32,33]	Determined by the area of different types of forests within the buffer zones of industrial enterprises and absorption indexes for American cities
Recreation and educational development	Area of public green space, per 1000 people Area of public green space within walking distance, %	Remote Sensing Data Recreational Standards	Determined by the area of different types of green infrastructure, its distance from residential areas, and recreational capacity

2.3.1. Removal of Air Pollution

Generally, the service of air purification is assessed using data on the tree absorption potential. We adopted the method suggested by [32,33] for Canadian and American cities. In those studies, the authors made use of the iTree software application to calculate the absorbed volume of pollutants by trees. Since there are no examples for Russian vegetation in the database, we applied the results of [32,33] for Canadian and American vegetation to similar forms of vegetation in Russia. In particular, [32,33] assessed the volume of gaseous pollutants removed by urban woodland, regardless of the proximity to roads.

For the current study, we assessed the ES volume of pollutant removal provided by tree vegetation using the forest type area and the related average value of air pollutant removal for each city. Further, we assumed that all woodland inside one city belongs to one corresponding forest type. Therefore, the provided volume (i.e., the amount of pollutants that can be absorbed by all urban trees) was calculated by multiplying the area of tree cover by the average value of air pollutant removal indicated in Table 3.

Table 3. Average values for the removal of air pollutants by various forest types, basing on the [32,33] results.

Forest Type	CO, t/ha/yr	SO_2, t/ha/yr	NO_x, t/ha/yr	CO + SO_2 + NO_x, t/ha/yr	Total, t/ha/yr
Dark coniferous	0.0002	0.0022	0.0072	0.0096	0.0124
Light coniferous	0.0002	0.0025	0.0078	0.0105	0.019
Broadleaf	0.0006	0.0033	0.0081	0.012	0.0171
Mixed	0.0004	0.001	0.0055	0.0069	0.0136
Small leaf	0.0002	0.0007	0.0047	0.0056	0.0144

The demanded ES volume is simply the total emissions from automobiles in the city. Data on vehicle emissions was taken from the annual report issued by the Federal Service for Hydrometeorology and Environmental Monitoring of Russia (Rosgidromet) [34].

The consumed volume is equal to the amount of pollutants that trees actually absorb. As will be shown below, the absorption capacity of trees (provided ES) is lower than the pollution emissions (demanded ESs) in all investigated cities; therefore, in all cases, the consumed volume is equal to the provided volume, i.e., the absorption capacity is fully used. The ratio between the demanded and provided ES volume indicates the extent to which a population's needs for this ES are satisfied.

Compared to automobiles, emissions from point sources present a smaller share of total emissions in Russia's most populated cities. To mitigate the negative impact of industrial zones on the environment and human health, special sanitary buffers are created around industrial areas. These must be of sufficient size to ensure that chemical, biological, and physical air pollution does not exceed established safety guidelines. Regarding industries classified in danger categories I and II, buffers must act to decrease concentration values to an acceptable level of health risk. In this study, we adopted the 300 m buffer (danger class III), under the assumption that the most populated cities do not have many dangerous industries within their borders. While the Sanitary Norms and Regulations of Russia (SanPin) [35] do not elaborate on the type of vegetation inside the buffers, it is stated that they must decrease the negative impact on the environment (including chemical and physical pollution), present a barrier between residential and industrial areas, absorb pollutants, and regulate the microclimate. In this study, we only considered the extent of woodland inside the buffer, as this is the most effective type of vegetation to mitigate negative chemical and physical impacts and to purify the air. Specifically, we selected all areas in the OSM land-use layer with the tag "industrial" and created merged buffers of a width of 300 m around these. Then, we calculated the area of tree vegetation inside these buffers and multiplied it by the values from Table 3 to obtain the provided ES volume.

At this stage, the demanded volume for the total emissions from point sources inside the city was set at the same level as vehicle pollution (the data was also taken from the Rosgidromet report). This assumption can be justified by the fact that there is no freely available geospatial data on identified point sources and emissions volume in Russia. This lack of information makes it impossible to calculate the demanded ES volume for each industrial area individually. In fact, green infrastructure inside the sanitary buffers is not required to remove all air pollutants but merely to decrease the level to the maximum permissible concentration (MPC).

2.3.2. Urban Microclimate Regulation

Urban green infrastructure plays a crucial role in regulating the urban microclimate, particularly in reducing the heat island effect [36,37]. An urban heat island is a built-up area with higher temperatures than nearby open space. This difference in temperatures is caused by a specific heat exchange between the atmosphere and the built-up area [18].

The intensity of an urban heat island depends on local meteorological factors and the ratio between built-up and unsealed green areas in the urban land-use system [13]. A case study of Berlin showed that large elements of green infrastructure have a more efficient cooling effect on the surrounding area than small elements [38]. Specifically, the temperature in the center of a large green element (>500 ha) is 5 °C lower than in adjacent territories (ibid.). The cooling effect in Moscow ranges from −4.3 to +0.3 °C (with an average of −1.3 °C) [39]. Depending on the size of the green element, the area benefiting from the cooling effect may vary between 500 and 1500 m [14,31].

Assuming that in cities with similar geographical features, large green elements (>500 ha) perform the most efficient cooling effect at a distance up to 1500 m, we determined the provided ES volume as the average area cooled by parks. Assuming a near circular form for the cooled area, we calculated the area using the following equation: $S3 = S1 + S2$, where S1 is 500 ha, S2 is the cooled area, and S3 is the total area of park and the area that it cools. The resulting sum of a park's area and its cooled area is 1893 ha. This is 3.8 times bigger than the park itself. We took this value as a coefficient to calculate the nominal area influenced by similar parks and their tree vegetation in each city.

The demanded ES volume is simply the urban area (including the area of tree cover).

2.3.3. Food Production

ES volume provided by green infrastructure is expressed as an indirect indicator, namely the gross yield of household croplands during the year. Typical cultivated crops are potatoes and other vegetables and fruits. Drawing on official statistical data for the municipalities, we additionally calculated the share of croplands and permanent croplands provided by the total urban area.

Households mostly grow potatoes and other vegetables as well as fruits for trade or private use. According to the recommendations of the World Health Organization, a healthy diet should include 400 g of fruits and vegetables every day (in addition to potatoes and other root vegetables). Considering that this recommendation also includes crops that cannot be cultivated in the temperate zone, such as bananas, oranges, grapefruits, etc., we halved this norm to 200 g/per day (or 73 kg/per year) to calculate the demanded volume. According to recommendations of the Ministry of Health of the Russian Federation, the annual rational norm of potato is 90 kg/per year. Thus, a total standard value for vegetable, fruits, and potato consumption can be set at 163 kg/per year. The demanded volume was calculated by multiplying this value by the total population of a city.

2.3.4. Recreational Services

ES volume provided by green infrastructure is here calculated as the maximum permissible number of people who can simultaneously and comfortably visit an element of urban green infrastructure for recreational walking. Our assessment was done in several stages. First, it was necessary to define the types of green infrastructure suitable for recreation. Here, we selected elements from the following OSM categories: "wood", "forest", "orchard", "grass", "meadow", "village_green", "recreation_ground", "garden", "park", "allotments", and "cemetery". Generally, the borders of these elements correspond with the outlines of green infrastructure we obtained from a classified NDVI image and Hansen's tree-cover raster, thus proving their relevance. Areas suitable for recreation were divided into two categories: protected areas (PAs) and other recreational GI.

The provided ES of mass weekend recreation was estimated as the maximum permissible recreational carrying capacity of PAs and other GI. For Pas, we adopted minimum estimates of the permissible recreational load of 2 per/ha and for other GI medium estimates of 50 per/ha [5]. The estimate of the demanded ES volume was based on the Moscow Park standard recommended number of simultaneous park visitors, namely 5% of the urban population. The provided volume of the ES of daily recreation was estimated as the proportion of the city's residential zone (from OSM) that has GI within an 800 m buffer, i.e., the standard distance of a 10-min walk (Accessible Natural Greenspace Standard -

ANGSt) [40]. The demanded volume is 100%, since the ideal situation implies that all residents have a green zone within walking distance.

2.3.5. Indicators of the Urbanization Rate and GI

To determine the influence of urbanization and city development on the GI area and ES volume, we assessed the urbanization rate using four indicators:

- Percentage ratios of urban area occupied by tree vegetation and non-tree vegetation;
- Percentage transformation of GI composition in the period 2000–2020;
- Percentage change in the urban population;
- Percentage change in built-up area.

The data for the first two parameters were obtained by modeling in ArcGIS with the use of remote sensing (Landsat images, Hansen's tree-cover data, OSM materials). The latter two indicators were obtained from official statistical data on the Russian population and land use, including the reports on the national census [10], and use and condition of the land (The Federal Service for State Registration, Cadastre and Cartography—Rosreestr).

3. Results

3.1. Composition of Green Infrastructure

The main features of green infrastructure in Russia's most populated cities are shown in Table 4. Tree vegetation of varying density makes up 16% to 61% of the total expanse of green infrastructure. The cities Omsk and Volgograd, located in the steppe zone, have the smallest share of tree cover at 18% and 16%, respectively, while cities of the forest zone, such as Ekaterinburg and Perm, have the most tree cover at 58.8% and 61.3%, respectively. (Figure 2). Although Voronezh is situated in an ecological zone with climate conditions similar to that of Omsk and Volgograd, its tree cover is about two times greater. The same disparity can be seen when comparing the lower extent of tree vegetation of Kazan with that of Perm and Ekaterinburg, which have similarly favorable climate conditions. Although Krasnoyarsk and Rostov-on-Don lie in different ecological zones, their extent of tree cover is similar. Our analysis showed that cities in the steppe zones often incorporate large forests (both natural and artificial) outside the main urban core into the administrative borders, generally to be used for recreation. In contrast, areas of woodland in the forest zone are usually not incorporated within the city.

Table 4. Percentage area of GI types in Russia's 16 largest cities.

GI Type	Max. Value	Min. Value	Average Value	Variance Value
Tree vegetation	61.3	16.0	36.7	207.2
non-tree vegetation	75.9	37.0	58.2	163.3
agricultural lands	11.8	1.4	5.2	9.6
protected areas	30.0	0.3	8.3	71.6

Of course, woodland is not the only element of urban GI, which encompasses other unsealed areas including farmland.

According to the General Plans for urban development, GI per capita varies from 4 to 135 m^2, depending on the city and district. The lowest figures are found in Chelyabinsk (2.1–5.8 m^2/per capita), Perm (4.0–10.0), Rostov-on-Don (6.7–10.0), and Volgograd (10.0), with the highest figures in Ekaterinburg (38.6) and Novosibirsk (88.0–135.0). Our research shows that three of the 16 cities have less than 100 m^2 of green infrastructure per capita; seven cities have about 100–200 m^2/per capita and six cities have more than 200 m^2 of GI per capita. If we consider that not all tree vegetation is captured in standard assessments, the actual per capita values will be much greater.

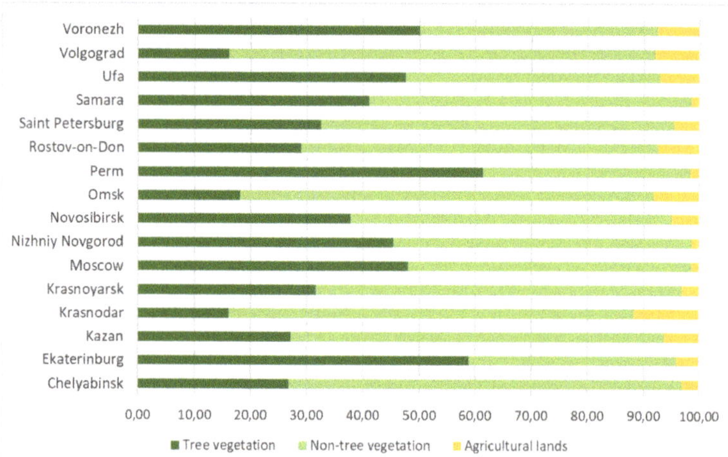

Figure 2. The composition of GI in Russia's most populated cities in 2016 (Moscow's GI was calculated for the city boundaries before 2012).

For our study, we also examined the relationship between the city's GI and the local climate (Figure 3). To describe climatic conditions, we used the Evaporation Precipitation Index (EP), which is a ratio between the average annual precipitation and evaporation. In general, values above 1 are more favorable for vegetation while values below 1 imply a harsh growing environment.

As no correlation was found between EP and the share of GI from total urban area (Figure 3a), we may assume that climate does not play a decisive role in determining the total GI area. While there is a tendency for cities with EP < 1 (steppe and forest-steppe cities, such as Omsk, Samara, Volgograd) to contain less green space than cities with higher EP values, at the same time, several cities with the most favorable climatic conditions (EP > 1) display a share of GI similar to cities in the steppe, e.g., Ekaterinburg. Saint Petersburg has the highest EP value (1.7) but simultaneously the smallest share of GI. Almost all cities except Krasnodar still meet the recommended 40% share of urban green [5].

In contrast, we found a certain correlation between the extent of tree vegetation (the most valuable form of GI) and climatic conditions (Figure 3b). While low EP (<1) implies a small share of tree vegetation (e.g., steppe zone), high EP (>1) does not always imply a large share of tree cover. Therefore, the degradation of GI in cities with EP above 1.0 but with a share of tree vegetation below 40% cannot be attributed to climate. In most cases, the processes of urbanization, economic development, and human activities influence GI to a much greater extent than natural conditions in the urban environment of Russia's largest cities.

3.2. *Removal of Air Pollution from Vehicles and Point Sources*

Evaluation of this ES in physical terms showed a very negative situation in all analyzed cities [28]. Motor vehicles are the main source of air pollution in most of the case studies. The share of vehicle emissions from total air emissions varies from 34% in Krasnoyarsk to 94% in Moscow. Gaseous pollutants contained in the exhaust fumes circulate in the air under the influence of atmospheric diffusion. Their concentration gradually decreases at greater heights and distances from roads. Such gases can circulate to other city districts where they undermine the air quality. Trees and shrubs in buffers around the roads are most affected by the pollutants; at the same time, they act as barriers to the further diffusion of pollutants. The provided volume depends on the area of woodland inside the city. This value varies greatly between the studied cities: the highest values are found in Moscow, Perm, and Voronezh, the smallest in Chelyabinsk, Omsk, Krasnodar, and Rostov-on-Don.

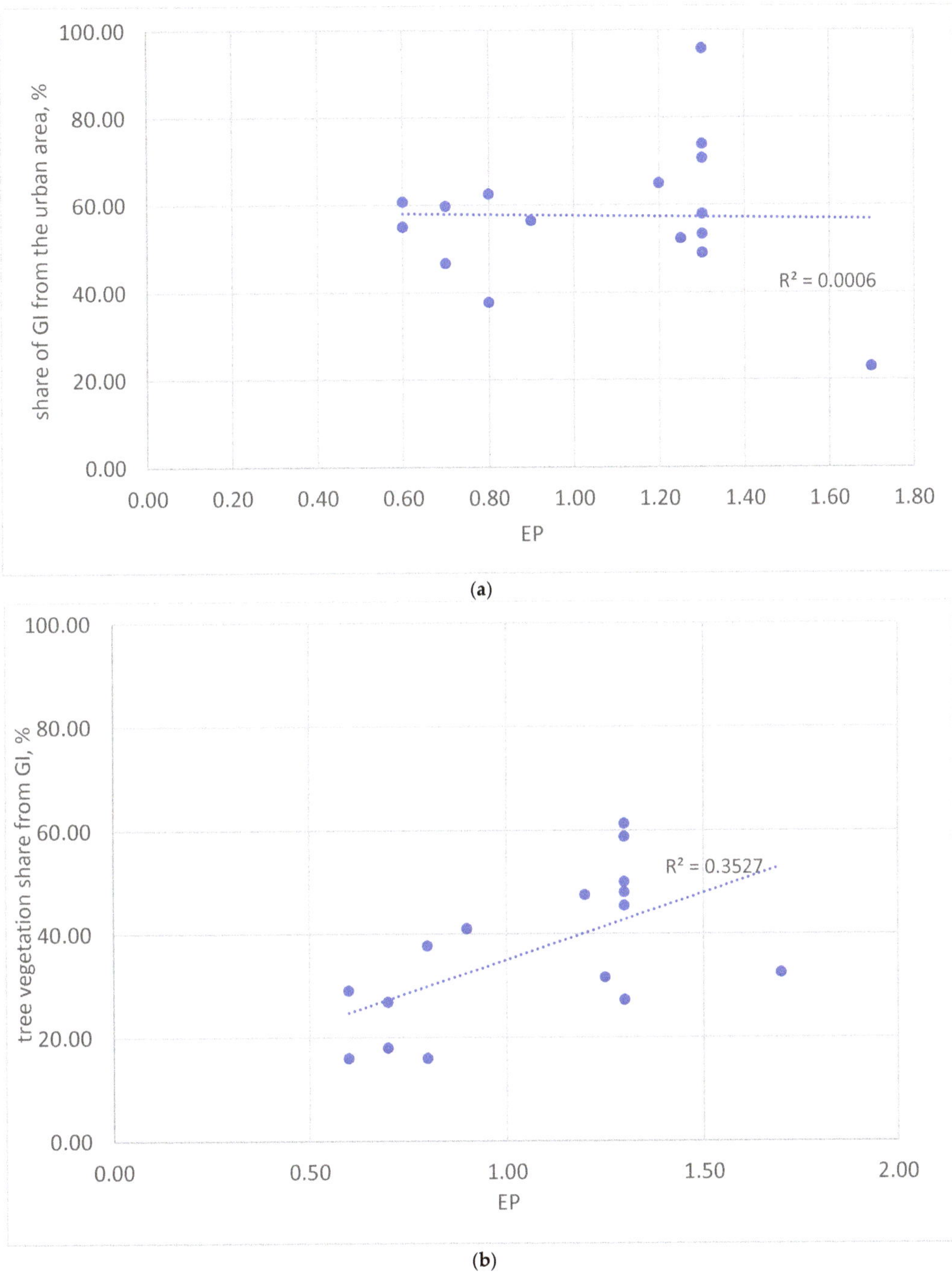

Figure 3. Relationships between the Evaporation Precipitation Index (EP) and (**a**) the share of urban GI area from total urban area; (**b**) share of tree vegetation from total urban area in Russian most populated cities.

There is a significant difference in the demanded volume between the cities: those with the greatest loads are Moscow, Saint Petersburg, and Kazan, while Volgograd, Krasnoyarsk, and Rostov-on-Don produce less than 70,000 tons of emissions per year.

The ratio of required to provided ES volume does not exceed 1% in the studied cities, i.e., this service is practically unavailable to the local population. Slightly more sulphur dioxide (SO_2) is absorbed than the other two investigated pollutants, carbon monoxide (CO) and nitrogen oxides (NOx). In this case, the demanded volume for removal is smaller than the consumed volume, meaning that more SO_2 emissions can be absorbed by the existing woodland. In contrast, CO is removed in smaller volumes than other pollutants due to its intense emissions and the low absorption capacity of CO by trees.

The results show that today's urban green infrastructure is unable to remove most emissions. The best results were found in Voronezh and Perm, which are rather green and have comparatively low emissions. It should be also stressed that Moscow, despite having the largest green area, shows the worst results for all studied gases due to the enormous volume of vehicle emissions. The correlation between the proportion of removed pollutants and the relative extent of green area is extremely low, also indicating the inadequate capacity of this ecosystem service.

Regarding the removal of air pollutants from stationary sources, the supplied volume of this ES was found to vary between 6.2 and 104.1 tons/year, depending on the area of green infrastructure inside the buffers.

As before, the ratio of required to provided ES volumes is less than 1% in all case studies. Considering that a large proportion of pollutants is transported away from the point source, we can assume that the major role of green infrastructure inside the sanitary buffers is to partly decrease the demanded volume of air purification provided by other urban and suburban forests, which also perform this ecosystem service.

The provided ES volume depends on the deforestation level in the sanitary buffers. In our study, we could not identify a single city where tree vegetation covered more than 30% of a buffer. The maximum proportion was found in Perm (30%), with the lowest values in Chelyabinsk (7%) and Volgograd (9%).

The extent of woodland in the sanitary buffers reflects the zone in which a city lies: for instance, values for this indicator are higher in the forest zone than in the steppe zone. At the same time, the sanitary buffers are much more deforested than the cities themselves. Further, the largest industrial zones were found in the biggest cities, namely Moscow and Saint Petersburg. Despite the large number of industrial zones, these two cities have the most highly vegetated sanitary buffers.

3.3. *Urban Microclimate Regulation: Cooling Effect*

Several of the studied cities do not possess sufficient green infrastructure to perform an optimal climate regulation, i.e., less than 100% of the urban precincts enjoys a cooling effect. These include cities in the steppe zone, such as Volgograd, Omsk, Krasnodar, Chelyabinsk, and Samara. Krasnoyarsk is also severely lacking in this service: only 23% of the urban area is cooled. Considering that this city has a humid continental climate (Dfb type, according to the Köppen climate classification) with rather hot summers (the average July temperature is about 20 °C), heat mitigation is thus crucial to ensure a comfortable urban environment. While Saint Petersburg was also found to be suboptimal in this ecosystem service, our adopted method is probably less suitable here as it does not take account of the numerous canals and water bodies in the city, which also provide a significant cooling effect. In contrast, the cooled area in Perm, Moscow, and Nizhniy Novgorod was found to be two times bigger than the urban precincts. This can be attributed to the high shares of developed green infrastructure and tree vegetation in these cities. It is important, however, to remember that green elements are not evenly distributed across most of Russia's large cities; rather, the outskirts are often close to large urban forests while urban cores may lack any vegetation. This situation is true of Perm, Rostov-on-Don, and Samara, where the largest shares of GI (up to 50–70%) are concentrated outside the urban core. Particularly in the case of Rostov-on-Don, forest massifs are situated on the almost undeveloped and unsettled riverbank, while the city center and residential areas are severely deficient in green areas. For more details on the evaluation of this ES, see [9].

3.4. Provisioning Services: Food Production

Agricultural lands make up less than 2% of the total urban area of Nizhniy Novgorod, Moscow (within the old pre-2012 boundaries, i.e., excluding the so-called New Moscow) and Samara. In contrast, about 12% of Krasnodar is made up of farmland (Figure 2). The share depends not only on the ecological zone and climate conditions but also on the historical formation of the city. Volgograd, Voronezh, Rostov-on-Don, Omsk, and Krasnodar are the only case studies where we found a considerable area of croplands belonging to agricultural businesses. In fact, most permanent croplands, such as fruit orchards, belong to private households. These private households (so-called dachas, a seasonal second home) occupy a significant area of suburban territories and play an important role in recreation and food-provisioning for many city-dwellers.

While modern large cities are generally not self-sufficient in food, our analysis showed considerable variation in the ratio of food production, from 5.5% in Samara to 31.3% in Omsk. It should be noted that this potential input is bigger than the percentage of agricultural lands in the city, implying that household croplands are highly productive and important for the food security of many of Russia's large cities. Notably, cities of the steppe zone with more favorable agroclimatic conditions and fertile soils (including chernozems) also have the highest values of this ES service, e.g., Rostov-on-Don, Krasnodar, Volgograd, and Omsk. However, these farmlands situated within the administrative borders mostly belong to agricultural holdings, which often export the yield to other regions. For more details on the evaluation of this ES, see [9].

3.5. Recreational Services: Forming Natural Conditions for Recreation

One of the most important ESs provided by urban green infrastructure is to establish conditions for recreational walking and sports. Unfortunately, it is technically difficult to assess this kind of ES in Russian cities: firstly, due to the variety of green elements with diverse conditions for recreation; and secondly, because of the wide range of recreational activities and a lack of standards to regulate the permissible recreational carrying capacity.

When considering the provided ES for weekend recreation based on the minimum value of permissible recreational carrying capacity (2 per/ha) for urban PAs and the medium value (50 per/ha) for other GI, the results show that all cities (except Rostov-on-Don) can supply a considerably larger volume of recreational service than demanded by local residents (the ratio is above 100%). The degree of ES provision for citizens changes when applying the minimum value (2 per/ha) for the whole GI. In this case, the ES supply exceeds the demand in only two cities: Voronezh and Perm. In other cases, the demand is greater than the supply. The average ES volume for daily (rather than weekend) recreation supply by urban GI is about 90% in most cities, i.e., more than 90% of the residential areas of Moscow, Ekaterinburg, Kazan, Nizhniy Novgorod, Voronezh, and Perm have GI within walking distance. This result can be expected since these cities also have the largest area of GI in general. More surprising, however, is the fact that GI in cities, such as Saint Petersburg, Samara, Volgograd, and Chelyabinsk, with relatively small green areas is sufficient to meet the needs of citizens for daily recreation. In contrast, Rostov-on-Don and Ufa show the worst results for this ecosystem service despite their relatively large green areas.

On the whole, the results for recreational services are better than for a number of other studied ESs in Russian cities. Together with microclimate regulation, recreational services can be called the most efficient.

3.6. Patterns of ES Provision in Russia's Most Populated Cities

For the present study, we analyzed the extent to which residents are provided with ESs (i.e., the ratio between provided and demanded ESs) in Russia's 16 largest cities. Our results show that the most efficient service in a number of cities is microclimate regulation and daily recreation. The red line in Figure 4 shows that in most cases, the volume of climate regulation is even excess to demands, i.e., the existing green infrastructure can provide more heat mitigation than required by the local population. Daily recreation is also a well-performed service, although the provided volume of weekend recreation (when considering the minimum recreational carrying capacity) is smaller. The most poorly provided services are food production and air purification. The latter service is only able to remove less than 0.01% of the airborne pollutants. Clearly, we cannot expect plants to absorb 100% of urban emissions; therefore, we set a value of 10% of the total emissions as the demanded volume. Yet even under this assumption, the best result for this service (in Voronezh) is still only 4.4% of the demand.

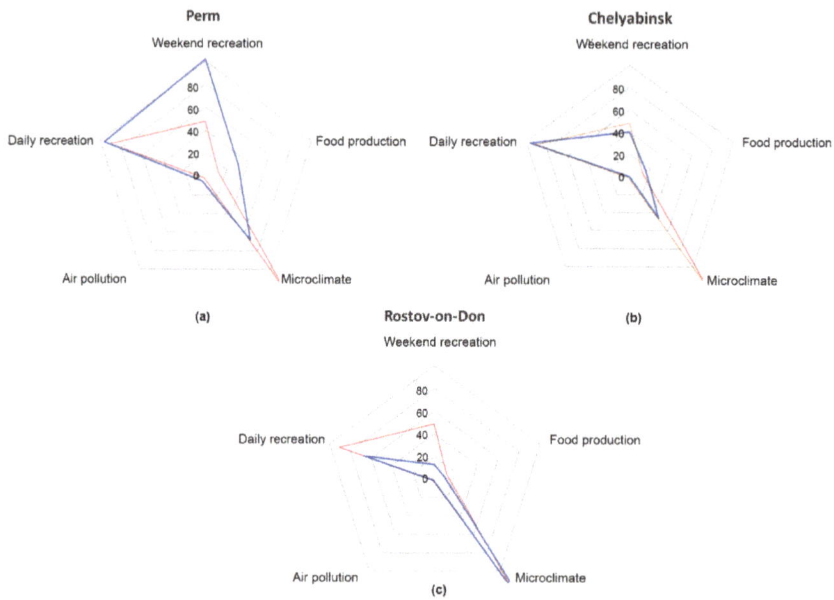

Figure 4. The extent to which residents are provided with ESs; the red lines show the average distribution of ESs in the 16 largest cities; the blue lines indicate the pattern in three selected cities: Perm (**a**), Chelyabinsk (**b**), and Rostov-on-Don (**c**).

Seven of the studied cities, namely Moscow, Novosibirsk, Saint Petersburg, Kazan, Omsk, Volgograd, and Ufa, have similar patterns of ES provision. However, we can identify three other types of patterns amongst the studied cities. In Perm (Figure 4a) and Voronezh, the most efficient service is weekend recreation; while all other services are also fairly efficient, microclimate regulation is not excessive as is the case in most of the investigated cities. This pattern can be called the most balanced, since the disparities between the volumes of the five services is less striking. Another pattern can be identified in Chelyabinsk (Figure 4b) and Krasnoyarsk. Here, the structure of ES provision indicates a sufficient supply of daily recreation but a deficiency in all other services, including climate regulation. The third pattern is associated with the cities Rostov-on-Don (Figure 4c), Samara, and Nizhniy Novgorod. Here, all ESs apart from heat mitigation are seen to be less efficient than in other cities. The second and third patterns of ES provision are the least balanced. Most of the cities with these patterns have some industrial specialization, such as Chelyabinsk, Samara, Rostov-on-Don, and Nizhniy Novgorod, which can partially explain this deficit in ESs, even though the green infrastructure parameters are acceptable. This ES pattern demonstrates that an abundance of green space does not necessarily imply that ecosystem services are well provided. In fact, the quality and location of green elements will also determine the volume and variety of ESs. We believe that this is an important finding with practical relevance for Russia's cities, which, in comparison to many large European and Asian cities, enjoy extensive GI within the urban precincts. These large shares of green space give the misleading impression that ES volumes are equally large.

3.7. Urbanization and Green Infrastructure Transformation: Case Studies of the Five Largest Cities

Russia's most heavily populated cities, i.e., Moscow, Saint Petersburg, Novosibirsk, Ekaterinburg, and Nizhniy Novgorod, all enjoy a particular status. Acting as pioneers and innovators, these cities define the GI development for other administrative centers in the country. Here, it is also important to note that the administrative boundaries have repeatedly changed in most of these cities over the last 20 years (for this reason, we used the most recent boundaries). The major changes to urban boundaries occurred after 2011. Several cities have followed the example of Moscow, whose area has enlarged by a factor of 2.5, with many small towns being incorporated into large neighboring cities. The recent study for TEEB-Russia [25] mostly made use of old boundaries.

Our assessment of GI transformation revealed a loss of green space in all cities (Figures 5 and 6). The greatest change was found in Moscow and Nizhniy Novgorod (a loss of 9.2% or 235 km^2 and 19% or 95 km^2, respectively), with the smallest in Ekaterinburg (a loss of only 0.3%). However, Ekaterinburg is also the only case where we found a gain in non-tree vegetation, against a loss in tree vegetation of about 0.8%. From our NDVI images and tree-cover raster, we could determine that most tree vegetation was transformed into non-tree vegetation, which is less valuable in terms of ES volume and variety.

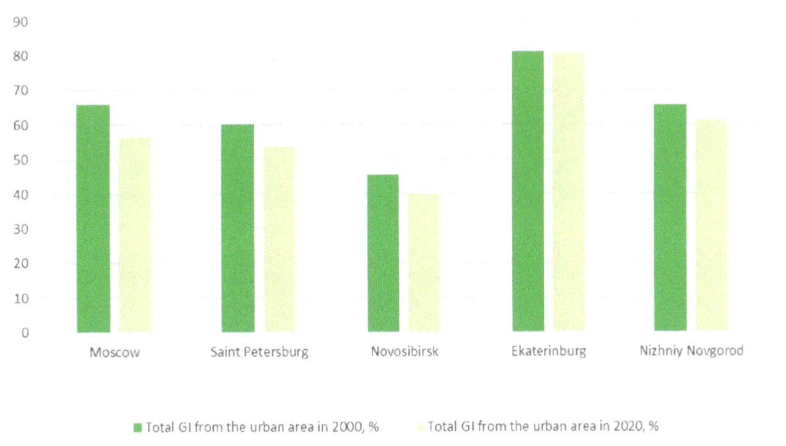

Figure 5. Loss of GI in Russia's largest cities in the period 2000–2020, %.

In other cities, the major transformation of GI concerns non-tree vegetation, while tree vegetation loss is little affected. The loss of this type of GI is about 3% in Nizhniy Novgorod and 7% in Moscow, generally due to the incursion of new residential districts at the cities' outskirts. In Moscow, the majority of green area has been replaced by roads and other infrastructure buildings as part of the development of "New Moscow" since 2012. Much less GI has been lost in the inner-city areas. It is important to note that our assessment at the city level may contains some errors, especially concerning small local changes inside the urban core, due to the spatial resolution of 30 x 30 m offered by the Landsat images and Hansen's tree cover raster. In particular, we assume that our results may be somewhat overstated in cities with a large share of agricultural lands within urban borders, since NDVI values can in reality reflect plowed fields rather than built-up areas. Although we attempted to deal with this problem by distinguishing agricultural lands by visual interpretation and OSM materials, the particular resolution of the remote sensing data still leaves some room for error.

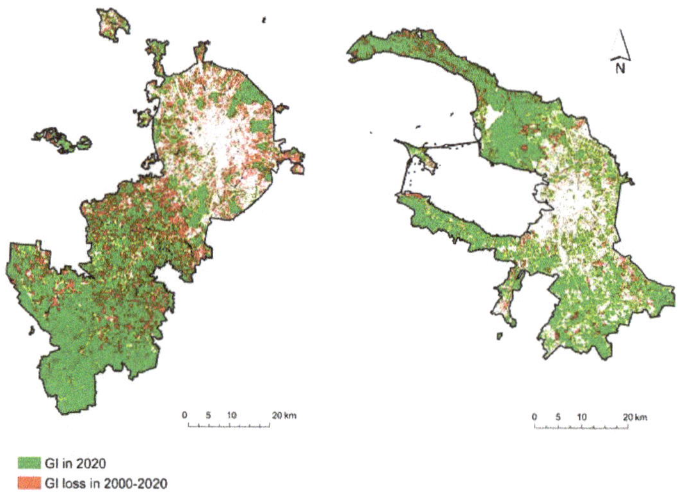

Figure 6. Loss of GI in the period 2000–2020, % in Moscow (**left**) and Saint Petersburg (**right**).

In general, and with the exception of Nizhniy Novgorod, the relative change in green area may seem quite insignificant in the investigated cities. Yet the absolute losses in green area may be fairly large: for example, Saint Petersburg lost about 100 km^2 of GI and Novosibirsk 28 km^2. While most transformations occur on the urban outskirts, so that the most densely populated inner cities have preserved their green areas and local residents can still enjoy their recreational ES, the depletion of the green belt brings with it a reduction in regulating and supporting services. In general, suburban forests perform most air and water purification. Agricultural lands have not changed much during our period of investigation, so that this provisioning service has also remained stable. According to our modeling, the most common land cover transition is when non-tree vegetation makes way for built-up areas. This is the worst form of transformation according to [41], since it leads to a total loss of ESs. The transition from tree vegetation to non-tree vegetation is less common and less damaging for the variety and volume of ESs. Regarding the transition from woodlands to built-up areas, although this change does take place in some central districts, it is not widespread in the studied cities.

4. Discussion

While the methods of ES modeling used here are applicable at the city level, they lack a certain accuracy and level of detail. For instance, the volume and variety of provided ESs depend not only on the total area of GI but also its quality. In this study, we simplified the quality assessment to consider only three types of GI: tree vegetation, non-tree vegetation, and agricultural land. Although this simplification is permissible for our scale and target results, a more complex assessment for the purposes of urban planning and development will require more indices and parameters. Vegetation should be classified by its health and condition [42], while the functions and value of green elements should be determined not only by their size but also form and location [43]. Moreover, fragmentation is an important factor to consider as it can determine the quality of ecosystems, various supporting services, and GI sustainability [44].

Ignoring intracity differences, our results reflect the average situation for an entire city. However, the supply and demand of ESs may vary between districts, depending on the population density, the location of industrial and residential areas, the volume of emissions from particular sources, as well as geographical and spatial features (prevailing wind directions and other atmospheric conditions, built-up structures, etc.) [45].

The results of this survey demonstrated that most cities with a large share of GI are sufficiently provided with climate regulating services. However, we stress that GI ele-ments, which supply this and other services, are usually concentrated outside the popu-lated urban core. In Russia, the administrative boundaries of most cities encompass not only built-up areas but vast territories of forests, grasslands, and agricultural lands so that the actual urban core can be two to three times smaller than the administrative boundaries. This means that despite the huge volume of supplied ESs, the real amount of consumed ESs can be much smaller, since consumers' homes are distant from the GI. We uncovered this problem in almost all the studied cities. The expansion of formal urban boundaries at the expense of vast suburban areas, as in the case of Moscow (see Section 3.7), can significantly improve the average indicators of ES provision, while in the city center, the contraction of GI can cause a decrease in ES supply to residents there.

For instance, the efficient performance of heat mitigation services depends first and foremost on the location of green elements, which ideally should be situated within the densely populated urban center and not only at the outskirts. This built-up center forms the core of the heat island, exposing the majority of city dwellers to excessive heat [46]. While our data only allows us to assess the configuration of green infrastructure qualitatively, the results show that the main challenge to climate change adaptation in all of Russia's largest cities is a lack of green infrastructure in the urban center. To improve our assessment methodology of climate regulation, it would be necessary to consider not just the scale of heat islands but also the size of existing green elements. Better knowledge of the form, size, and location of parks can help to calculate the affected area with greater accuracy.

To assess more precisely the input of urban agricultural lands to food provisioning, it would be necessary to take account of the particular features of the local fruit and vegetable markets as well as the consumption of crops (statistical data on different types of crops) and people's preferences for local products [38].

Results on recreational services can be made more accurate by, firstly, using the most reliable information on the extent of green infrastructure capable of performing recreational services and its current state; secondly, by defining the types of recreational activities in order to specify the norms of carrying capacity for each one individually; and thirdly, by detailing the norms for the permissible recreational carrying capacity that does not deteriorate natural ecosystems in consideration of the specific environmental and geographical features of the area, i.e., slopes, soil and vegetation cover, factors that disturb fauna, etc. [28]. Although it is too early to make predictions about changes to urban ESs in this regard, we can speculate that recreational services will gradually become the prevailing component in the pattern of ESs.

Regarding the assessment of air purification by trees, currently, there is no such tool in Russia, such as iTree, to show absorption rates and limits for different species native to the country [33]. As an alternative, we used average values for US and Canadian cities of approximately the same populations and with similar vegetation zones. Obviously, the results would be more accurate if dedicated data for Russian cities and vegetation can be used.

To calculate the mentioned parameters, you need well-structured and constantly updated materials at the local level. Currently, open data is collected for the whole city at best and not for the constituent districts. While economic and demographic data are more or less exhaustive, diverse environmental datasets on species, weather conditions, protected areas, and vegetation state are not presented in a single well-organized form. Moreover, as there is no official data on emissions from point sources, it is impossible to locate districts with the highest demand for air purification services. Detailed studies cannot be carried out at the city level due to the lack of data on visitors to parks or their recreational preferences. The literature contains a number of studies that have investigated the ESs of individual green elements, urban districts, or even entire cities in Russia; yet, very few methods have been proposed to assess the ESs in several cities simultaneously. Therefore, the improved availability of more detailed databases on cities could ensure not only an accurate ES calculation but also the assessment of a greater variety of ESs.

Some may argue that it is not useful or feasible to compare the volume and variety of ES in cities that are situated in completely different natural zones. In our opinion, this is generally true for small towns in Russia. It is obvious that towns in the north of the country are less green than in the south. Towns of medium size do not greatly transform the environment so that natural conditions still remain the governing factor behind GI features. However, large cities are a completely different case. They form their own urban environment with transformed or reduced natural processes, and the quality as well as the quantity of GI largely depend on urban planning [25]. In turn, improvement measures are stimulated or hampered by investments, public policy, as well as the general needs and trends of the population.

Regarding especially valuable GI elements, these include not only urban PAs but also riparian green zones. Unlike most urban forests, rivers usually run through urban centers. They are of exceptional importance for several reasons. First, they perform extra heat mitigation services. Second, natural riparian ecosystems are of the utmost importance due to the supporting and regulating services they provide. Like wetlands, these ecosystems are responsible for the majority of water purification, erosion, and flood control services. Third, areas adjacent to rivers and other water bodies are popular for the various recreational opportunities they provide. Fourth and finally, being unique intrazonal ecosystems, they are a natural habitat for a number of species. Rivers are natural corridors between different green elements in the city as well as between urban and suburban ecosystems. Hence, they support the stability and health of GI. In this way, riparian areas can be multifunctional elements of GI [47]. Yet having provided water and energy to industries for many decades, riparian areas in many large Russian cities are burdened with abandoned plants and factories, old embankments, and docks. Nonetheless, rivers and riparian zones still constitute valuable ES assets in many cities, and thus must be improved and developed.

Russian cities can be said to be fairly unique in regard to GI and related ESs. On the one hand, their population growth rates over the last 20 years have been relatively high, and urban sprawl is quite evident in many cities. On the other hand, and unlike many Asian or American cities of similar populations, many large cities in Russia have preserved extensive green areas from the Soviet era, which are now often being transformed by new developments and upgrading plans or because of the encroachment of built-up areas. The remaining parts of the Soviet green networks are actually natural ecosystems with sedimentary forests, which function just like any other natural forest outside the urban area. Regarding the provision of ESs, these GI elements are of particular value as they are highly efficient suppliers of regulating and supporting services in comparison to other artificial or semi-natural green elements. However, to preserve these urban forests in their current state, it is necessary to protect them from intense recreational or other use, which for the city authorities is an inefficient use of land. It is crucial to further develop plans and schemes on GI structuring in urban and suburban planning and GI assessment by using more indicators like biodiversity richness, connectivity and fragmentation, and zoning the studied area to define the most vulnerable or valuable territories [22].

Today, the green infrastructure of many of Russia's large cities is gradually transforming. Although extensive woodlands still make up the biggest part of urban green infrastructure, new districts and infrastructure are slowly swallowing these up; in their place, new artificial green areas, such as modern parks and squares, are being constructed. The main function of these elements is recreation. They also perform other services, such as heat mitigation or water purification, but at much lower volumes. On the positive side, these artificial elements can provide desperately needed green infrastructure in densely built-up inner-city districts; in contrast, most urban forests are situated in low-density areas on the outskirts, which already possess much street vegetation.

In Russia, the largest cities often function as role models to others. This is particularly true of Saint Petersburg and Moscow. In both of these cities, an ecological strategy has been announced to encourage an environmental-oriented lifestyle that includes a healthy urban environment featuring GI. In recent years, however, we note not only a loss of GI (see Section 3.7, Figure 6) but also a deterioration in its quality due to the unfavorable distribution of natural areas [9]. Such trends in real GI and ES change serve as dangerous examples to other cities, despite the declared strategies to improve the urban environment. Nevertheless, for most cases, there was an insignificant decline in green area. In the context of urbanization, GI intactness can be interpreted as a positive result. There is not a lot of undeveloped urban area to increase GI, but it is necessary to preserve existing green elements.

The question of the most efficient greening scheme for urban areas remains unanswered for many cities [48,49]. Moreover, with their specific structure and development history, Russia's cities also require specific planning solutions.

5. Conclusions

Our analysis of the current state of urban GI and the most important associated ESs in Russia's 16 largest cities showed that both GI and ESs are vital factors in determining the quality of life of city-dwellers. However, there is great variation between the examined cities in their stocks of GI and supply of ESs. Some were found to have very low ES provision. This suggests that indicators for GI and ESs should be included in the decision-making processes of large cities.

In particular, we analyzed the key regulating services (i.e., the removal of air pollutants from point sources and automobiles, regulation of the urban microclimate), recreational ESs (i.e., the formation of natural conditions for recreation), and provisioning ESs (i.e., food production). By comparing the provided and demanded ESs, we found that all of Russia's largest cities are deficient in the ES of air purification. Reflecting minimum standards for the supply of green space for recreation, it can be stated that the availability of recreational ES is sufficient in most cases. These results, however, need careful elaboration based on the maximum recreational carrying capacity of ecosystems, which varies for different types of ecosystems, climatic conditions, and recreational activities.

Urbanization processes, mass construction, increasing population densities, and the environmentally inadequate development of transport infrastructure in Russia's most populated cities are reducing both the area and quality of GI, and consequently the supply of ESs, especially in districts undergoing breakneck development. As a result, environmental indicators for the quality of life of the urban population are decreasing.

Our case study of the five most populated cities in the period 2000–2020 showed that densely built-up areas continued to expand at the expense of open spaces. The development programs of these cities will further reduce the stock of green space, in particular destroying mature woodland aged 40–60 years, which is especially efficient in the provision of ESs.

In Moscow and Saint Petersburg, transport infrastructure is being developed without any consideration of ecosystem preservation. This is causing fragmentation of suburban ecosystems and destroying links between suburban and urban ecosystems. The resulting isolated ecosystems have low sustainability, thus further endangering biodiversity.

Considering the key importance of ESs for the well-being of citizens and the trends of increasing urbanization and reductions in the stock of urban GI, it is urgently necessary to establish systems for the accounting and monitoring of GI and ESs in all Russian cities, especially its largest settlements. Evaluation methods should consider the diversity of GI elements, the full range of ESs, and the intracity differences of these indicators.

Major revisions to urban planning policy are required to safeguard the constitutional rights of citizens to a healthy environment. These should fully reflect scientifically backed standards for green infrastructure and its quality, including indicators of biodiversity and urban ESs.

Author Contributions: Conceptualization and methodology, O.K., O.I. and E.B.; investigation, O.K. and O.I.; writing-original draft preparation, O.K, O.I.; writing—review and editing, E.B., K.G., O.K. and O.I. All authors have read and agreed to the published version of the manuscript.

Funding: This study was a part of the TEEB Russia research, commissioned by the German Federal Agency for Nature Conservation (BfN) with funds from the German Federal Ministry for the Environment, Nature Conservation, and Nuclear Safety (BMU) and were supported by the Ministry of Natural Resources and Environment of the Russian Federation.

Acknowledgments: We would like to thank Derek Henderson for the language polishing of the manuscript. Also, we appreciate the help of Eugeny Kolbowsky in completing discussion and introduction.

Conflicts of Interest: The authors declare no conflict of interest.

References

1. Elmqvist, T. *Urbanisation, Biodiversity and Ecosystem Services: Challenges and Opportunities;* Springer: Dordrecht, The Netherlands, 2013; pp. 31–52.
2. Federal State Statistics Service of Russia (Rosstat). Data Base on the Municipal Parameters. Available online: https://rosstat.gov.ru/storage/mediabank/munst.htm (accessed on 13 November 2020).
3. Atlas of Urban Expansion. Available online: http://www.atlasofurbanexpansion.org/data (accessed on 10 October 2019).
4. Millennium Ecosystem Assessment (MEA). Available online: https://www.millenniumassessment.org/en/index.html (accessed on 10 October 2019).
5. USSR State Committee Forestry. *Temporal Guidelines for Recreational Capacity Assessment for Excursions, Tourism Management, Mass Tourism and Temporal Standards for These Capacities;* USSR State Committee Forestry: Moscow, Russia, 1987; pp. 1–33.
6. Mach, M.E.; Martone, R.G.; Chan, K.M.A. Human impacts and ecosystem services: Insufficient research for trade-off evaluation. *Ecosys. Serv.* **2015**, *16*, 112–120. [CrossRef]
7. Wang, J. A multiscale analysis of urbanization effects on ecosystem services supply in an urban megaregion. *Sci Total Environ.* **2019**, *662*, 824–833. [CrossRef]
8. Dear, M.; Scott, A.J. *Urbanization and Urban Planning in Capitalist Society;* Routledge: Oxfordshire, UK, 2018; Volume 7, pp. 3–20.
9. Klimanova, O.A. (Ed.) Ecosystem Services of Russia: Prototype National Report. In *Green Infrastructure and Ecosystem Services of the Largest Cities in Russia;* BCC Press: Moscow, Russia, 2021; Volume 3, pp. 14–42.
10. Sbitnev, A.V. Methodological aspects of the assessment of phytotoxic properties of ice-melter reagents. *Hyg. Sanit.* **2016**, *95*, 773–778. [CrossRef]
11. Lafortezza, R.; Davies, C.; Sanesi, G.; Konijnendijk, C.C. Green infrastructure as a tool to support spatial planning in European urban regions. *IForest* **2013**, *6*, 102–108. [CrossRef]
12. Grunewald, K.; Junxiang, L.; Gaodi, X.; Lennart, K.S. *Towards Green Cities—Urban Biodiversity and Ecosystem Services in China and Germany;* Springer: Cham, Switzerland, 2018; pp. 105–173.
13. Grunewald, K.; Bastian, O.; Louda, J.; Arcidiacono, A.; Brzoska, P.; Bue, M.; Cetin, N.I.; Dworczyk, C.; Dubova, L.; Fitch, A.; et al. Lessons learned from implementing the ecosystem services concept in urban planning. *Ecosys. Serv.* **2021**, *49*, 101273. [CrossRef]
14. Von Stulpnagel, A.; Horbert, M.; Sukopp, H. The importance of vegetation for the urban climate. In *Urban Ecology;* Sukopp, H., Hejny, S., Kowarik, I., Eds.; SPB Academic Publishing: Hague, The Netherlands, 1990; pp. 175–193.
15. Bolund, P.; Hunhammer, S. Ecosystem Services in Urban Areas. *Ecol. Econ.* **1999**, *29*, 293–301. [CrossRef]
16. Gómez-Baggethun, E.; Gren, D.; Barton, J.; Langemeyer, T.; McPhearson, P.; O'Farrell, E.; Andersson, Z. Urban ecosystem services. In *Urbanization;* Elmqvist, T., Fragkias, M., Goodness, J., Güneralp, B., Marcotullio, P., McDonald, R.I., Eds.; Springer: Dordrecht, The Netherlands, 2013; pp. 175–251.
17. Gómez-Baggethun, E.; Barton, D.N. Classifying and valuing ecosystem services for urban planning. *Ecol. Econ.* **2013**, *86*, 235–245. [CrossRef]
18. Forman, R.T. *Urban Ecology: Science of Cities;* Cambridge University Press: Cambridge, UK, 2014; pp. 126–129.
19. Salata, K.D.; Yiannakou, A. Green Infrastructure and climate change adaptation. *TeMA J. Land Use* **2016**, *9*, 7–24.
20. TEEB. The Economics of Ecosystems and Biodiversity: Ecological and Economic Foundations. Available online: http://teebweb.org/our-work/country-studies/teeb-inspired-studies/russia/ (accessed on 10 October 2019).
21. Honeck, E.; Moilanen, A. Implementing Green Infrastructure for the Spatial Planning of Peri-Urban Areas in Geneva, Switzerland. *Sustainability* **2020**, *12*, 1387. [CrossRef]
22. Bukvareva, E.N.; Zamolodchikov, D.G. *Ecosystem services of Russia: Prototype National Report Vol. 1. Terrestrial Ecosystems Services;* BCC Press: Moscow, Russia, 2018; pp. 5–36.
23. Chaparro, L.; Terradas, J. *Report on Ecological Services of Urban Forest in Barcelona;* Barcelona City Council, Department of Environment: Barcelona, Spain, 2009; pp. 1–96.

24. Pickett, S.T.A.; Cadenasso, M.; Grove, M.; Nilon, C.; Pouyat, R.; Zipperer, W.; Costanza, R. Urban Ecological Systems: Linking Terrestrial Ecological, Physical, and Socioeconomic Components of Metropolitan Areas. *Annu. Rev. Ecol. and Syst.* **2003**, *32*. [CrossRef]
25. Meerow, S.; Newell, J.P. Spatial planning for multifunctional green infrastructure: Growing resilience in Detroit. *Landscape Urban Plan.* **2017**, *159*, 62–75. [CrossRef]
26. Vladimirov, V.V. *City and Landscape*; Mysl: Moscow, Russia, 1986; pp. 3–26.
27. Bityukova, V.R. Environmental Rating of Cities of Russia. *Ecol. Ind. Russ.* **2015**, *19*, 34–39.
28. Kim, H. Understanding recreation demands and visitor characteristics of urban green spaces: A use of the zero-inflated negative binomial model. *Urban For. Urban Gree.* **2021**, *201*, 127332. [CrossRef]
29. System of Environmental-Economic Accounting—Ecosystem Accounting (SEEA EA). Available online: https://seea.un.org/ecosystem-accounting (accessed on 1 September 2021).
30. Hansen, M.C.; Potapov, P.V.; Moore, R.; Hancher, M.; Turubanova, S.A.; Tyukavina, A.; Thau, D.; Stehman, S.V.; Goetz, S.J.; Loveland, T.R.; et al. High-Resolution Global Maps of 21st-Century Forest Cover Change. *Science* **2013**, *342*, 850–853. [CrossRef] [PubMed]
31. Tyrväinen, L.; Pauleit, S.; Seeland, K.; De Vries, S. Benefits and uses of urban forests and trees. In *Urban Forests and Trees*; Reference Book; Springer: Berlin, Germany, 2005; pp. 81–114.
32. Nowak, D.J.; Crane, D.E.; Stevens, J.C. Air pollution removal by urban trees and shrubs in the United States. *Urban For. Urban Gree.* **2016**, *4*, 115–123. [CrossRef]
33. Nowak, D.J.; Hirabayashi, S.; Doyle, M.; McGovern, M.; Pasher, J. Air pollution removal by urban forests in Canada and its effect on air quality and human health. *Urban For. Urban Gree.* **2018**, *29*, 40–48. [CrossRef]
34. Federal Service for Hydrometeorology and Environmental Monitoring of Russia (Rosgidromet). *State of Air pollution in the Cities in Russia*; Annual Report; FGBU: Saint Petersburg, Russia, 2017; pp. 4–227.
35. Sanitary Regulations and Standards. *SANPIN 2.2.1/2.1.1.1200-03 Sanitary Protection Zones and Sanitary Classification of Industrial, Residential and Other Objects*; Standartinform: Moscow, Russia, 2003; pp. 1–53.
36. Gill, S.; Handley, J.; Ennos, R.; Pauleit, S. Adapting Cities for Climate Change: The Role of the Green Infrastructure. *Built Environ.* **2007**, *33*, 115–133. [CrossRef]
37. Höppe, P. The physiological equivalent temperature—a universal index for the biometeorological assessment of the thermal environment. *Int. J. Biometeorol.* **1999**, *43*, 71–75. [CrossRef]
38. Yacamán, O.; Carolina, D.F.; Rafael, M.O. Green infrastructure planning in metropolitan regions to improve the connectivity of agricultural landscapes and food security. *Land* **2020**, *9*, 414. [CrossRef]
39. Sorokina, E.A.; Lokoshhenko, M.A. Cold Islands in Moscow. In *Proceedings of the V International Ecological Congress ELPIT-2015*; SNC: Samara, Russia, 2015; pp. 16–20.
40. Ma, J. Green Infrastructure and Urban Liveability: Measuring Accessibility and Equity. Doctoral dissertation, University of Auckland Research, Auckland, New Zealand, May 2016.
41. Jopke, C. Interactions among ecosystem services across Europe: Bagplots and cumulative correlation coefficients reveal synergies, trade-offs, and regional patterns. *Ecol. Ind.* **2015**, *49*, 46–52. [CrossRef]
42. Kureel, N. Modelling vegetation health and stress using hyperspectral remote sensing data. *Modeling Earth Sys. Environ.* **2021**, *2*, 1–16.
43. Seppelt, R.; Dormann, C.; Eppink, F.; Lautenbach, S.; Schmidt, S. A quantitative review of ecosystem service studies: Approaches, shortcomings and the road ahead. *J. Appl. Ecol.* **2011**, *48*, 630–636. [CrossRef]
44. Kowe, P.; Mutanga, O.; Dube, T. Advancements in the remote sensing of landscape pattern of urban green spaces and vegetation fragmentation. *Int. J. Remote Sens.* **2021**, *42*, 3797–3832. [CrossRef]
45. Zulian, G.; Raynal, J.; Hauser, R.; Maes, J. Urban Green Infrastructure: Opportunities and Challenges at the European Scale. In *Ecosystem Services and Green Infrastructure. Cities and Nature*; Arcidiacono, A., Ronchi, S., Eds.; Springer: Cham, Switzerland, 2021; pp. 17–28.
46. Seiwert, A.; Stefanie, R. Understanding the term green infrastructure: Origins, rationales, semantic content and purposes as well as its relevance for application in spatial planning. *Land Use Policy* **2020**, *97*, 104785. [CrossRef]
47. Hermida, M.A.; Cabrera, N.; Osorio, P.; Cabrera, S. Methodology for the assessment of connectivity and comfort of urban rivers. *Cities* **2019**, *95*, 102376. [CrossRef]
48. Haines-Young, R.; Potschin, M.B. *Common International Classification of Ecosystem Services (CICES)and Guidance on the Application of the Revised Structure, Vol. 5.1.*; The Paddocks: Nottingham, Great Britain, 2018; pp. 7–12.
49. Von Haaren, C.; Lovett, A.A.; Albert, C. *Landscape Planning with Ecosystem Services: Theories and Methods for Application in Europe*; Landscape Series 24; Springer: Berlin/Heidelberg, Germany, 2019; pp. 19–42.

land

MDPI

Article

High Resolution Land Cover Integrating Copernicus Products: A 2012–2020 Map of Italy

Paolo De Fioravante [1,2,*], Andrea Strollo [1], Francesca Assennato [1], Ines Marinosci [1], Luca Congedo [1] and Michele Munafò [1]

1 Italian Institute for Environmental Protection and Research (ISPRA), Via Vitaliano Brancati 48, 00144 Rome, Italy; andrea.strollo@isprambiente.it (A.S.); francesca.assennato@isprambiente.it (F.A.); ines.marinosci@isprambiente.it (I.M.); luca.congedo@isprambiente.it (L.C.); michele.munafo@isprambiente.it (M.M.)

2 Department of Innovation in Biology, Agri-Food and Forest Systems (DIBAF), University of Tuscia, Via San Camillo de Lellis SNC, 01100 Viterbo, Italy

* Correspondence: pdefioravante@unitus.it

Abstract: The study involved an in-depth analysis of the main land cover and land use data available nationwide for the Italian territory, in order to produce a reliable cartography for the evaluation of ecosystem services. In detail, data from the land monitoring service of the Copernicus Programme were taken into consideration, while at national level the National Land Consumption Map and some regional land cover and land use maps were analysed. The classification systems were standardized with respect to the European specifications of the EAGLE Group and the data were integrated to produce a land cover map in raster format with a spatial resolution of 10 m. The map was validated and compared with the CORINE Land Cover, showing a significant geometric and thematic improvement, useful for a more detailed and reliable evaluation of ecosystem services. In detail, the map was used to estimate the variation in carbon storage capacity in Italy for the period 2012–2020, linked to the increase in land consumption

Keywords: land cover; Copernicus; land monitoring; ecosystem services; CORINE land cover; carbon storage capacity; EAGLE matrix

Citation: De Fioravante, P.; Strollo, A.; Assennato, F.; Marinosci, I.; Congedo, L.; Munafò, M. High Resolution Land Cover Integrating Copernicus Products: A 2012–2020 Map of Italy. *Land* **2022**, *11*, 35. https://doi.org/10.3390/land11010035

Academic Editor: Adrianos Retalis

Received: 28 November 2021
Accepted: 20 December 2021
Published: 26 December 2021

1. Introduction

Territory is a source of resources such as food, biomass and raw materials and provides essential ecosystem services that support production functions, regulate natural cycles, provide cultural and spiritual benefits [1].

The European Environment Agency introduced the concept of "land system", which defines the territory as a set of terrestrial components that includes all the processes and activities related to its anthropic use [2,3]. This concept considers the territory as an integrated system [4] that combines everything related to land use and land cover. The study of land cover and land use is essential to understand the causes and effects of human-determined changes [5], seeing as now changes and transformations that occur within the land system lead to consequences for the well-being of humans and the environment at the local, regional and global levels.

1.1. Estimation and Monitoring of Ecosystem Services

The management of territory is fundamental, since its transformation alters the environmental processes and related ecosystem services [6], which are the benefits that humans obtain directly or indirectly from terrestrial ecosystems [7].

Since 2005, the concept of ecosystem services has been placed on the political agenda [8–10] thanks also to the Millennium Ecosystem Assessment (MEA). The MEA is an international research project launched in 2001 with the aim of evaluating the consequences of ecosystem changes on human well-being and establishing actions to improve the conservation and sustainable use of ecosystems and their contribution to health.

The Millennium Ecosystem Assessment [11] provides a classification of ecosystem services which consist of four groups:

- *Provisioning services*, which provide products obtained from ecosystems, such as food, raw materials and water.
- *Regulating services*, i.e., the benefits provided through ecosystem processes, such as carbon storage, erosion control.
- *Cultural services*, which represent the nonmaterial benefits that people obtain through spiritual enrichment, cognitive development and aesthetic experience.
- *Supporting services*, which constitute a "transversal" category that supports the production of other services, providing living space for plants and animals or maintaining genetic diversity. They differ from other categories since their impact on people is indirect or is visible after a very long period.

The Economics of Ecosystems and Biodiversity [12] provides a classification aligned with MEA, while the Common International Classification of Ecosystem Services (CICES) excludes "support services" and renames the category of "regulation services" as "regulation and maintenance services", including habitat maintenance [13,14].

In 1997, Costanza [7] published one of the first assessments of ecosystem services, which was then updated in 2014 [15]. Other application examples for more limited areas are readily available in the scientific literature: assessment on pollination can be found both from a biophysical [16,17] and economic [18–20] point of view. The United Kingdom was among the first European country to draw up a complete official report in line with the Millennium Ecosystem Assessment [21]. Rabe et al. [22] developed a network of national indicators on ecosystem services in Germany using CORINE Land Cover data for the analysis of nine ecosystem services divided into three classification categories (according to CICES). Spain published an official report in 2016 [23] related to twelve ecosystem services. Outside the European context, there are applications on a national scale in China [24], South Africa [25] and South Korea [26].

Soil has fundamental functions for nature and humankind and is the source of many ecosystem services [27–29]:

- *Fertility*: the nutrient cycle ensures fertility in the soil and, at the same time, the release of nutrients necessary for plant growth;
- *Filter and reserve*: the soil can act as a filter against pollutants and can store large quantities of water, useful for plants and for the mitigation of floods;
- *Structural*: soils represent the support for plants, animals and infrastructures;
- *Climate regulation*: the soil, in addition to being the largest carbon sink, regulates the emission of important greenhouse gases (N_2O and CH_4);
- *Conservation of biodiversity*: soils are an immense reservoir of biodiversity. They represent the habitat for thousands of species capable of preventing the action of parasites or facilitating waste disposal;
- *Resource*: soils can be an important source of supply of raw materials.

All soils perform their functions at the same time (food production, water purification, carbon sequestration, etc.) in a different way according to land use and pedogenetic characteristics. For example, the rate of carbon sequestration and water purification is higher in a natural area than in an agricultural one, which however has greater production capacity.

Carbon sequestration and storage is an important regulatory service linked to the attitude of ecosystems to fix greenhouse gases. This service contributes to climate regulation and is fundamental in defining adaptation strategies to climate change [30,31]. The capacity to store carbon depends, among other things, on land use and cover and on the climate [32]. Land use, land use change and forestry (LULUCF) activities can act as sources of emissions or store carbon by acting as sinks. In particular, natural and seminatural forest ecosystems have the highest carbon sequestration potential. Once natural land is urbanized or degraded, it loses its ability to retain carbon which, as a result, is emitted into the atmosphere [33]. Urban expansion, land consumption, deforestation and forest degradation limit the ability of natural areas to store carbon and have contributed to these emissions by releasing carbon stored in forests, vegetation and soil [6,34,35].

1.2. Land Monitoring

Land cover and land use are strongly related and for many applications both information categories are required [36]. To meet different monitoring needs, data with different characteristics from a spatial, temporal and thematic point of view were introduced.

In this respect, different initiatives have been developed. The purpose of the Copernicus program is to collect information on the earth's surface and organize it according to criteria that allow to compare different data, to exchange data between EU countries and to increase the number of users. The Copernicus Land Monitoring Service (CLMS) allows researchers to obtain geographic information on soils and on numerous variables related to them (such as the state of the vegetation or the water cycle), supporting applications in a wide variety of sectors, such as territorial planning, management of water resources and forests, agriculture and food security. CORINE Land Cover is one of the main products belonging to CLMS. It has guaranteed information for the whole European territory since 1990, with 44 land cover and land use classes and geometric detail of 25 hectares.

Recently, data with higher spatial and thematic details have been introduced in the context of the CLMS Local Component. It aims at providing detailed information on critical areas from an environmental point of view, which require specific and detailed monitoring. Currently, this Copernicus component offers land cover and land use maps in vector format, with high spatial resolution and a 6-year update frequency for four categories of areas. Urban Atlas refers to the CLC classification system, describing with higher detail the land cover and land use characteristics of urban areas, while Riparian Zone and Natura 2000 use the ecosystem types defined in the Mapping Assessment of Ecosystems and their Services (MAES) [37], which are based on the CLC classes too.

These aforementioned data adopt classification systems based on different combinations of land cover and land use classes that are difficult to compare and integrate with those of other data. In order to coordinate data flows from a thematic point of view, the EAGLE group (EIONET Action Group on Land monitoring in Europe) was created. It aims at defining a conceptual methodology to describe land cover and land use information in a consistent data model. EAGLE is not a classification system but a tool to describe classes of a given classification system by tracing them to the segments related to the three categories. This allows to better understand the characteristics, the overlaps and the possible conversions between different classification systems and provides a basis to define new ones. The EAGLE model aims at separating the land cover and land use components through data modelling systems applicable at different scales and in different contexts, while maintaining compatibility with existing databases.

The EAGLE data model is based on the definition of three blocks, called "categories":

- *Land cover components (LCC),* which refer to the definition of "land cover" provided by the INSPIRE directive 2007/2/CE. The LCCs are mutually exclusive and exhaustive and can be used as a modelling element to semantically describe a class definition or to map landscape;
- *Land use attributes (LUA),* that follow in principle the Hierarchical INSPIRE Land Use Classification System (HILUCS), with some changes to fit the purpose of the EAGLE concept. The LUA are attached to the land cover unit;
- *Landscape characteristics (CH),* which describe further details of the land cover components. The first level distinguishes "land management", "spatial pattern", "crop type", "mining product type", "ecosystem types", "height zone", "(bio-)physical characteristics", "general parameters", "status" and "temporal" parameters. This enhances the integration between national activities and European land monitoring initiatives encouraging a bottom-up approach in data production.

The problem of interoperability and non-homogeneity between data is also evident at a national level. The National Land Consumption Map offers national coverage, with annual update and EAGLE compliant classification system, while most of the data available at the regional level are inconsistent, not updated and difficult to relate to each other.

Despite the large amount and variety of land cover and land use data available at national and European level, currently CLC is the only product capable of supporting an assessment of ecosystem services on a national scale [38], since it guarantees the mapping of the entire national territory and has a thematic detail suitable for the purpose. However, the low spatial resolution and the presence of mixed classes reduce the reliability of the assessments based on them.

In this sense, the first objective of this research concerns the development of a methodology that makes the main Copernicus and national land cover and land use data comparable and integrable, in order to obtain a product with national coverage that allows to overcome the limits of the CLC in terms of classification system and geometric detail.

Furthermore, the activity refers to an EAGLE compliant classification system with a thematic detail useful for conducting an assessment of ecosystem services, with particular reference to the variation of carbon stocks. This change was assessed with respect to theincrease in land consumption between 2012 and 2020.

2. Materials and Methods

2.1. Overview

The methodology presented in this study integrates Copernicus and national data for the production of a land cover map capable of supporting the ecosystem services assessment. Data were reclassified according to an EAGLE compliant classification system and merged into a 10 m resolution land cover map of Italy. The map was used to assess the loss of carbon storage capacity for the period 2012–2020, associated with land consumption (Figure 1).

2.2. Study Area

The analysis was carried out for the entire Italian territory (Figure 2), which covers 301,338 km^2. The country is composed mainly of hills (41.0%) and mountains (35.0%), while the remaining 23.0% of the territory is covered by plains. To the north is the mountain range of the Alps, which exceeds 4,000 m in altitude. In this area, the alpine climate prevails, with high rainfall with a maximum of 2,500–3,500 mm. In the peninsular area there is the Apennine mountain range, reaching its highest peak in Abruzzo with Gran Sasso (2,912 m) and characterized by a continental climate. The coastal area has a Mediterranean climate, with average annual rainfall that reaches a minimum of 500 mm in Apulia and Molise.

Land cover is characterized by forest in mountain areas, with conifer concentration in alpine areas. Crops and most of the urbanized areas are concentrated in the plains and along the coast.

2.3. Land Cover Classification System

The activities described in this paper refers to a sixteen-class classification system. The classes are defined in accordance with the EAGLE group specifications [39–41] and are organized into five levels (Table 1).

The classification system is based on previous activities of the working group [41] and improved to maintain the thematic detail offered by the Copernicus and national input data. The first three classes coincide up to the third level of detail with Eagle concept land cover components. Wetland class and the fourth and fifth classification levels are based on EAGLE characteristics (LCH) definitions.

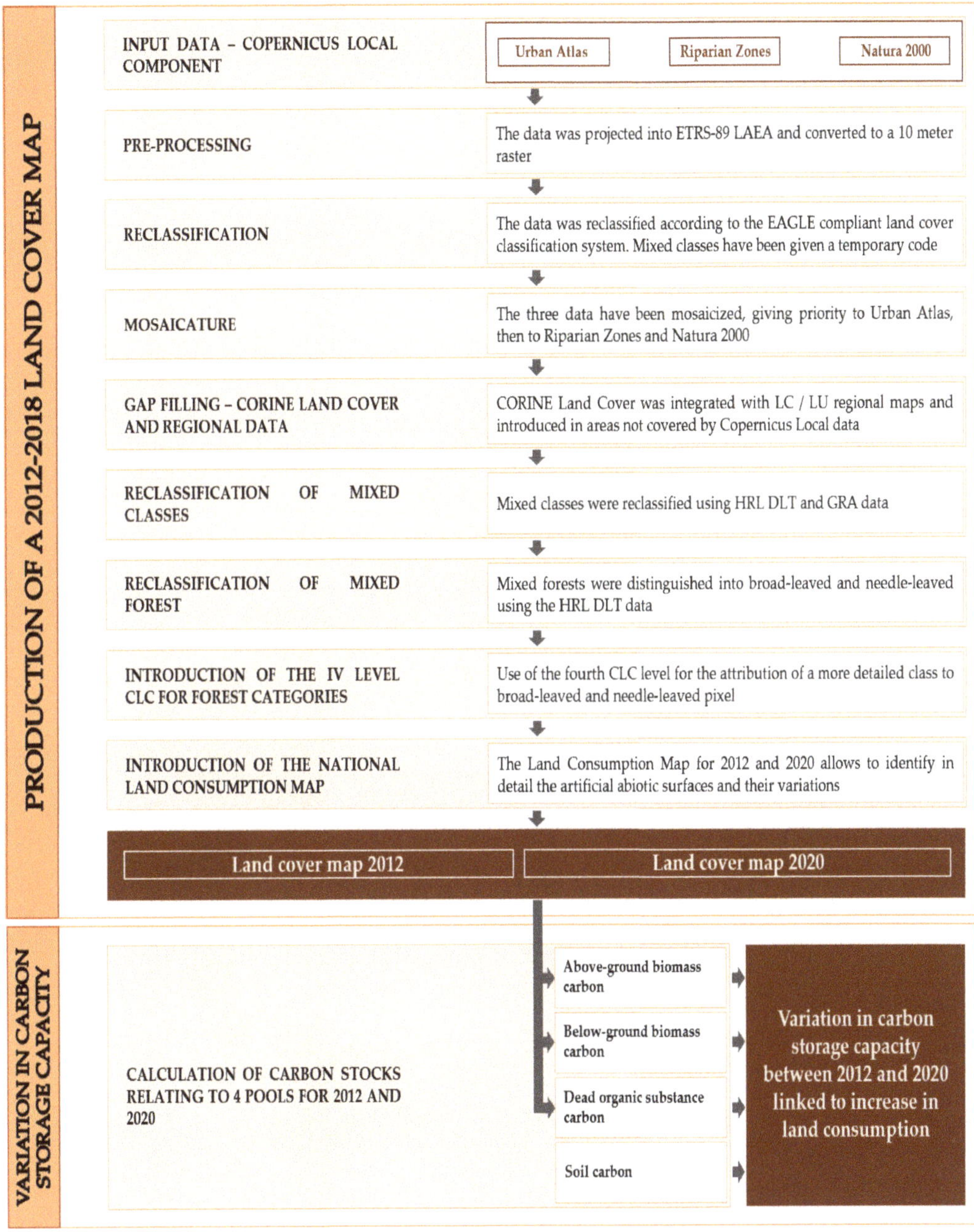

Figure 1. Workflow of the methodology for the production of the land cover map and evaluation of ecosystem services. The mosaic between Urban Atlas, Riparian Zones, and Natura 2000 was created (projected and rasterized at 10 m), then supplemented with the regional maps and CORINE Land Cover in the areas not covered by these three data. The distinction of forest categories and mixed classes was based on HRLs, while the National Land Consumption Map for 2012 and 2020 made it possible to identify artificial abiotic surfaces and their variation. The map allowed the calculation of three of the four pools considered for the estimation of the variation in carbon storage capacity.

Figure 2. Study area—Italy.

Table 1. Land cover classification system.

Land Cover									
I Level		II Level		III Level		IV Level		V Level	
1	Abiotic Non-vegetated surfaces	11	Artificial abiotic						
		12	Natural abiotic	121	Consolidated (bare rocks, cliffs)				
				122	Unconsolidated (beaches, dunes, sands)				
2	Biotic vegetated surfaces	21	Woody vegetation	211	Trees	2111	Broad-leaved		
						2112	Needle-leaved		
						2113	Permanent crops	21131	Orchards
								21132	Olive groves
								21133	Wood plantations
				212	Shrubs	2121	Vineyards		
						2122	Shrubland		
		22	Herbaceous vegetation	221	Periodically	2211	Pastures		
						2212	Arable land		
				222	Permanent				
3	Water surfaces	31	Water bodies						
		32	Permanent snow and ice						
4	Wetlands								

1. **Abiotic non-vegetated:** The class includes any unvegetated surfaces. At the second classification level, the class is subdivided between man-made artificial structures (artificial abiotic surfaces) and natural material surfaces (natural abiotic surfaces).

Artificial abiotic surfaces include impervious surfaces and reversible land consumption, according to the definition of ISPRA National Land Consumption Map [42]. Reversible land consumption indicates areas where the natural cover has been removed or replaced due to anthropogenic interventions, such as soil compaction or excavation, forming a non-impermeable and undeveloped surface. The main difference from the EAGLE model is the inclusion of quarries and extraction sites in this class, as detailed by De Fioravante et al. [41]. Natural abiotic surfaces are any kind of material that remains in its natural consistence or form, either with or without anthropogenic influence. The latter class presents a third-level distinction between consolidated unvegetated natural surfaces (bare rocks, cliffs) and unconsolidated unvegetated natural surfaces (beaches, dunes, sands).

2. **Biotic vegetated:** The class includes any vegetated surfaces, with or without anthropogenic influence. At the second classification level, woody and herbaceous vegetation are distinguished.

The third classification level of woody vegetation divides trees and shrubs. Trees are then classified at the fourth level among broad-leaved, needle-leaved and permanent crops. Broad-leaved and needle-leaved have a fifth level based on CORINE Land Cover, while permanent crops distinguish olive groves, orchards and woody plantation. Shrubs are distinguished on the fourth level in vineyards and natural areas.

Herbaceous vegetation is divided into periodic and permanent. The periodic herbaceous class corresponds to the managed areas, subdivided at the fourth level into pastures and arable land, while the natural herbaceous is related to natural unmanaged grassland.

3. **Water surfaces:** The class includes natural or artificial solid and liquid water. The second classification level distinguishes water bodies from permanent snow and ice. Water bodies includes liquid water regardless of shape, position, salinity and origin. Permanent snow and ice includes accumulations that persist throughout the year regardless of seasonal variations.

4. **Wetlands:** The class does not have a direct correspondence with the EAGLE LCC, as it is considered an LCH. The class was however included to maintain the information content offered by the input data. In detail, a definition was adopted aligned with the CORINE Land Cover, including in the class the inland wetlands (inland marshes and peat bogs) and coastal wetlands (salt marshes, salines, intertidal flats) while lagoons and estuaries are associated to water bodies.

2.4. Selection of Input Data

The map is based on the integration of national and European data. At the European level, the main CLMS data for 2012 were considered, which is the reference year of most of the available data.

At the national level, the National Land Consumption Map (LCM) was used for the identification of artificial abiotic surfaces and some regional land cover and land use maps supported the characterization of agricultural areas.

2.4.1. CLMS Data

Data part of the local and pan-European components of the CLMS were considered. Regarding the local component, Urban Atlas (UA), Riparian Zones (RZ) and Natura 2000 (N2000) data were used (Table 2), while for the pan-European component, reference was made to CORINE Land Cover (CLC) and the high resolution layers (HRL) (Table 3).

Table 2. Copernicus Land Monitoring Service—local component data.

	Urban Atlas	**Riparian Zones**	**Natura 2000**
Data type	Vector	Vector	Vector
Classes	27	56	55
MMU	0.25 ha (class 1) 1 ha (class 2–5)	0.5 ha	0.5 ha

Table 3. Copernicus Land Monitoring Service—Pan-European component data.

	High Resolution Layers	**CORINE Land Cover**
Data type	Raster	Vector
Classes	4	44
MMU	Pixel 10 × 10 m	25 ha (status) 5 ha (changes)

2.4.2. National Data

LCM was used as national data [41,43–45]. It is a 10 m raster available for the entire Italian territory. In addition, regional LC/LU maps were selected for Apulia, Latium, Abruzzo, Veneto, Liguria, Lombardy and Basilicata, in order to increase the spatial and thematic detail in areas not covered by the Copernicus local component data.

2.5. Production of a National Land Cover Map Based on Copernius and National Data

2.5.1. Data Pre-Processing

Data were projected in the European Terrestrial Reference System 1989 (ETRS-89) and in Lambert Azimuthal Equal Area (LAEA) projection. Then they were converted into 10 m resolution raster, aligned with LCM and HRLs data (the only two data already in raster format).

2.5.2. Map Production

Map production involves the steps described below. From a geometric point of view, priority was given to Copernicus local products (UA, RZ, N2000), which offer better spatial resolution than other available land cover/land use data. In the areas not covered by the

Local data, the CLC and the regional maps available for 2012 (derived from CLC) have been inserted.

From a thematic point of view, the input data were homogenized with respect to the classification system of Table 1. For the mixed classes, Copernicus HRL data were used to distinguish the woody component from herbaceous vegetation, then reference was made to the definitions of such mixed classes to distinguish natural areas from agricultural ones. The CLC data made it possible to attribute a detailed prevalent forest categories to areas classified as broad-leaved and needle-leaved, while the LCM locates the consumed land.

Reclassification of Copernicus UA, RZ, N2000 Data and Creation of the Basic Mosaic

Local component data show the highest thematic and geometric detail, therefore they were used as a basis for the map. They were first reclassified according with the classification system defined in Table 1, while temporary codes have been introduced for the classes without a direct correspondence (Table 4) and for the artificial surfaces and mixed forests. These areas have been assigned a land cover class according to the procedure described below, starting from the information provided by the HRL, LCM and CLC data.

Table 4. Temporary codes for mixed land cover classes, artificial surfaces and mixed forests.

Temporary Codes		UA	RZ	N2000	CLC
91	Green urban areas	14100			141
92	Annual crops associated with permanent crops		23100	2310	241
	Complex cultivation patterns	24000	23200	2320	242
93	Land principally occupied by agriculture with significant areas of natural vegetation		23300	2330	243
94	Agroforestry		23400	2340	244
95	Transitional woodland and scrub		34100	3410	324
	Damaged forest			3500	
96	Sparsely vegetated areas	33000	61000	6100	333
97	Burnt areas		63200	6320	334
99	Permanent crops (vineyards, fruit trees, olive groves)	22000	22100	2210	
998	Urban areas and artificial surfaces	11100–13400, 14200	11110–14000	1110–1400	111–133, 142
999	Mixed forest	31000	33100		3131–3132

Reclassification was carried out considering the correspondences of Table A1 for N2000, of Table A2 for UA and of Table A3 for RZ, which are reported in Appendix A.

The three reclassified data were merged. Where more data were present at the same time, priority was given to UA, as the higher spatial resolution allows for a more detailed description of the complex pattern that characterizes urban areas. Regarding RZ and N2000, the comparison between the two data in the overlapping areas showed greater detail in RZ's description of the territory, which was considered a priority over N2000.

Reclassification and Introduction of CLC Data and Regional Maps

The CLC in 10 m raster format has been reclassified considering the correspondences of Table A4 (see Appendix A). CLC was then integrated with regional data. In detail, the polygons related to the following land cover classes (Table 1) were exported from the regional maps: 121, 122, 21131, 21132, 2121, 2122, 2212 and 222. They have been converted to raster and superimposed on the CLC in order to allow the detection of patches smaller than the CLC MMU. These data were then added to the mosaic in the areas not covered by UA, RZ and N2000.

Use of HRLs for the Classification of Uncertain Classes (Temporary Code 91–99)

The HRLs DLT, GRA and WAW related to 2012 have been mosaicized and used to assign a land cover class to the uncertain classes of Table 4.

In detail, the mixed land cover classes were compared with the HRLs DLT and GRA from the geometric point of view and with reference to the class definitions, in order to identify the data matches (Table 5). This comparison made it possible to distinguish the arboreal component (identified by HRL Forest) from the herbaceous and the unvegetated areas (identified by HRL Grassland) in the mixed classes. To distinguish natural vegetation from the agricultural areas, reference was made to the initial definition of mixed classes.

Table 5. Attribution of classes to temporary codes using HRL. DLT = dominant leaf type, GRA = grassland.

Temporary Code	Area Covered by DLT		Area Covered by GRA		No GRA and No DLT Area	
91	21131	Orchards	2211	Pastures	2121	Vineyards
92			2212	Arable land		
93	2111 or 2112	Broad-leaved or Needle-leaved	2211	Pastures	2212	Arable land
94			2212	Arable land		
95			2122	Shrubland		
96			222	Permanent		
97			222	Permanent	121	Natural abiotic surfaces
99	21131	Orchards	2121	Vineyards		

HRL WAW was used to refine water bodies. In areas covered by high resolution data (UA, RZ, N2000) the water bodies of the HRL WAW are classified as 122 "unconsolidated natural abiotic surfaces". In areas covered only with CLC, water bodies mapped by the HRL were added, as they are smaller than the CLC minimum mapping unit (MMU).

Use of HRLs to Classify Broad-Leaved and Needle-Leaved in the Mixed Forest (Temporary Code 999)

The mixed forest pixels coming from UA, RZ and N2000 were distinguished in broad-leaved and needle-leaved through HRL DLT. In the mixed forest pixels where there was no correspondence with the HRL, the class was attributed according to a proximity criterion, starting from the Euclidean allocation made on HRL DLT.

Use of the Fourth CLC Level for the Attribution of a More Detailed Prevalent Class to Broad-Leaved and Needle-Leaved Pixel

Broad-leaved and needle-leaved woodlands were reclassified to the fifth classification level (not shown in Table 1) on the basis of the fourth CLC classification level available for the Italian territory. In the forest pixels without direct correspondence with the fourth CLC level, the more detailed class was assigned according to a proximity criterion, starting from the Euclidean allocation conducted on broad-leaved and needle-leaved CLC polygons.

Inclusion of LCM and Assignment of the Class to Temporary Codes 998

The LCM for 2012 and 2020 were superimposed on the map obtained in the previous steps, since it is the most detailed available data. Pixels classified as 998 that do not fall within the LCM are related to urban areas of the Copernicus data without artificial land cover. They have therefore been attributed to the "permanent herbaceous" class.

2.5.3. Accuracy Assessment

The 16-class map for 2012 was validated. Accuracy assessment consists of a first phase of quality control conducted through a systematic visual search for macroscopic errors. A quantitative accuracy assessment was then performed through the photointerpretation of a sample of points, which were then compared with the values of the land cover map at the same locations. The sample size was assessed using the methodology proposed by Olofsson [46], which is widely adopted in literature [47–49].

The sample size (n) is calculated starting from the areas of each class and from the definition of a first attempt user accuracy, using the following equation [47–50]:

$$\frac{(\sum W_i S_i)^2}{[S(\hat{O})]^2}$$

where:

W_i—is area proportion of each classes in the considered map
U_i—user accuracy of class i. A conservative scenario was assumed, considering $U_i = 0.6$ for all classes.
S_i—standard deviation of stratum i, $S_i = \sqrt{(U_i(1 - U_i))}$ [50]. Considering $U_i = 0.6$, it turns out $S_i = 0.49$ for all classes.
$S(\hat{O})$—is the target standard error for overall accuracy. It was assumed to be 0.01 as suggested by Olofsson [46], which corresponds to a confidence interval of 1%.
A sample of size 2400 was obtained (Table 6).

Table 6. Calculation of sample size.

Land Cover Class	Area (ha)	W_i	$W_i{*}S_i$
Artificial abiotic surfaces	2,102,288	0.070	0.034
Consolidated (bare rocks, cliffs)	802,539	0.027	0.013
Unconsolidated (beaches, dunes, sands)	59,325	0.002	0.001
Broad-leaved	7,805,217	0.259	0.127
Needle-leaved	2,167,261	0.072	0.035
Orchards	619,759	0.021	0.010
Olive plantations	1,088,099	0.036	0.018
Wood plantations	46,356	0.002	0.001
Vineyards	689,746	0.023	0.011
Shrubland	1,375,125	0.046	0.022
Pastures	1,372,790	0.046	0.022
Arable land	9,183,519	0.305	0.149
Permanent herbaceous	2,337,826	0.078	0.038
Water bodies	402,830	0.013	0.007
Permanent snow and ice	37,000	0.001	0.001
Wetlands	50,295	0.002	0.001
Total	30,139,975		1
Total number of samples		**2400**	

The 2400 points were distributed among the classes considering the average value between equal and area-proportional distribution (see https://fromgistors.blogspot.com/2019/09/Accuracy-Assessment-of-Land-Cover-Classification.html, accessed on 25 December 2021) (Table 7).

Table 7. Sample allocation.

Classes	Allocation		
	Equal	Proportional	Final
Artificial abiotic surfaces	150	167	159
Consolidated(bare rocks, cliffs)	150	64	107
Unconsolidated (beaches, dunes, sands)	150	5	100
Broad-leaved	150	622	386
Needle-leaved	150	173	162
Orchards	150	49	100
Olive plantations	150	87	119
Wood plantations	150	4	77
Vineyards	150	55	103
Shrubland	150	109	130
Pastures	150	109	130
Arable land	150	731	441
Permanent herbaceous	150	186	168
Water bodies	150	32	91
Permanent snow and ice	150	3	77
Wetlands	150	4	77
Total	**2400**	**2400**	**2427**

A stratified random sampling was conducted to identify on each class the corresponding number of points calculated in Table 7. The points were photointerpreted with very high resolution images, considering 2012 as the reference year.

2.5.4. Ecosystem Services—Carbon Storage Capacity Assessment

The land cover map was used as input data for the assessment of ecosystem services, and in particular for the carbon storage capacity. The analysis is based on a simplified scheme that considers the amount of carbon stock constant over time. The variation of this ecosystem service refers to two versions of the land cover map, which exploit the National Land Consumption Map for 2012 and 2020. In this sense, the variation in carbon storage capacity must be understood as a reduction linked to the increase in land consumption. Actually, while urbanization improves human social and economic well-being, on the other hand it has a negative impact on human ecological well-being, which is closely related to the level of ecosystem services and which puts urban development and human well-being at risk in the future [6].

As input data for the carbon stock, many different bibliographic sources were used and integrated, such as the National Inventory of Forests and Forest Carbon Tanks (INFC) and the recent map of organic carbon created as part of the activities of the Global Soil Partnership [51]. Specific coefficients were used to identify the contribution deriving from the different pools [52–54]. There are four main pools of carbon in nature [55], recognized and classified by the Intergovernmental Panel on Climate Change [56], which are analyzed for each portion of the territory and each type of land cover:

Above-Ground Biomass (AGB)

It includes all the tissues of plant organisms outside the soil (such as stems, branches, leaves, seeds, etc.). The volume is calculated as:

$$AGB = a * GSV + b * GSV * e - c * GSV \quad (1)$$

where:

GSV = growing stock volume

a, b, c = specific coefficients for each forest type [53]

To switch from biomass to the fraction of stored carbon, the values are multiplied by 0.5

Below-Ground Biomass (BGB)

It includes the root system of plants. The volume is calculated as [54]:

$$BGB = GSV * BEF * WBD * R \quad (2)$$

where:

GSV = growing stock volume
BEF = biomass expansion factor
WBD = wood basic density
R = crown/roots ratio, tabulated for the different species [52,54].

To switch from biomass to the fraction of stored carbon, the values are multiplied by 0.5

The Carbon Contained in the Dead Organic Substance (DOS)

The pool includes the necromass, the woody plant residues, the litter, the finer residues not yet decomposed. As regards the epigeal biomass, specific multiplicative coefficients are considered to be applied to the values obtained from the calculation shown above, for example 0.20 for evergreen plants and 0.14 for deciduous trees [56].

Specific formulas for each species present in the bibliography were used for the litter [52,54].

Soil Carbon

The pool includes organic and mineral layers including up to a depth of 30 cm. It is evaluated starting from the data produced by CREA-ABP, CNR-Ibimet, regions and some universities as part of the Global Soil Partnership/FAO initiative [51] as an Italian contribution to the Global Soil Organic Carbon map. The map offers the values of the carbon contained in the soil in raster format with a resolution of 1 km.

In detail, for the forest cover areas, data from the National Inventory of Forests and Forest Carbon Tanks (INFC) were used [57]. The inventory provides differentiated values both by region and according to the different plant species, with reference to the classes of the CLC fourth classification level.

For the other land cover classes, estimates from the literature were used: the pool values for artificial areas were considered zero while for the other natural and agricultural surfaces the literature values reported in Table 8 were used [34].

Table 8. Carbon content for land cover classes.

LC Class	AGB	BGB	DOS	Total
	Mg C/ha			
Agricultural areas	5	-	-	58.1
Orchards	10	-	-	62.1
Woody plantations	28.55	5.25	1.75	99.45
Pastures	-	-	-	78.9
Other natural areas	3.05	-	-	69.95
Urban areas	-	-	-	-
Sparsely vegetated areas	-	-	-	-

With reference to the carbon values of permanent crops, the values of Table 9 were considered [58].

Table 9. Carbon stock values for some agricultural classes.

LC Class	AGB	BGB
	Mg C/ha	
Olive trees	9.1	2.6
Vineyards	5.5	4.4
Fruit trees	8.3	5.6

3. Results

3.1. Land Cover Map and Accuracy Assessment

The map of Figure 3 was obtained by applying the procedure described in the previous chapter.

Accuracy assessment was conducted on the map of Figure 3 and provides the results shown in Figure 4 and summarized in Table 10. The map has an overall accuracy of 90%. The omission error is less than 20% in all classes and slightly higher than this value for permanent snow and ice and wetlands.

Table 10. Overall, user and producer accuracy of the 2012 land cover map.

Overall Accuracy		
0.83		
Class name	**User's accuracy**	**Producer's accuracy**
Artificial abiotic surfaces	0.98	0.83
Consolidated (bare rocks scree, cliffs)	0.89	0.82
Unconsolidated (beaches, dunes, sands)	0.97	0.85
Broad-leaved	0.92	0.91
Needle-leaved	0.88	0.96
Orchards	0.82	0.86
Olive plantation	0.98	0.94
Wood plantations	0.90	0.84
Vineyards	0.79	0.89
Shrubland	0.84	0.83
Pastures	0.88	0.93
Arable land	0.92	0.93
Permanent herbaceous	0.76	0.83
Water bodies	0.90	0.95
Permanent snow and ice	0.99	0.79
Wetlands	0.88	0.79

The commission error is just over 20% in two of the 16 classes, while it is less than 3% for the artificial abiotic surfaces, beaches, dunes and sands, olive groves and permanent snow and ice.

Analyzing the first classification level of the 2012 land cover map, over 88% of the national surface is vegetated, followed by abiotic surfaces (9.83%) and water bodies and wetlands (1.46 and 0.16%). At the second classification level, in the abiotic class the artificial component prevails (6.98%) followed by bare rocks. Regarding vegetation, woody and herbaceous vegetation show comparable surfaces (45.7% and 42.7% respectively). In the woody class the arboreal component prevails, which occupies 38.91% of the national surface (Table 11).

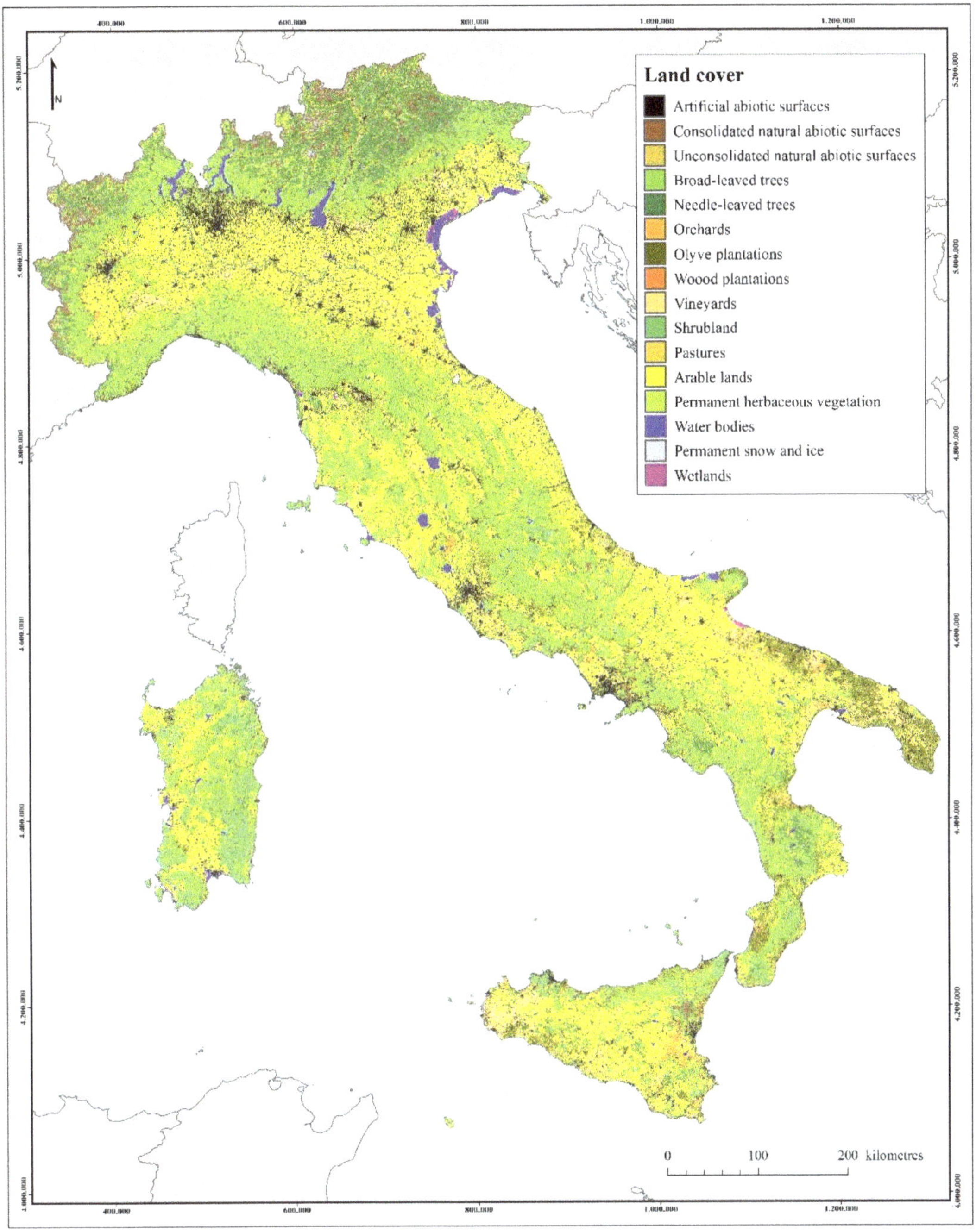

Figure 3. Land cover map of Italy for 2012.

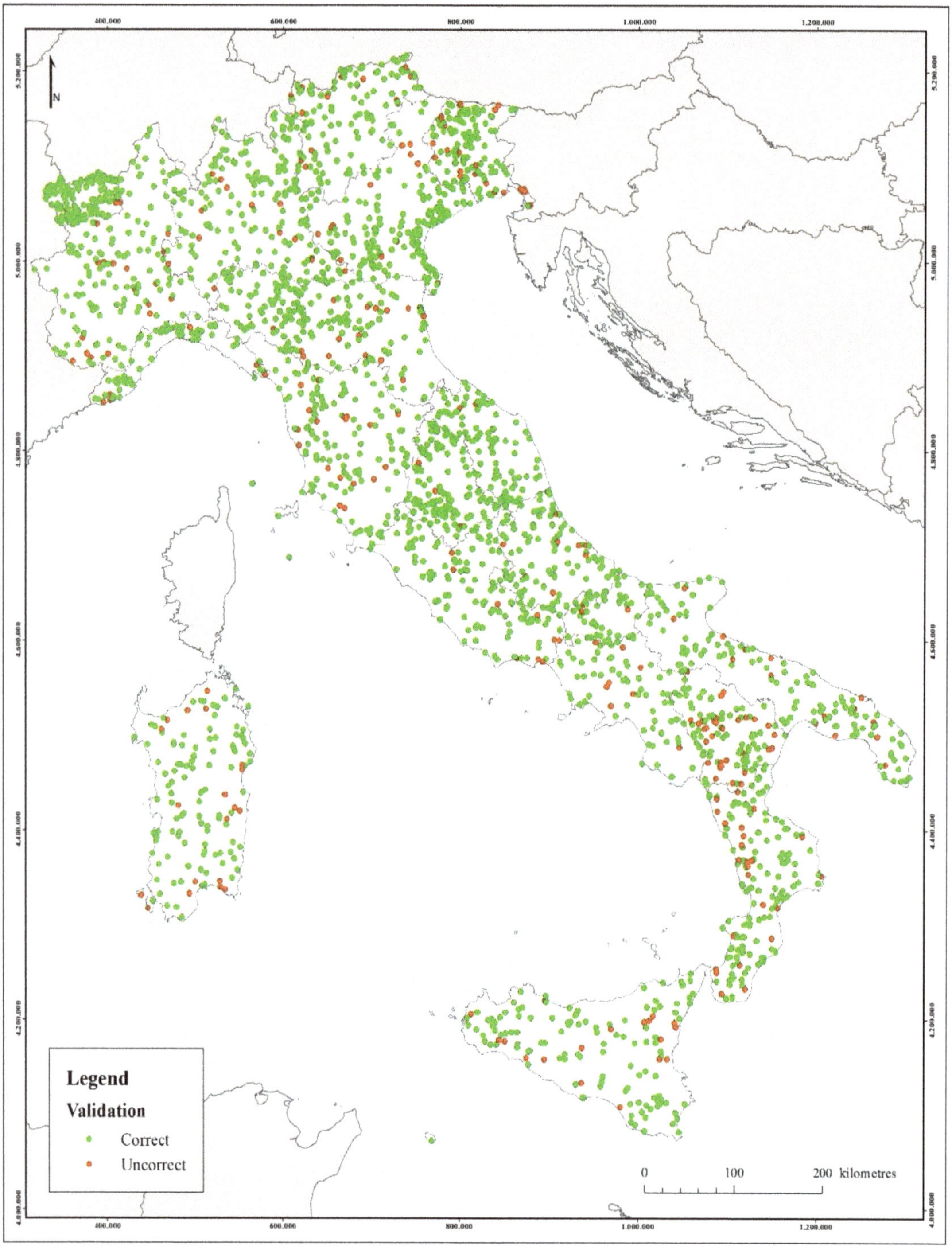

Figure 4. Result of the accuracy assessment with reference to the sample of photo-interpreted points.

Table 11. Land cover classes 2012 (first and second classification level).

	ha	% Total	% Class
Abiotic Surfaces	**2,964,151**	**9.83**	
Artificial abiotic surfaces	802,539	2.66	27.07
Natural abiotic surfaces	59,325	0.20	2.00
Bioticvegetated	**26,685,696**	**88.54**	
Woody vegetation	13,791,561	45.76	51.68
Herbaceous vegetation	12,894,135	42.78	48.32
Water surfaces	**439,830**	**1.46**	
Water bodies	402,830	1.34	91.59
Permanent snow and ice	37,000	0.12	8.41
Wetlands	**50,295**	**0.17**	

Considering the maximum thematic detail, class 2212 (arable land) prevail, which occupies 30.47% of the national territory, followed by broad-leaved trees (25.90%). All other classes occupy less than 10% of the territory and 11 out of 16 classes less than 5% (Table 12).

Table 12. Land cover classes 2012 (up to fifth classification level).

LC Code	Class Name	ha	%
11000	Artificial abiotic surfaces	2,102,288	6.98
12100	Consolidated (bare rocks scree, cliffs)	802,539	2.66
12200	Unconsolidated (beaches, dunes, sands)	59,325	0.20
21110	Broad-leaved	7,805,217	25.90
21120	Needle-leaved	2,167,261	7.19
21131	Orchards	619,759	2.06
21132	Olive plantations	1,088,099	3.61
21133	Wood plantations	46,356	0.15
21210	Vineyards	689,746	2.29
21220	Shrubland	1,375,125	4.56
22110	Pastures	1,372,790	4.55
22120	Arable land	9,183,519	30.47
22200	Permanent herbaceous	2,337,826	7.76
31000	Water bodies	402,830	1.34
32000	Permanent snow and ice	37,000	0.12
40000	Wetlands	50,295	0.17
Total		**30,139,972**	**100.00**

3.2. Estimation of the Carbon Storage Capacity

The application to the 2012 map of the methodology for calculating the carbon stocks obtained the results shown in Figure 5, which show a strong concentration of the carbon stocks in the alpine and mountain areas, while the value is significantly reduced in the agricultural areas of the Padana plain and in particular in Sicily and in the Tavoliere delle Puglie.

In Italy, 2,898,672 tons of stored carbon (stock) were lost due to land consumption between 2012 and 2020 (Table 13). In detail, this value relates to transformations from natural to artificial land cover, excluding restorations and changes between other different land cover classes. Analysing the results on a regional scale, almost a quarter of the total losses are concentrated in the Veneto (384,537 tons, equal to 13.27% of the total) and Lombardy (319,666 tons, 11.03% of the total), while each of the other regions is affected by less than 10% of the changes. The minor losses are in Aosta Valley (13,206 tons), Molise (21,242 tons) and Liguria (25,237 tons), which together host less than 2% of the changes.

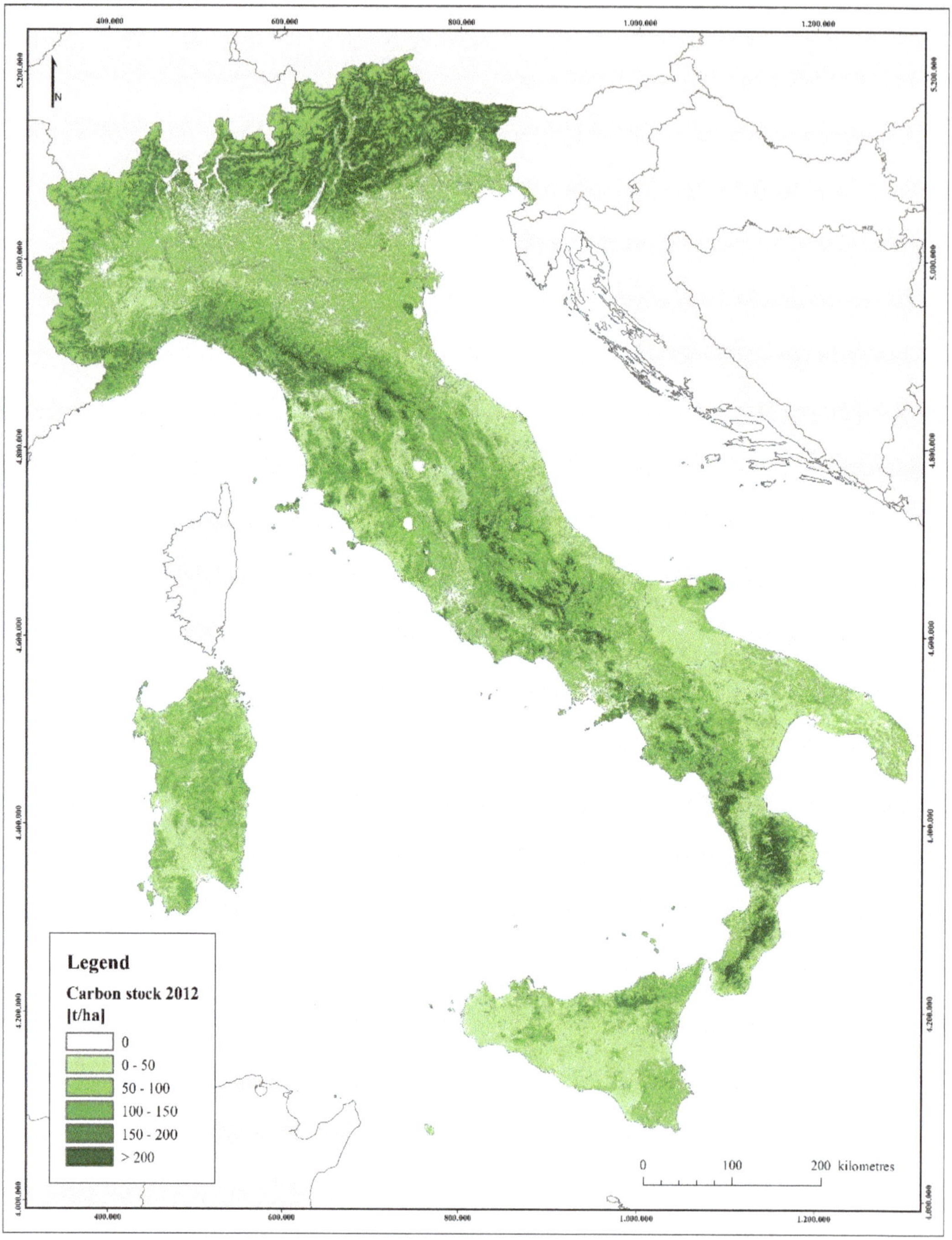

Figure 5. Carbon stock (2012).

Table 13. Variation in carbon storage capacity (2012–2020).

Region	C Storage Variation (2012–2020)	
	(Tons)	(%)
Abruzzo	86.034	2.97
Basilicata	52.086	1.80
Calabria	102.138	3.52
Campania	200.661	6.92
Emilia-Romagna	274.205	9.46
Friuli-Venezia Giulia	91.222	3.15
Latium	203.818	7.03
Liguria	25.237	0.87
Lombardy	319.667	11.03
Marche	81.746	2.82
Molise	21.243	0.73
Piedmont	203.725	7.03
Apulia	210.723	7.27
Sardinia	88.891	3.07
Sicily	225.972	7.80
Tuscany	120.143	4.14
Trentino-South Tyrol	134.470	4.64
Umbria	58.947	2.03
Aosta Valley	13.207	0.46
Veneto	384.537	13.27
Italy	2.898.672	100.00

4. Discussion

The effectiveness of ecosystem services assessment methodologies is linked to the availability of spatial data that allow an accurate and reliable description of the territory [59,60].

The CLC is still one of the few data able to guarantee information on land cover and land use for the entire national territory and is a reference for studies on a national scale. However, the new monitoring needs in terms of updating frequency and spatial resolution have revealed the limits of the project. The increase in geometric detail guaranteed by the introduction of high resolution data from the Copernicus local component and national data made it possible to describe the territory more accurately than the CLC in critical areas, such as urban areas, riparian areas and protected areas, although important portions of the national territory are still covered only by CLC.

Figure 6 shows a more detailed representation of green urban areas, which is relevant if we consider that the ecosystem functions and services that urban soils are able to offer are often neglected [61].

The increase in geometric detail also makes it possible to improve the representation of small urban agglomerations, roads and riparian areas (Figures 6–9).

Anyway, all the Copernicus data used for this research present critical issues from a thematic point of view, since their classification system is based on land cover and land use attributes with numerous mixed classes, such as the CLC class 2.4 (heterogeneous agricultural areas) and 3.2 (shrub and/or herbaceous vegetation associations).

The proposed methodology allows a more detailed description of the mixed classes, (Figures 8 and 9) and makes it possible to distinguish the arboreal component from the herbaceous one, the agricultural areas from the natural ones, arable land from permanent crops. That allows the carbon storage capacity of the different classes to be analysed separately and more accurately. Although aspects such as the diameter of the trunk, the density of the trees or the characteristics of the undergrowth are not considered, the map distinguishes the tree species in a detailed way, maintaining the fourth classification level of CLC, which is available only for the Italian territory [62].

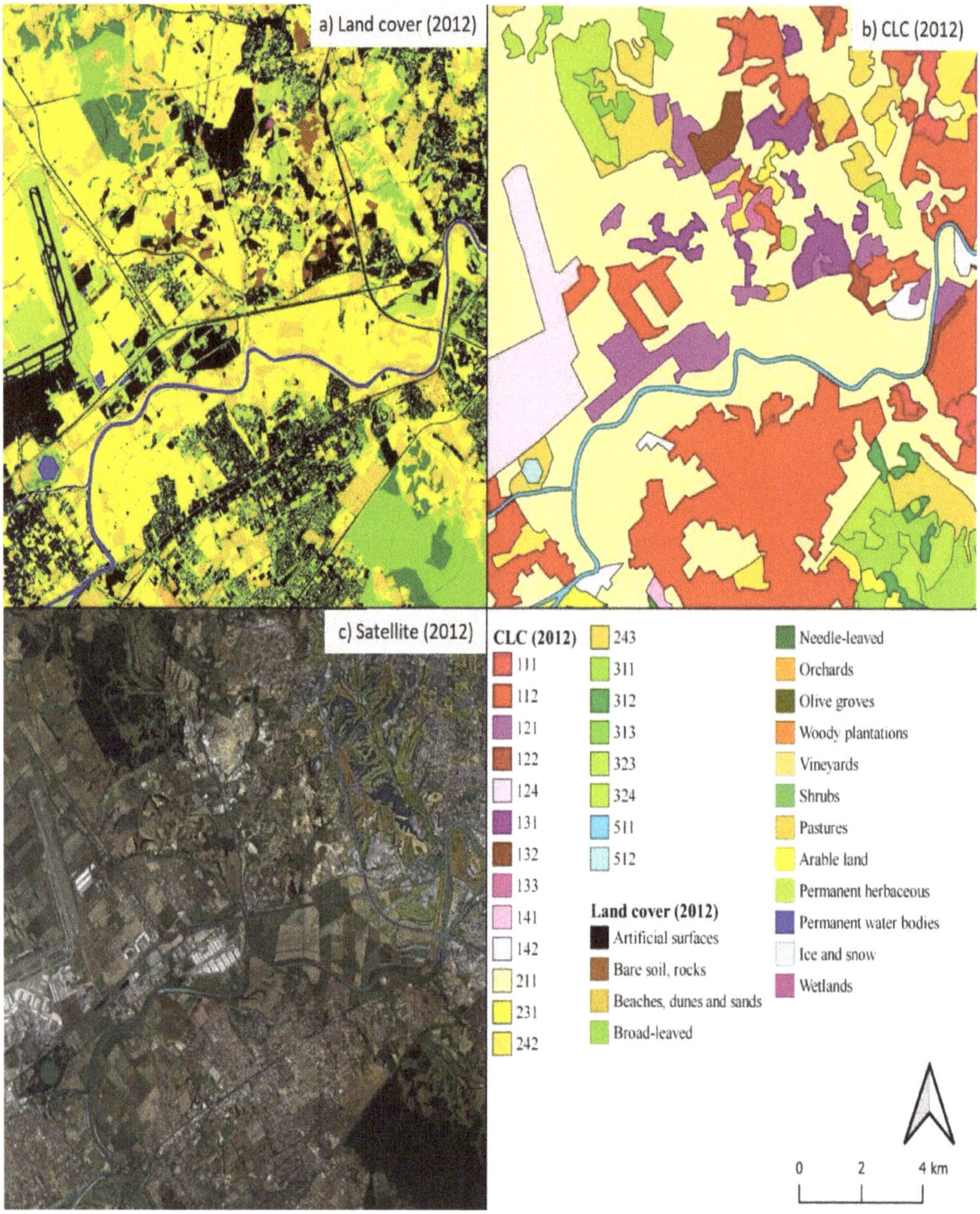

Figure 6. Comparison between land cover map for 2012 (**a**), CORINE Land Cover (**b**) and satellite image (**c**), focusing on urban area.

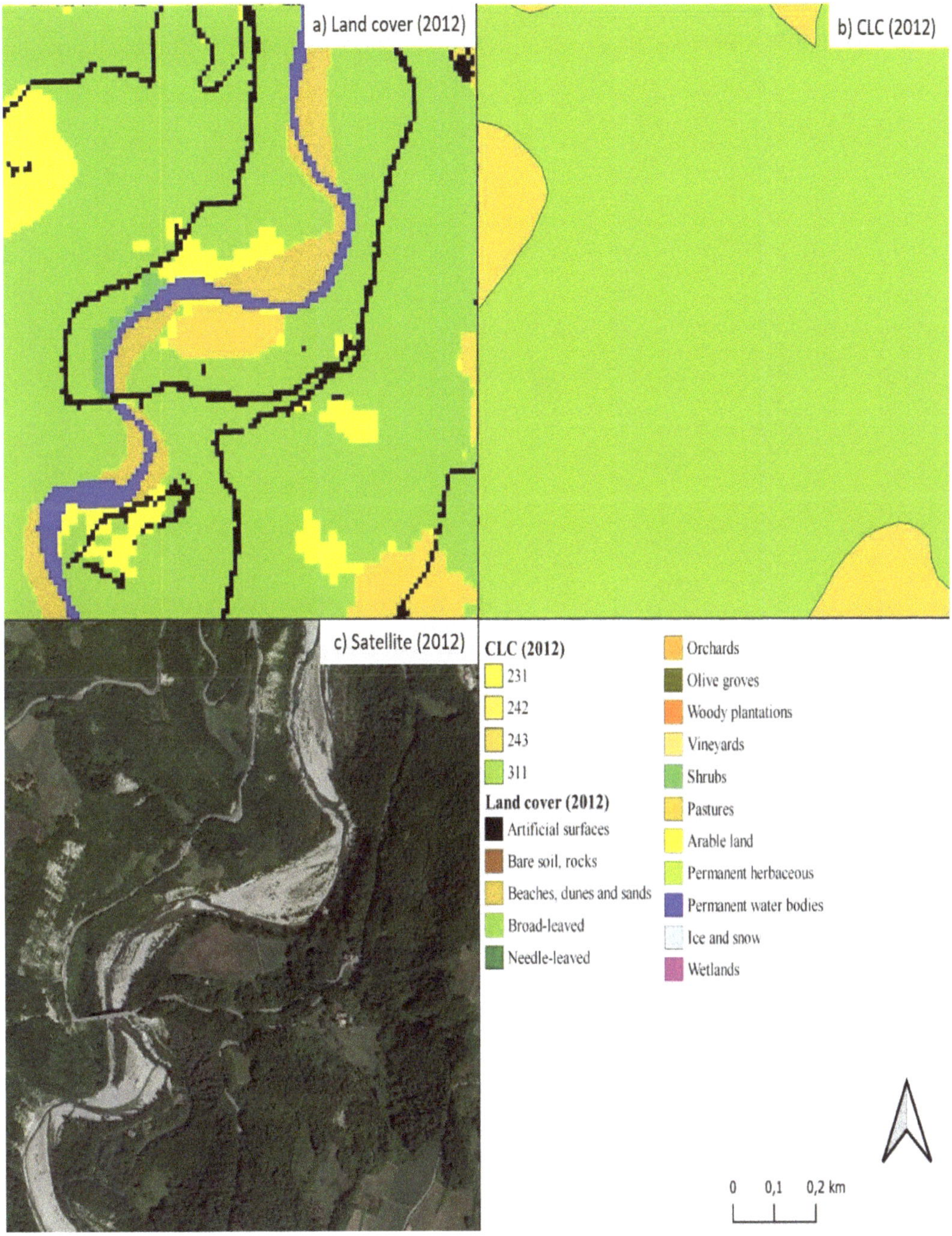

Figure 7. Comparison between land cover map for 2012 (**a**), CORINE Land Cover (**b**) and satellite image (**c**), focusing on water courses.

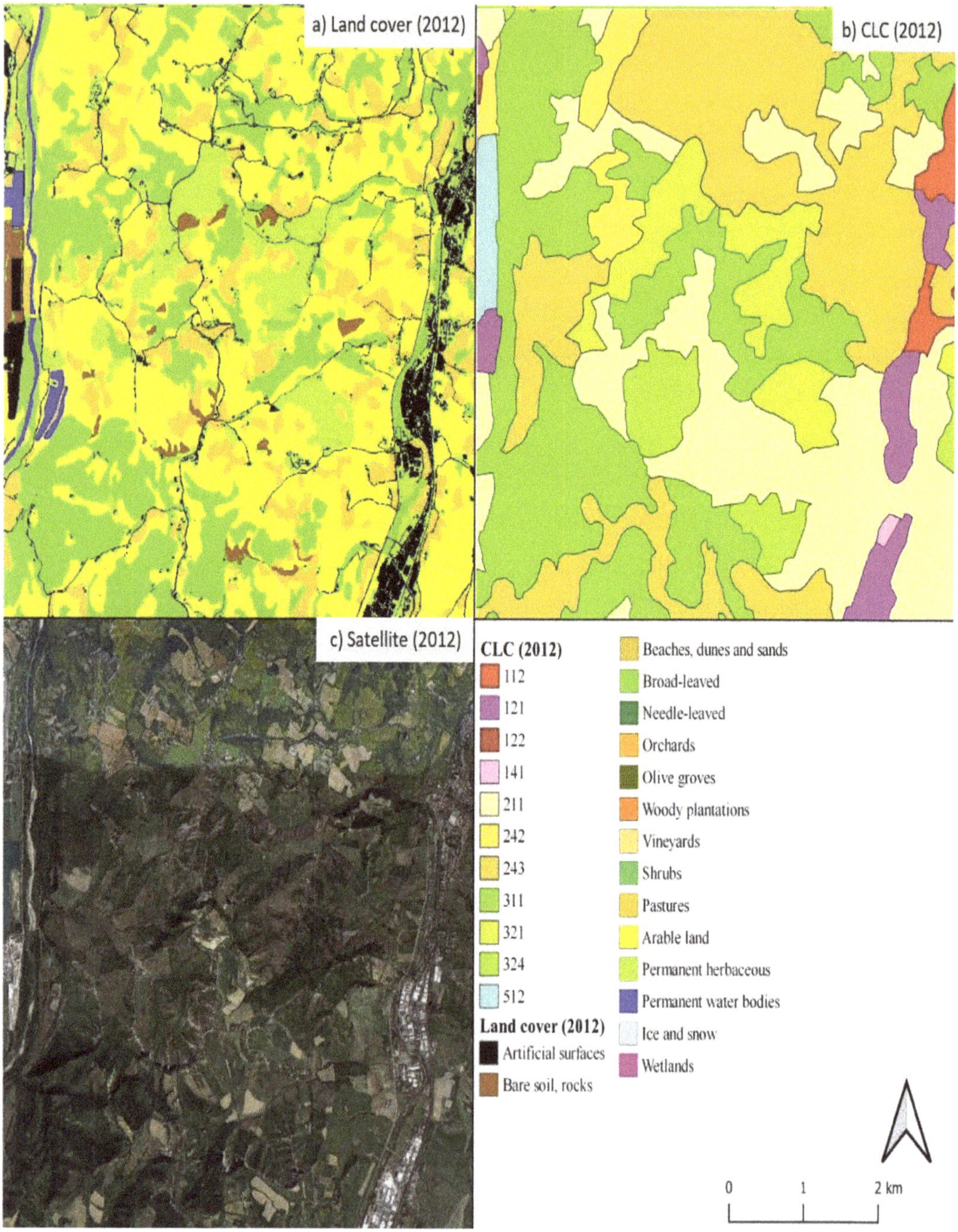

Figure 8. Comparison between land cover map for 2012 (**a**), CORINE Land Cover (**b**) and satellite image (**c**), focusing on small patches of natural vegetation and agricultural areas.

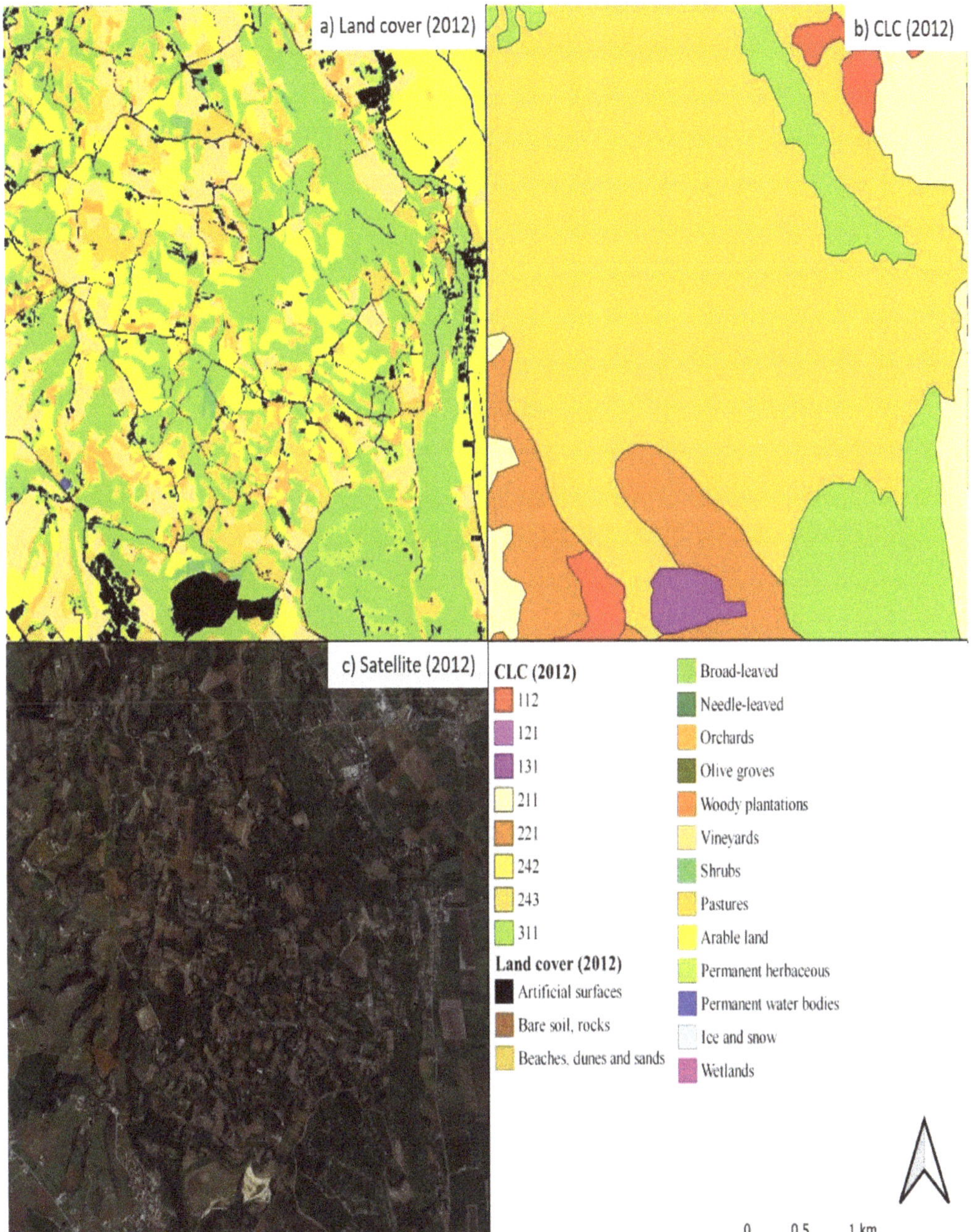

Figure 9. Comparison between land cover map for 2012 (**a**), CORINE Land Cover (**b**) and satellite image (**c**), focusing on heterogeneous agricultural areas.

An aspect that will require further development is the possibility of defining the correspondence between the classes of Copernicus data and a classification system oriented to the description of habitats, which would be more functional for conducting studies on ecosystem services. In this sense it would be necessary to integrate ancillary data not easily available on a national scale [63].

The data available at the time of the research limited the study on carbon stocks only to variations caused by increased land consumption. The availability of new Copernicus data updated to 2018 will allow the production of land cover maps capable of evaluating the variations of ecosystem services associated with other land cover changes occurred between 2012 and 2018 [60]. However the update frequency remains too low for numerous monitoring activities. Actually, the LCM is updated annually, while Copernicus data every 6 years and the maps available at regional level in Italy are often based on CLC data and are updated in a few cases (Lombardy, updated to 2018, Tuscany 2016, Liguria 2018, Latium 2016), while other maps are less up to date (2012 for Sicily and Apulia, 2013 for Abruzzo and Basilicata, 2006 for Calabria and 2009 for Campania).

Initiatives such as the next "CLC Plus" are expected to be decisive in the near future, guaranteeing the introduction of updated and interoperable products, more suitable for carrying out the monitoring activities necessary to meet institutional needs. ISPRA is conducting other research activities in this direction [41,43–45,64], through the definition of a land cover classification methodology for the production of maps with Sentinel resolution, annual update frequency and EAGLE compliant classification system, capable of providing updated and reliable products for monitoring on a national scale, which can be integrated with the activities of the "CLC Plus" and the National Strategic Plan for the Space Economy.

5. Conclusions

Since the 19th century, anthropogenic activities have led to a significant increase in the level of carbon dioxide in the atmosphere [35] and negatively affected the regeneration capacity and balance of ecosystems. Urban expansion, deforestation and forest degradation have contributed to these emissions by releasing the carbon stored naturally in forests, vegetation and soil [34,35]. This type of carbon is added to greenhouse gas emissions related to industries and energy production and to the products of impure combustion. Terrestrial ecosystems are able to sequester as much carbon as is currently in the atmosphere but over the course of the century terrestrial biosphere is likely to become a net source of carbon due to factors connected with climate change, pollution and the over-exploitation of resources that will alter the structure, reduce biodiversity and perturb functioning of most ecosystems, and compromise the services they currently provide.

The land cover and land use changes involve landscape fragmentation, reduction of biodiversity and loss of green areas important for carbon accumulation and more generally for the provision of ecosystem services. Current conservation practices are generally poorly prepared to adapt to this level of change, and effective adaptation responses are likely to be costly to implement.

The monitoring of carbon stock accounting is an institutional duty enshrined in the Kyoto protocol and the Paris agreements and an important driver in defining adaptation strategies to climate change [33]. Effective monitoring strategies of land cover and land use changes are essential for studying the phenomenon. In this sense, the methodology presented in this paper represents a step forward for large-scale assessments of ecosystem services more relevant to reality, since compared to the CLC it provides products for the entire national territory with greater geometric detail and a better description of mixed areas.

The methodology is easily applicable in other territorial areas, since it is based on Copernicus data available for many European countries, furthermore the use of an EAGLE compliant classification system makes the methodology easily adaptable to the specific availability of national data.

The main limitation of the methodology concerns the low update frequency of the input data, which limited the monitoring of ecosystem services to variations related to land consumption, which is the only data updated annually for the Italian territory.

A first future development concerns the updating of the map using the new Copernicus Local and Pan-European products for 2018. This implementation will allow the evaluation of the variations in the carbon stocks associated not only with land consumption but also with other land cover changes.

The products of the application of the methodology are also in continuity with other activities carried out by the working group and can constitute a useful support tool for the development of land cover classification methodologies with high update frequency for the satisfaction of the institutional needs envisaged by the new "Space Economy" Strategic Plan and for the creation of "Istances" in the new CLC Plus Project. In this sense, an important added value of this research is linked to the suitability of products with respect to present and future national and European initiatives and standards in the field of remote sensing.

Author Contributions: Conceptualization, P.D.F., A.S., L.C. and M.M.; methodology P.D.F. and A.S.; software, P.D.F., L.C. and A.S.; validation, A.S., L.C. and M.M.; formal analysis, P.D.F. and A.S.; investigation, P.D.F. and A.S.; resources, F.A. and M.M.; data curation, P.D.F. and A.S.; writing—original draft preparation P.D.F., A.S. and L.C.; writing—review and editing, P.D.F., A.S., F.A., M.M. I.M. and L.C.; visualization, P.D.F. and A.S.; supervision, F.A. and M.M.; project administration, F.A. and M.M.; funding acquisition, F.A. and M.M. All authors have read and agreed to the published version of the manuscript.

Funding: This research was funded by the Italian Institute for Environmental Protection and Research (ISPRA) structural funds.

Institutional Review Board Statement: Not applicable.

Informed Consent Statement: Not applicable.

Data Availability Statement: Data presented in this study are available on request from the corresponding author. The data are not publicly available because they are part of ongoing research.

Conflicts of Interest: The authors declare no conflict of interest.

Appendix A

Below are the tables containing the correspondences referred to for the first reclassification of the data N2000 (Table A1), UA (Table A2), RZ (Table A3) and CORINE Land Cover (Table A4). In this first reclassification there are the temporary codes of Table 3.

Table A1. Reclassification table for N2000.

N2000	Class Name	LC	N2000	Class Name	LC
1110	Urban fabric (predominantly public and private units)	998	4212	Seminatural grassland without woody plants (C.C.D. $\leq$ 30%)	93
1120	Industrial, commercial and military units	998	4220	Alpine and subalpine natural grassland	22200
1210	Road networks and associated land	998	5110	Heathland and Moorland	22200
1220	Railways and associated land	998	5120	Other scrub land	21220
1230	Port areas and associated land	998	5200	Sclerophyllous vegetation	21220
1240	Airports and associated land	998	6100	Sparsely vegetated areas	96
1310	Mineral extraction, dump and construction sites	998	6210	Beaches and dunes	12200
1320	Land without current use	998	6220	River banks	12200
1400	Green urban, sports and leisure facilities	998	6310	Bare rocks and rock debris	12100
2110	Arable land	22120	6320	Burnt areas (except burnt forest)	97

Table A1. *Cont.*

N2000	Class Name	LC	N2000	Class Name	LC
2120	Greenhouses	22120	6330	Glaciers and perpetual snow	32000
2210	Vineyards, fruit trees and berry plantations	99	7100	Inland marshes	40000
2220	Olive groves	21132	7210	Exploited peat bog	40000
2310	Annual crops associated with permanent crops	92	7220	Unexploited peat bog	40000
2320	Complex cultivation patterns	92	8110	Coastal salt marshes	40000
2330	Land principally occupied by agriculture with significant areas of natural vegetation	93	8120	Salines	40000
2340	Agroforestry	94	8130	Intertidal flats	40000
3110	Natural and seminatural broad-leaved forest	2111	8210	Coastal lagoons	31000
3120	Highly artificial broad-leaved plantations	2111	8220	Estuaries	31000
3210	Natural and seminatural coniferous forest	2112	9110	Interconnected water courses	31000
3220	Highly artificial coniferous plantations	2112	9120	Highly modified water courses and canals	31000
3310	Natural and seminatural mixed forest	999	9130	Separated water bodies belonging to the river system	31000
3320	Highly artificial mixed plantations	99	9210	Natural water bodies	31000
3410	Transitional woodland and scrub	95	9220	Artificial standing water bodies	31000
3420	Lines of trees and scrub	95	9230	Intensively managed fish ponds	31000
3500	Damaged forest	95	9240	Standing water bodies of extractive industrial sites	31000
4100	Managed grassland	22110	10000	Sea and ocean	31000
4211	Seminatural grassland with woody plants (C.C.D. $\geq$ 30%)	93			

Table A2. Reclassification table for UA.

UA	Class Name	LC	UA	Class Name	LC
11100	Continuous Urban Fabric (>80%)	998	13400	Land without current use	998
11210	Discontin. Dense Urban Fabric (50–80%)	998	14100	Green urban areas	91
11220	Discontin. Medium Density Urban Fabric (30–50%)	998	14200	Sports and leisure facilities	998
11230	Discontinuous Low Density Urban Fabric (10–30%)	998	21000	Arable land (annual crops)	22120
11240	Discontinuous Very Low Density Urban Fabric (<10%)	998	22000	Permanent crops (vineyards, fruit trees, olive groves)	99
11300	Isolated Structures	998	23000	Pastures	22110
12100	Industrial, commercial, public, military and private units	998	24000	Complex and mixed cultivation patterns	92
12210	Fast transit roads and associated land	998	25000	Orchards at the fringe of urban classes	21131
12220	Other roads and associated land	998	31000	Forests	999
12230	Railways and associated land	998	32000	Herbaceous vegetation associations (natural grassland, moors...)	22200
12300	Port areas	998	33000	Open spaces with little or no vegetations (beaches, dunes, bare rocks, glaciers)	96
12400	Airports	998	40000	Wetland	40000
13100	Mineral extraction and dump site	998	50000	Water bodies	31000
13300	Construction sites	998			

Table A3. Reclassification table for RZ.

RZ	Class Name	LC	RZ	Class Name	LC
11110	Continuous Urban Fabric (IMD ≥ 80%)	998	41000	Managed grassland	22110
11120	Dense Urban Fabric (IMD ≥ 30–80%)	998	42100	Seminatural grassland	93
11130	Low Density Fabric (IMD < 30%)	998	42200	Alpine and subalpine natural grassland	22200
11200	Industrial, commercial and military units	998	51100	Heathland and Moorland	22200
12100	Road networks and associated land	998	51200	Other scrub land	21220
12200	Railways and associated land	998	52000	Sclerophyllous vegetation	21220
12300	Port areas and associated land	998	61000	Sparsely vegetated areas	96
12400	Airports and associated land	998	62100	Beaches and dunes	12200
13100	Mineral extraction, dump and construction sites	998	62200	River banks	12200
13200	Land without current use	998	63100	Bare rocks and rock debris	12100
14000	Green urban, sports and leisure facilities	998	63200	Burnt areas (except burnt forest)	97
21100	Arable land	22120	63300	Glaciers and perpetual snow	32000
21200	Greenhouses	22120	71000	Inland marshes	40000
22100	Vineyards, fruit trees and berry plantations	99	72100	Exploited peat bog	40000
22200	Olive groves	21132	72200	Unexploited peat bog	40000
23100	Annual crops associated with permanent crops	92	81100	Coastal salt marshes	40000
23200	Complex cultivation patterns	92	81200	Salines	40000
23300	Land principally occupied by agriculture with significant areas of natural vegetation	93	81300	Intertidal flats	40000
23400	Agroforestry	94	82100	Coastal lagoons	31000
31100	Natural and seminatural broad-leaved forest	2111	82200	Estuaries	31000
31200	Highly artificial broad-leaved plantations	2111	91100	Interconnected water courses	31000
32100	Natural and seminatural coniferous forest	2112	91200	Highly modified water courses and canals	31000
32200	Highly artificial coniferous plantations	2112	91300	Separated water bodies belonging to the river system	31000
33100	Natural and seminatural mixed forest	999	92100	Natural water bodies	31000
33200	Highly artificial mixed plantations	999	92200	Artificial standing water bodies	31000
34100	Transitional woodland and scrub	95	92300	Intensively managed fish ponds	31000
34200	Lines of trees and scrub	95	92400	Standing water bodies of extractive industrial sites	31000
35000	Damaged forest	95	10000	Sea and ocean	31000

Table A4. Reclassification table for CLC.

CLC	Class Name	LC	CLC	Class Name	LC
111	Continuous urban fabric	998	31113117	Broad-leaved forest	2111
112	Discontinuous urban fabric	998	31213125	Needle-leaved forest	2112
121	Industrial or commercial units	998	31313132	Mixed forest	999
1211	Photovoltaic fiends	998	3211	Continuous natural grasslands	22200
122	Road and rail networks and associated land	998	3212	Discontinuous natural grasslands	22200
123	Port areas	998	322	Moors and heathland	21220
124	Airports	998	3231	High Mediterranean scrub	21220
131	Mineral extraction sites	998	3232	Low Mediterranean scrub and the garrigue	21220
132	Dump sites	998	324	Transitional woodland shrub	95

Table A4. *Cont.*

CLC	Class Name	LC	CLC	Class Name	LC
133	Construction sites	998	3241	Forest harvesting	95
141	Green urban areas	91	331	Beaches, dunes, sands	12200
142	Sport and leisure facilities	998	332	Bare rocks	12100
2111	Intensive nonirrigated arable land	22120	333	Sparsely vegetated areas	96
2112	Extensive nonirrigated arable land	22120	334	Burnt areas	97
212	Permanently irrigated land	22120	335	Glaciers and perpetual snow	32000
213	Rice fields	22120	411	Inland marshes	40000
221	Vineyards	21210	412	Peat bogs	40000
222	Fruit trees and berry plantations	21131	421	Salt marshes	40000
223	Olive groves	21132	422	Salines	40000
224	Woody plantation	21133	423	Intertidal flats	40000
2241	New woody plantation	21133	511	Water courses	31000
231	Pastures	22110	512	Water bodies	31000
241	Annual crops associated with permanent crops	92	521	Coastal lagoons	31000
242	Complex cultivation patterns	92	522	Estuaries	31000
243	Land principally occupied by agriculture, with significant areas of natural vegetation	93	523	Sea and ocean	31000
244	Agroforestry	94			

References

1. Munafò, M. *Consumo di Suolo, Dinamiche Territoriali e Servizi Ecosistemici—Edizione 2018*; ISPRA: Rome, Italy, 2018.
2. Verburg, P.H.; Erb, K.-H.; Mertz, O.; Espindola, G. Land System Science: Between global challenges and local realities. *Curr. Opin. Environ. Sustain.* **2013**, *5*, 433–437. [CrossRef] [PubMed]
3. Pedroli, G.B.M.; Meiner, A. *Landscapes in Transition: An Account of 25 Years of Land Cover Change in Europe*; European Environment Agency: København, Denmark, 2017.
4. EEA. Land Systems at European Level—Analytical Assessment Framework. 2018. Briefing no. 10/2018. pp. 1–9. Available online: https://www.eea.europa.eu/themes/landuse/land-systems (accessed on 28 November 2021).
5. Anderson, J.R. Land use and land cover changes. A framework for monitoring. *J. Res. Geol. Surv.* **1977**, *5*, 143–153.
6. Liu, R.; Dong, X.; Wang, X.; Zhang, P.; Liu, M.; Zhang, Y. Study on the relationship among the urbanization process, ecosystem services and human well-being in an arid region in the context of carbon flow: Taking the Manas river basin as an example. *Ecol. Indic.* **2021**, *132*, 108248. [CrossRef]
7. Costanza, R.; D'Arge, R.; de Groot, R.; Farberll, S.; Grassot, M.; Hannon, B.; Limburg, K.; Naeem, S.; O'Neilltt, R.V.; Paruelo, J.; et al. The value of the world's ecosystem services and natural capital. *Nature* **1997**, *387115*, 253–260. [CrossRef]
8. Gómez-Baggethun, E.; De Groot, R.; Lomas, P.L.; Montes, C. The history of ecosystem services in economic theory and practice: From early notions to markets and payment schemes. *Ecol. Econ.* **2010**, *69*, 1209–1218. [CrossRef]
9. Danley, B.; Widmark, C. Evaluating conceptual definitions of ecosystem services and their implications. *Ecol. Econ.* **2016**, *126*, 132–138. [CrossRef]
10. Teague, A.; Russell, M.; Harvey, J.; Dantin, D.; Nestlerode, J.; Alvarez, F. A spatially-explicit technique for evaluation of alternative scenarios in the context of ecosystem goods and services. *Ecosyst. Serv.* **2016**, *20*, 15–29. [CrossRef]
11. Corvalan, C.; Hales, S.; McMichael, A.; Butler, C.; Campbell-Lendrum, D.; Confalonieri, U.; Leitner, K.; Lewis, N.; Patz, J.; Polson, K.; et al. A Report of the Millennium Ecosystem Assessment. In *Ecosystems and Human Well-Being*; Island Press: Washington, DC, USA, 2005; Volume 5.
12. Xepapadeas, A. *The Economics of Ecosystems and Biodiversity: Ecological and Economic Foundations*; Kumar, P., Ed.; Earthscan: London, UK; Washington, DC, USA, 2011; Volume 16, pp. 239–242. ISBN 978-1-84971-212-5 (HB).
13. La Notte, A.; D'Amato, D.; Mäkinen, H.; Paracchini, M.L.; Liquete, C.; Egoh, B.; Geneletti, D.; Crossman, N.D. Ecosystem services classification: A systems ecology perspective of the cascade framework. *Ecol. Indic.* **2017**, *74*, 392–402. [CrossRef]
14. Potschin, M.; Haines-Young, R. Defining and measuring ecosystem services. In *Routledge Handbook of Ecosystem Services*; Potschin, M., Haines-Young, R., Fish, R., Turner, R.K., Eds.; Routledge: London, UK; New York, NY, USA, 2016; pp. 25–44.
15. Costanza, R.; de Groot, R.; Sutton, P.; van der Ploeg, S.; Anderson, S.J.; Kubiszewski, I.; Farber, S.; Turner, R.K. Changes in the global value of ecosystem services. *Glob. Environ. Chang.* **2014**, *26*, 152–158. [CrossRef]
16. Zulian, G.; Maes, J.; Paracchini, M.L. Linking land cover data and crop yields for mapping and assessment of pollination services in Europe. *Land* **2013**, *2*, 472–492. [CrossRef]

17. Groff, S.C.; Loftin, C.S.; Drummond, F.; Bushmann, S.; McGill, B. Parameterization of the InVEST crop pollination model to spatially predict abundance of wild blueberry (Vaccinium angustifolium Aiton) native bee pollinators in Maine, USA. *Environ. Model. Softw.* **2016**, *79*, 1–9. [CrossRef]
18. Leonhardt, S.D.; Gallai, N.; Garibaldi, L.A.; Kuhlmann, M.; Klein, A.-M. Economic gain, stability of pollination and bee diversity decrease from southern to northern Europe. *Basic Appl. Ecol.* **2013**, *14*, 461–471. [CrossRef]
19. Lautenbach, S.; Seppelt, R.; Liebscher, J.; Dormann, C.F. Spatial and temporal trends of global pollination benefit. *PLoS ONE* **2012**, *7*, e35954. [CrossRef] [PubMed]
20. Hanley, N.; Breeze, T.D.; Ellis, C.; Goulson, D. Measuring the economic value of pollination services: Principles, evidence and knowledge gaps. *Ecosyst. Serv.* **2015**, *14*, 124–132. [CrossRef]
21. UK National Ecosystem Assessment. *The UK National Ecosystem Assessment: Synthesis of the Key Findings*; UNEP-WCMC: Cambridge, UK, 2011.
22. Rabe, S.-E.; Koellner, T.; Marzelli, S.; Schumacher, P.; Grêt-Regamey, A. National ecosystem services mapping at multiple scales. *The German exemplar. Ecol. Indic.* **2016**, *70*, 357–372. [CrossRef]
23. Santos-Martín, F.; García Llorente, M.; Quintas-Soriano, C.; Zorrilla-Miras, P.; Martín-Lopez, B.; Loureiro, M.; Benayas, J.; Montes, M. *Spanish National Ecosystem Assessment: Socio-economic valuation of ecosystem services in Spain. Synthesis of the key findings*; Biodiversity Foundation of the Spanish Ministry of Agriculture, Food and Environment: Madrid, Spain, 2016.
24. Xie, G.; Zhang, C.; Zhen, L.; Zhang, L. Dynamic changes in the value of China's ecosystem services. *Ecosyst. Serv.* **2017**, *26*, 146–154. [CrossRef]
25. Turpie, J.K.; Forsythe, K.J.; Knowles, A.; Blignaut, J.; Letley, G. Mapping and valuation of South Africa's ecosystem services: A local perspective. *Ecosyst. Serv.* **2017**, *27*, 179–192. [CrossRef]
26. Choi, H.-A.; Song, C.; Lee, W.-K.; Jeon, S.; Gu, J.H. Integrated approaches for national ecosystem assessment in South Korea. *KSCE J. Civ. Eng.* **2018**, *22*, 1634–1641. [CrossRef]
27. Daily, G.C.; Matson, P.A.; Vitousek, P.M. Ecosystem services supplied by soil. In *Nature's Services: Societal Dependence on Natural Ecosystems*; Daily, G.C., Ed.; Island Press: Washington, DC, USA, 1997; pp. 113–132.
28. Wall, D.H.; Bardgett, R.D.; Covich, A.P.; Snelgrove, P.V.R. *The Need for Understanding How Biodiversity and Ecosystem Functioning Affect Ecosystem Services in Soils and Sediments*; Island Press: Washington, DC, USA, 2004.
29. Dominati, E.; Patterson, M.; Mackay, A. A framework for classifying and quantifying the natural capital and ecosystem services of soils. *Ecol. Econ.* **2010**, *69*, 1858–1868. [CrossRef]
30. Fryer, J.; Williams, I.D. Regional carbon stock assessment and the potential effects of land cover change. *Sci. Total Environ.* **2021**, *775*, 145815. [CrossRef] [PubMed]
31. Mathew, I.; Shimelis, H.; Mutema, M.; Minasny, B.; Chaplot, V. Crops for increasing soil organic carbon stocks–A global meta analysis. *Geoderma* **2020**, *367*, 114230. [CrossRef]
32. Clerici, N.; Cote-Navarro, F.; Escobedo, F.J.; Rubiano, K.; Villegas, J.C. Spatio-temporal and cumulative effects of land use-land cover and climate change on two ecosystem services in the Colombian Andes. *Sci. Total Environ.* **2019**, *685*, 1181–1192. [CrossRef]
33. Savaresi, A.; Perugini, L.; Chiriacò, M.V. Making sense of the LULUCF Regulation: Much ado about nothing? *Rev. Eur. Comp. Int. Environ. Law* **2020**, *29*, 212–220. [CrossRef]
34. Sallustio, L.; Quatrini, V.; Geneletti, D.; Corona, P.; Marchetti, M. Assessing land take by urban development and its impact on carbon storage: Findings from two case studies in Italy. *Environ. Impact Assess. Rev.* **2015**, *54*, 80–90. [CrossRef]
35. Vashum, K.T.; Jayakumar, S. Methods to estimate above-ground biomass and carbon stock in natural forests-a review. *J. Ecosyst. Ecography* **2012**, *2*, 1–7. [CrossRef]
36. Sallustio, L.; Munafò, M.; Riitano, N.; Lasserre, B.; Fattorini, L.; Marchetti, M. Integration of land use and land cover inventories for landscape management and planning in Italy. *Environ. Monit. Assess.* **2016**, *188*, 48. [CrossRef]
37. Maes, J.; Teller, A.; Erhard, M.; Liquete, C.; Braat, L.; Berry, P.; Egoh, B.; Puydarrieus, P.; Fiorina, C.; Santos, F.; et al. *An Analytical Framework for Ecosystem Assessments under Action 5 of the EU Biodiversity Strategy to 2020*; Publications Office of the European Union: Luxembourg, 2013; ISBN 9789279293696.
38. Cabral, P.; Feger, C.; Levrel, H.; Chambolle, M.; Basque, D. Assessing the impact of land-cover changes on ecosystem services: A first step toward integrative planning in Bordeaux, France. *Ecosyst. Serv.* **2016**, *22*, 318–327. [CrossRef]
39. Arnold, S.; Kosztra, B.; Banko, G.; Smith, G.; Hazeu, G.; Bock, M. The EAGLE concept—A vision of a future European Land Monitoring Framework. In Proceedings of the 33th EARSeL Symposium towards Horizon 2020, Matera, Italy, 3–6 June 2013; pp. 551–568.
40. Kleeschulte, S. *Technical Specifications for Implementation of a New Land-Monitoring Concept Based on EAGLE. D3: Draft Design Concept and CLC-Backbone, CLC-Core Technical Specifications, Including Requirements Review*, 3rd ed.; European Environment Agency: Copenhagen, Denmark, 2019; p. 78.
41. De Fioravante, P.; Luti, T.; Cavalli, A.; Giuliani, C.; Dichicco, P.; Marchetti, M.; Chirici, G.; Congedo, L.; Munafò, M. Multispectral Sentinel-2 and SAR Sentinel-1 Integration for Automatic Land Cover Classification. *Land* **2021**, *10*, 611. [CrossRef]
42. Munafò, M. *Consumo di Suolo, Dinamiche Territoriali e Servizi Ecosistemici. Edizione 2020*; ISPRA: Rome, Italy, 2020.
43. Munafò, M. *Consumo di Suolo, Dinamiche Territoriali e Servizi Ecosistemici Edizione 2021 Rapporto ISPRA SNPA*; ISPRA: Rome, Italy, 2021; ISBN 9788844810597.

44. Luti, T.; De Fioravante, P.; Marinosci, I.; Strollo, A.; Riitano, N.; Falanga, V.; Mariani, L.; Congedo, L.; Munafò, M. Land Consumption Monitoring with SAR Data and Multispectral Indices. *Remote Sens.* **2021**, *13*, 1586. [CrossRef]
45. Strollo, A.; Smiraglia, D.; Bruno, R.; Assennato, F.; Congedo, L.; De Fioravante, P.; Giuliani, C.; Marinosci, I.; Riitano, N.; Munafò, M. Land consumption in Italy. *J. Maps* **2020**, *16*, 113–123. [CrossRef]
46. Olofsson, P.; Foody, G.M.; Herold, M.; Stehman, S.V.; Woodcock, C.E.; Wulder, M.A. Good practices for estimating area and assessing accuracy of land change. *Remote Sens. Environ.* **2014**, *148*, 42–57. [CrossRef]
47. FAO Map Accuracy Assessment and Area Estimation: A Practical Guide. *National Forest Monitoring and Assessment Working Paper*. 2016. 46/E. p. 69. Available online: https://www.fao.org/publications/card/fr/c/e5ea45b8-3fd7-4692-ba29-fae7b140d07e/ (accessed on 28 November 2021).
48. Stehman, S.V.; Czaplewski, R.L. Design and Analysis for Thematic Map Accuracy Assessment—An application of satellite imagery. *Remote Sens. Environ.* **1998**, *64*, 331–344. [CrossRef]
49. Stehman, S.V.; Foody, G.M. Key issues in rigorous accuracy assessment of land cover products. *Remote Sens. Environ.* **2019**, *231*, 111199. [CrossRef]
50. Cochran, W.G.; William, G. *Sampling Techniques*; John Wiley& Sons: New York, NY, USA, 1977.
51. Lupia, F. *MARSALa-A Model-Based Irrigation Water Consumption Estimation at Farm Level*; INEA: Brussels, Belgium, 2013.
52. Vitullo, M.; De Lauretis, R.; Federici, S. *La Contabilità del Carbonio Contenuto Nelle Foreste Italiane*; Agenzia per la Protezione dell'ambiente e per i Servizi Tecnici: Rome, Italy, 2007.
53. Di Cosmo, L.; Gasparini, P.; Tabacchi, G. A national-scale, stand-level model to predict total above-ground tree biomass from growing stock volume. *For. Ecol. Manag.* **2016**, *361*, 269–276. [CrossRef]
54. Romano, D.; Arcarese, C.; Bernetti, A.; Caputo, A.; Contaldi, M.; De Lauretis, R.; Di Cristofaro, E.; Gagna, A.; Gonella, B.; Taurino, E.; et al. *Italian Greenhouse Gas Inventory 1990–2015. National Inventory Report 2017*. 2017. Available online: https://www.isprambiente.gov.it/en/publications/reports/italian-greenhouse-gas-inventory-1990-2015.-national-inventory-report-2017 (accessed on 28 November 2021).
55. Li, S.; Liu, Y.; Yang, H.; Yu, X.; Zhang, Y.; Wang, C. Integrating ecosystem services modeling into effectiveness assessment of national protected areas in a typical arid region in China. *J. Environ. Manag.* **2021**, *297*, 113408. [CrossRef]
56. Penman, J.; Gytarsky, M.; Hiraishi, T.; Krug, T.; Kruger, D.; Pipatti, R.; Buendia, L.; Miwa, K.; Ngara, T.; Tanabe, K. Good practice guidance for land use, land-use change and forestry. *Good Pract. Guid. L. Use Land-Use Chang. For.* **2003**. Available online: https://www.ipcc.ch/publication/good-practice-guidance-for-land-use-land-use-change-and-forestry/ (accessed on 28 November 2021).
57. Marchetti, M.; Sallustio, L.; Ottaviano, M.; Barbati, A.; Corona, P.; Tognetti, R.; Zavattero, L.; Capotorti, G. Carbon sequestration by forests in the National Parks of Italy. *Plant Biosyst. Int. J. Deal. Asp. Plant Biol.* **2012**, *146*, 1001–1011. [CrossRef]
58. Canaveira, P.; Manso, S.; Pellis, G.; Perugini, L.; De Angelis, P.; Neves, R.; Papale, D.; Paulino, J.; Pereira, T.; Pina, A. Biomass Data on Cropland and Grassland in the Mediterranean Region. Final Report for Action A4 of Project MediNet. 2018. Available online: http://www.lifemedinet.com/ (accessed on 28 November 2021).
59. Burke, T.; Whyatt, J.D.; Rowland, C.; Blackburn, G.A.; Abbatt, J. The influence of land cover data on farm-scale valuations of natural capital. *Ecosyst. Serv.* **2020**, *42*, 101065. [CrossRef]
60. Tomlinson, S.J.; Dragosits, U.; Levy, P.E.; Thomson, A.M.; Moxley, J. Quantifying gross vs. net agricultural land use change in Great Britain using the Integrated Administration and Control System. *Sci. Total Environ.* **2018**, *628*, 1234–1248. [PubMed]
61. Morel, J.L.; Chenu, C.; Lorenz, K. Ecosystem services provided by soils of urban, industrial, traffic, mining, and military areas (SUITMAs). *J. Soils Sediments* **2015**, *15*, 1659–1666. [CrossRef]
62. Sambucini, V.; Marinosci, I.; Bonora, N.; Chirci, G. *La Realizzazione in Italia del Progetto CORINE Land Cover 2006*; ISPRA: Rome, Italy, 2010; ISBN 9788844804770.
63. Tomaselli, V.; Dimopoulos, P.; Marangi, C.; Kallimanis, A.S.; Adamo, M.; Tarantino, C.; Panitsa, M.; Terzi, M.; Veronico, G.; Lovergine, F. Translating land cover/land use classifications to habitat taxonomies for landscape monitoring: A Mediterranean assessment. *Landsc. Ecol.* **2013**, *28*, 905–930. [CrossRef]
64. Spadoni, G.L.; Cavalli, A.; Congedo, L.; Munafò, M. Analysis of Normalized Difference Vegetation Index (NDVI) multi-temporal series for the production of forest cartography. *Remote Sens. Appl. Soc. Environ.* **2020**, *20*, 100419. [CrossRef]

land

MDPI

Article

Ecosystem Services Valuation for the Sustainable Land Use Management by Nature-Based Solution (NbS) in the Common Agricultural Policy Actions: A Case Study on the Foglia River Basin (Marche Region, Italy)

Elisa Morri * and Riccardo Santolini

Department of Humanistic Studies (DISTUM), Campus Scientifico Enrico Mattei, Carlo Bo University of Urbino, 61029 Urbino, Italy; riccardo.santolini@uniurb.it
* Correspondence: elisa.morri@uniurb.it

Abstract: Agricultural land is a very important ecosystem that provides a range of services like food, maintenance of soil structure, and hydrological services with high ecological value to human wellbeing Ecosystem Services (ESs). Understanding the contribution of different agricultural practices to supply ESs would help inform choices about the most beneficial land use management. Nature-based Solutions (NbS) are defined by IUCN (International Union for Conservation of Nature) as actions to protect, sustainably manage and restore natural or modified ecosystems, which address societal challenges (e.g., climate change, food and water security, or natural disasters) effectively and adaptively, while simultaneously providing human wellbeing and biodiversity benefits. Some actions farmers can implement in the new Rural Development Programs (RDP 2021–2022 and 2023–2027) can be considered as NbS and could affect the quantity, quality, and time of some ESs related to water regulation and supply, N adsorption and erosion protection. This study aims to evaluate these ESs in different scenarios in the upper Foglia river basin (Italy) and at a local scale (farming), and to highlight the issue to compensate farmers for the production of public goods which benefit the whole society (ESs) by the implementation of RDP's actions. These scenarios highlight how actions have positive effects on ecosystem services and their economic value related to land use management, on maintaining agricultural practices by integrating Water Frame Directive (2000/60/EC), Directive 2007/60/EC on the management of flood risks and highlighting the potential role of farmers in a high diversity landscape. This study highlights a new way to evaluate the processes of natural capital in the production of public goods, which benefits the whole society (ESs), by emphasizing the economic and environmental role of farmers in producing them and putting on the table data to trigger a PES (Payment for Ecosystem Services) mechanism. To facilitate decision making, robust decision support tools are needed, underpinned by comprehensive cost-benefit analyses and spatially modeling in which agriculture can be a strategic sector to optimize food production and environmental protection in harmony with the Farm to Fork (F2F) strategy.

Keywords: ecosystem services; nature-based solutions; conservative agriculture; erosion protection; water regulation and supply; N balance

Citation: Morri, E.; Santolini, R. Ecosystem Services Valuation for the Sustainable Land Use Management by Nature-Based Solution (NbS) in the Common Agricultural Policy Actions: A Case Study on the Foglia River Basin (Marche Region, Italy). *Land* **2022**, *11*, 57. https://doi.org/10.3390/land11010057

Academic Editor: Luca Congedo

Received: 3 December 2021
Accepted: 27 December 2021
Published: 31 December 2021

1. Introduction

Agriculture is a dominant form of land management worldwide, and agricultural ecosystems account for almost 40% of land area [1]. In the European Union (EU) territory, rural areas account for over 77% (47% farmland, 30% forests) and their inhabitants, farming communities and other residents, representing approximately half of the entire population. The EU's Common Agricultural Policy (CAP) aims to support agriculture to ensure food security (in the context of climate change) and to promote sustainable and balanced development throughout the European rural areas, including those where production conditions are difficult [2]. Furthermore, there is a growing need for rapid and extensive actions to avoid the serious impacts of climate change, whose risks are at the top of the rankings in terms of impact and probability over the next 10 years [3]. The most worrying events are the floods triggered by extreme storms with possible landslides, especially in rural areas.

Even for hill and mountain areas, agricultural and forestry activities have been the main territorial modeling factors, in some cases creating very suggestive landscapes, but also affecting the territories often arranged for the degradation of soils and hydrogeomorphological instability [4]. Particularly in Italy, with a man's widespread presence in a differentiated landscape, specific agricultural and forestry practices have built an extensive water regulation network and stabilization of slopes contrasting erosion, especially in the past [5].

This work aimed to evaluate Ecosystem Services (ESs) as a benefit to the population provided by natural capital, associated with the water regulation (quality and quantity), soil erosion and N adsorption. The ESs were evaluated by comparing different scenarios provided by agriculture actions as derived from the Rural Development Programs (RDP) scheme (2014–2020), in an ecological Functional Unit (sub-basin of the Foglia river, Central Apennines) and at the local scale, to highlight the issue to economically recognize farmers for the production of public goods (ES), which benefits the whole society, emphasizing the economic and environmental role of farmers in producing them. Since the 1950s, the agriculture intensification with monoculture systems, the intensive tillage, and the high applications of fertilizers and pesticides maximizing crop and livestock production caused the loss of environmental quality and human well-being (landslide, flood, etc.) with higher costs recovery for the local communities [6–8].

Given this scenario of adaptation to climate change and improving the resilience of the landscape, agriculture is called upon to perform several functions: to meet the needs of citizens concerning the supply (availability, price, variety, quality, and safety); to protect the environment and ensure farmers a decent standard of living. At the same time, it is necessary to preserve rural communities and landscapes as a valuable part of Europe's heritage.

Since the mid-1980s, the European Commission introduced a range of policies. Over the last two decades, the CAP has been reformed and modified following three priorities: efficient food production, sustainable management of natural resources, and balanced development of the rural areas across the EU [9].

From CAP, the RDP are elaborated by the Member States to target funding from the European Agricultural Fund for Rural Development (EAFRD) to farmers [10]; it considers the agricultural sector as a key component in a sustainable economy [11] and Reg. (EU) 1305/2013 of the European Parliament and the Council on support for rural development by the EAFRD introduces Agri-Environmental-Climate (AEC) measures (art. 28) [12] and Agri-Environmental Schemes (AES) that play a crucial role for meeting society's demand for positive environmental outcomes from agriculture [13]. Whereas agricultural systems produce most of our food, they drive significant environmental degradation. This tradeoff between development and environmental conservation objectives is not an immutable outcome as agricultural systems are simultaneously dependents and providers of ecosystem services [14,15] depending on agricultural practices.

Furthermore, the enhancement of ESs needs to be addressed at different scales and should include the interaction between agriculture practices conserving biodiversity and stakeholders (farmers, managers, policymakers, etc.) to optimize service delivery [16]. AES represent one of the main tools of RDP: they are packages of actions implemented voluntarily by farmers who receive financial support in return for adopting environmentally-friendly practices [17], that could be optimized to reduce the adverse effects of agriculture, as improving water quality [18].

In particular, sustainable management of natural resources and climate action is one of the three objectives of the CAP post-2013 [19] and the new CAP (2023–27) eco-schemes, providing stronger incentives for climate-and environmentally-friendly farming practices and approaches (such as organic farming, agroecology, carbon farming, etc.) as tools to the environmental care and landscapes objectives. These actions can be considered Nature-based solutions (NbS), namely solutions inspired and supported by nature, which are cost-effective and simultaneously provide environmental, social, and economic benefits and help build resilience. Such solutions bring more and more diverse, natural features and processes into landscapes and agroecosystems through locally adapted, resource-efficient, and systemic interventions [20].

In addition, one of the most innovative aspects of CAP post-2013 was the introduction of payments for agricultural practices beneficial for the climate and the environment, which is generally referred to as "the greening component with measures as to comply with crop diversification, requirements, to maintain permanent grassland and to have ecological focus areas on farmland" [21]. The greening component has a prominent role in ensuring biodiversity objectives and the potential to protect and support ecosystems producing important ESs [11–21]. Consequently, the effect on the environment has been strengthened, providing payments to farmers who make commitments for making mandatory greening actions in order to get direct payments (1st pillar) and developing AEC measures in the EU's rural development policy (2nd pillar) through a set of actions relating to environmental protection and maintenance of the landscape, also as an adaptation to climate change. Regulation 1305/2013 stipulates that "restoring, preserving and enhancing eco-systems related to agriculture" and "promoting resource efficiency and supporting the shift towards a low carbon and climate-resilient economy" are two of the EU's six policy priorities for its rural development policy in the period 2014–2020 with payments to support afforestation, agroforestry, organic farming; compensating farmers for costs and income loss associated with restrictions introduced by the Wild Birds, Habitats and/or Water Framework Directives, etc. [21]. These aspects are increasingly topical and urgent in the next CAP (2023–2027) and are the target of objectives of agriculture and climate mitigation (obj. no. 4), efficient soil management (obj. no. 5), biodiversity, and farmed landscapes (obj. no. 6).

These measures are aimed to encourage farmers to protect and enhance the environment on their farmland by compensating them for the provision of ESs [9]. Whittingham [22] highlights the link between ESs and AEC payments and suggests that the compromise strategies combining patterns of biodiversity and ecosystem services may be the best solution for future initiatives, especially when operating at scale landscape [23,24]. In this context, agriculture can affect a wide range of ESs, including food and materials for human consumption, but also non-market services, such as water quality and quantity, soil and air quality, carbon sequestration, pollination services, seed dispersal, pest mitigation, biodiversity, and protection from disturbances [25–28].

These ESs are not usually considered in the development of agricultural management strategies [29] because they are not directly linked to the market as food or raw materials [4] or lack incentives for the provision that come with prices [15].

However, in addition to forest ecosystems [30,31], the contribution of agricultural land and practices on water supply and erosion control (regulating services) is mainly associated with the reduction of water losses and sediment by runoff or water evaporation that contribute to water retention and groundwater recharge.

When agronomic management practices are performed with respect to environmental quality, agroecosystems can produce high ecological and economical values and public goods, appreciated by people but that often are not recognized by market values [15]. Reed et al. [32] affirm that to link ecosystem service provision with payments to farmers, it is necessary to identify and put a price on each service, either directly by offering contracts for a set of outcomes or indirectly by offering contracts for a set of actions. Farmers' behavior and motivations play a central role in terms of overall measure uptake, and therefore policy success at the European [33] as well as at the Italian level [34,35], highlighting the importance of a positive attitude towards nature conservation by farmers as one motivation for participation in AEC measures [36].

Marche Region in the RDP 2014–2020 and for 2021 defines the Agro-Environmental Agreement (AEA) as "a set of commitments for farmers in a limited area, supported through a mix of RDP measures activated to reach specific environmental goals". In addition, other recent policies promote innovative management regimes, particularly for ecosystem services of regulation and water-related, which can provide potential funding mechanisms to integrate the actions of the Biodiversity Strategy 2030 [11] with the actions of the Water Framework Directive (2000/60/EC-WFD) and the Directive 2007/60/EC on the assessment and management of flood risks, by including in the public and/or private sectors actions aimed to preserve and enhance ecosystems and their services by the green and blue infrastructure and the restoration of degraded ecosystems.

Traditionally, agroecosystems were considered primarily as a source of provisioning services associated with food and raw materials production (fiber and biomass for bioenergy) [15,21,37], sometimes reducing other ecosystem services [38]. The efforts of farmers have focused on the provision of services as they are strongly linked to the market that generates income. At the same time, there are also non-market services while their contribution is reduced or absent due to unsustainable agricultural practices [29] that can cause serious instability problems.

Studies suggested that it may be possible to manage agroecosystems to support many ecosystem services while still maintaining or enhancing the provisioning services that agroecosystems were designed to produce [29,39,40] considering how agricultural practices can result in synergies between agricultural production and ESs [41].

In this context, regulation services are the most important related to the water cycle and soil instability, taking a different level of interest from other ESs [42,43]. The availability of water, both in terms of quantity and quality, is also heavily influenced by the functioning of ecosystems [38–44]. For example, the secondary drainage network may be an important sink for bioavailable nitrogen owing to its hydrological connections with terrestrial systems, high rates of biological activity, and streambed sediment environments that favor microbial denitrification [45].

The agroecosystems depend heavily on a cluster of regulation ESs, provided by natural ecosystems that can be measured in an Ecological Functional Unit in which is possible to determine levels of natural capital usability, as security of ecosystem function supporting other services, and also as tools to evaluate agriculture's dependence on natural capital or to address sustainable planning [43–46].

Many studies try to evaluate the value of agroecosystem services [8,15,47] also in Italy [29,48] and how agri-environment schemes under the Common Agricultural Policy could be adapted to derive a higher return of ecosystem services from agricultural land [32], but the incorporation of ecosystem services into agricultural land-use decision-making remains limited [7].

Extensive literature, starting from the paper of Costanza et al. [49], evaluates the ESs at a global or local scale [50–52]; however, there is a lack of detailed understanding of ESs in a different scenario of agricultural practices [53], especially in Italy where the focus was on mainly an in peri-urban context [54,55].

We conducted an ESs evaluation related to water regulation (quality and quantity) and soil erosion to show how agriculture actions as derived from the RDP scheme (2014–2020), can be a tool to highlight the role of farmers in protecting the landscape, contrasting climate change, producing public goods, and also economically recognizing their role and involve it in a PES (payment for ecosystem services) mechanism [56].

Concerning this, Jack [21] suggests that the private markets may increasingly develop to provide payments for landowners who help protect ecosystem services.

This study was supported by "Consorzio di Bonifica delle Marche" Institution, with the aim to involve the farmers in an AEA to solve problems of water quality and erosion of the study area that are also in accordance with the results required by the WFD (2000/60/EC) and the Directive on the assessment and management of flood risks (2007/60 /EC).

In this research, we have tried to make NbS solutions concrete with actions based on the management of natural dynamics, capitalizing on the accumulated knowledge of ESs. In that sense, we define nature-based solutions as any transition to the use of ESs with decreased input of non-renewable natural capital and increased investment in renewable natural processes.

This study starts with the identification of the ecological Functional Unit in which ESs as water regulation and supply, soil erosion, and N adsorption are evaluated in different scenarios of agricultural practices with the application of the AEA to understand how CAP's actions are able to produce a public benefit and to recognize the role of farmers in producing it. Subsequently, we evaluated the same ES at a farming scale to inform farmers on the value of some agricultural practices in maintaining ecosystem services and producing benefits to the local population and put this information on decision makers' attention to trigger PES mechanisms.

2. Study Area

The study area is the Foglia River Basin, located in Italy's northern region of the Apennine Mountains, the northern Marche Region (Figure 1). This river basin, with a torrential regime, is representative of many others reaching the Adriatic Sea in terms of geomorphology and socio-economic dynamics.

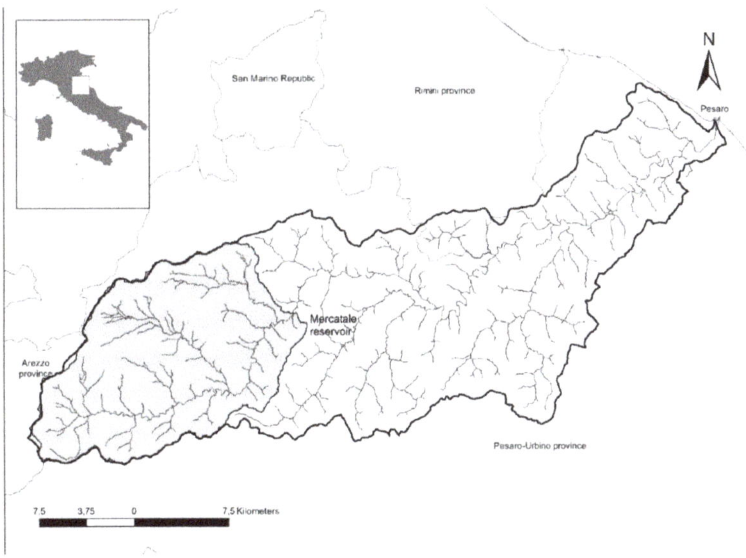

Figure 1. The study area of the upper Foglia river basin.

In particular, we focused on the upper part of the Foglia river basin that covers approximately 22,720 ha (total 70,000 ha). It is characterized by a combination of agricultural areas (43%), forests (40%), and natural and semi-natural grasslands (11%); urban areas cover 6% of the total area as derived by land use map [57].

The agricultural and zoo-technical activities in this area caused water degradation as indicated by the Regional Environment Protection Agency (ARPA) data on water quality monitoring (www.arpamarche.it accessed on 12 July 2015), and in the middle and lower part of the river basin, there are high irrigation, drinking and industry demands and water quality problems (namely, pollution by nitrates and eutrophication). Since the 1980s, progressive deterioration in groundwater quality and increasing demand has generated a greater use of surface water as the main source (56%), thus requiring physical and chemical treatments. In addition, several minor water systems (<2000 AE) are not able to remove all nutrients (e.g., organic load); many of these systems are not connected to drain networks.

Furthermore, a high percentage of the river basin risks soil erosion: 33% of land use types in the Foglia river basin are at a relatively high erosion risk (5–20 t ha^{-1} yr^{-1}) [58].

These problems are connected to the Mercatale reservoir located in the study area, with a water storage capacity of approximately 5.91 Mm^3, currently reduced to 4.8 Mm^3 caused by silting. Water stored in this basin is used for irrigation (3500 ha) and drinking water (1 Mm^3 for 8 municipalities) from the "Consorzio di Bonifica delle Marche" Institution.

The Mercatale reservoir represents an important reservoir to support water demand that significantly increases during summer because of 5.7 million tourists, especially along the coast, in addition to the 130,000 residents of the study area.

The timing of water demands and availability and erosion issues critically depend on land use in the upper regions of the basin. This area was one of those selected by the Marche region to apply an Agro-Environmental Scheme (AES) in which some agricultural practices can promote ecosystem services.

3. Materials and Methods

Considering the problems of the study area (water regulation/supply/quality and siltation connected to the Mercatale reservoir), 3 ESs were investigated, as described in the following paragraphs.

3.1. *Water Regulation and Supply*

The water regulation and supply services (WRSS), (Equation (3)), were estimated as composed by the volume of available water (*AW*) and considering the supply portion withheld by different land use as the contribution to water storage and natural release efficiency of water conservation (EW).

AW, (Equation (1)), was evaluated by considering each land use class extension's average surplus values (S). In particular, surplus values are approximated by the difference between precipitation input (P) and real evapotranspiration (ETR) [59], as described in Morri et al. [30].

$$AW = \sum S_{ij} \times Aj \tag{1}$$

where S_{ij} is surplus (mm) over *i*-pixel with *j*-land cover typology and Aj is the land cover extension (m^2).

The EW coefficient for different land cover typologies derived from Hümann et al. [60] for forest and was elaborated from Nedkkov and Burkhard [61] for the other land use classes (Table 1).

Table 1. Efficiency of water conservation provided by different land use classes (derived from Humann et al. [60] for forest and elaborated from Nedkkov and Burkhard [61] for the other land use classes).

Land Use Classes	Efficiency of Water Conservation EW (% of S)
Urban area (industrial, road, urban fabric); bare rocks	0
Arable land	17
Continue and discontinue grasslands; complex cultivation patterns; Sclerophyllous vegetation	33
Shrubby buffer strips; vineyards; transitional woodland-shrub	50
Broad-leaved forest, hygrophilous forest, mixed forest; arboreous buffer strips	67
Coniferous forest or mixed (mainly coniferous)	83
Water bodies, water courses	100

This coefficient was applied to the *AW* value for each land use class (Equation (3)) to evaluate water storage and release regulation. Urban areas have a value of 0 considering that available water is not stored in this kind of land cover and can contribute to runoff or overland flow.

For the S evaluation, meteo-climatic data were related to monthly precipitations data for the period 1982–2011 from Regional Information System Weather-Hydro-pluviometric-SIPMIP (http://84.38.48.145/sol/indexjs.php?lang=it, accessed on 12 July 2015).

The period 1982–2011 was selected to have an historical series of data and consider the start of global climatic change to understand if the hydrologic cycle can depend on it [62–64]. Maps of climatic average annual and monthly data were created (temperature and precipitation).

Real evapotranspiration (*ETr*) was calculated as described in Zhang et al. [65] that consider the *w* parameter (plant-available water coefficient) related to land cover to simulate different scenarios considering different land use in the study area (Equation (2)).

$$ETr = 1 + \frac{W\,\frac{E0}{P}}{1 + W\,\frac{E0}{P} + \left(\frac{EO}{P}\right)^{-1}} \quad (2)$$

where *P* is the precipitation, *E*0 is the potential evapotranspiration calculated as indicated in Zhang et al. [65], and *w* are the parameters for forests (2), grassland and arable lands (0.5), urban or bare rock (0.1). For the other land use classes not considered by Zhang et al. [65], for the w parameter evaluation, we constructed a regression line with the Kc parameter that describes the effect of both crop transpiration and soil evaporation as modeled by FAO [66] (Figure 2). Data are shown in Table 2.

Finally, the economic value of water regulation and supply service (*WRS* expressed in €) was calculated as described in Equation (3):

$$WRSS = CDW \times \sum j \sum i \times EWj \times AWij \quad (3)$$

where *CDW* is the price of a unit of bulk water (0.35 €/m^3) defined by the local water service company (Hera and Marche Multiservizi), *EWj* is the retaining coefficient for *j*-land cover typology, and *AWij* is available water over *i*-pixel with *j*-land cover typology.

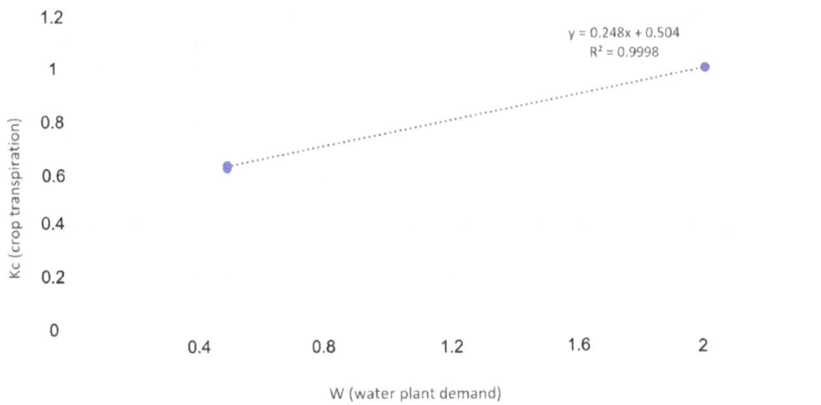

Figure 2. Regression line from Kc parameters (FAO) and W parameters for the land use classes not indicated in Zhang et al. [65].

Table 2. Parameters used for evaluation of soil loss by erosion and the real evapotranspiration.

Land Use Classes	C	Kc	W
Transitional woodland-shrub	0.01	0.8	1.2
Sclerophyllous vegetation	0.01	0.8	1.2
Industrial and commercial units	0	0.001	0.1
Sport and recreation ports	0.01	0.63	0.5
Green urban areas	0.01	0.63	0.5
Water bodies	0	1	0.1
Broad-leaved forest (Fagus)	0.01	1	2
Broad-leaved forest (Quercus)	0.01	1	2
Hygrophilous forest	0.01	1	2
Coniferous forest	0.01	1	2
Broad-leaved forest	0.01	1	2
Mixed forest (mainly coniferous)	0.01	1	2
Mixed forest (mainly broad-leaved)	0.01	1	2
Construction sites	0	0.001	0.1
Water courses	0	1	0.1
Shrubby buffer strips	0.01	0.63	0.5
Arboreous buffer strips	0.01	0.63	0.5
Continue grasslands	0	0.001	0.1
Discontinue grasslands	0.55	0.001	0.1
Road and rail networks	0.335	0.62	0.5
Bare rocks	0.55	0.001	0.3
Arable land	0.335	0.67	0.8
Arable land-bare surface	0	0.001	0.1
Complex cultivation patterns	0.55	0.7	0.2
Urban fabric	0.28	0.67	0.5
Vineyards	0.01	0.8	1.2
Grassing vineyards	0.01	1	2

3.2. *Erosion Protection*

The study area's erosion protection service (EP) emerges as a decrease in soil loss by erosion in different scenarios. Thus, the economic value of soil protection was estimated as the avoided cost of restoring soil where erosion might occur, similarly as described in Morri et al. [30]. Because conservative agricultural management is the most effective in soil protection, this value was estimated by the difference in the potential erosion between different scenarios applying conservation agriculture management.

The difference of soil loss expressed as ton ha^{-1} y^{-1} from different scenarios approximated the contribution to erosion reduction and was expressed as a benefit of the conservative agriculture measures for soil protection. Such a contribution was multiplied by the average cost for transporting and restoring a unit volume of soil and divided by the average soil density, as expressed in Equation (4), where C is the cost for transporting and locating a unit volume of soil (41 €/m^3) [67], $X_2 - X_1$ is the difference between the average soil loss (t ha^{-1} yr^{-1}) provided by different scenarios, and SD is the soil density (1.4 g/m^3) of the study area [68].

$$EP = C \times (X_2 - X_1)/SD \quad (4)$$

To evaluate the soil loss as erosion risk by water in the study area we used the Revised Universal Soil Loss Equation (RUSLE model) [69], as expressed by Equation (5):

$$A = R \times K \times LS \times C \times P \quad (5)$$

where:

A = estimated average soil loss (ton ha^{-1} y^{-1}).
R = rainfall-runoff erosivity factor (Jm^{-2} mmh^{-1}).
K = soil erodibility factor (t ha^{-1} R unit).

L = slope length factor (dimensionless).
S = slope steepness factor (dimensionless).
C = cover-management factor (dimensionless).
P = support practice factor (dimensionless).

The R factor quantifies the effect of raindrop impact and reflects the amount and rate of runoff likely to be associated with rain. For the study area, R-values were derived by Diodato [70] using regional relationships for estimating the erosion index from only three rainfall parameters as annual precipitation (a), annual maximum daily precipitation (b), and annual maximum hourly precipitation (c) that are reported in Italian Hydrographic Service newsletters as described in Equations (6) and (7):

$$EI_{30\text{-annual}} = 12.142 \times (abc)^{0.6446} \quad (6)$$

where $EI_{30\text{-annual}}$ (in MJ mm ha^{-1} y^{-1}) is the annual erosive empirical index to calculate R as:

$$R = \frac{1}{N} \times \sum_{1}^{N} EI_{30-\text{annual}} \quad (7)$$

where N is the year period.

K factor was provided by the Agency for Agro-food Sector Services of the Marche Region (ASSAM) [58]; LS factors were calculated by using the semiautomatic method for ARCGIS 10 of Pelton et al. [71] (http://gis4geomorphology.com/ls-factor-in-rusle/, accessed on 4 June 2015).

C factor depends on land use, and it varies from eroded cultivated soil (C = 0) to eroded non-cultivated soil (C = 1). In the study area, the C factor was defined from the land use map for each land cover class as indicated in Table 2 and derived from ASSAM.

P factor reflects the impact of support practices as the average annual erosion rate. It is the soil loss ratio with contouring and/or strip cropping to that with straight row farming up-and-down slope. *P*-values for different slope classes were used as indicated in Table 3 for the contouring practices that can be potentially developed in the study area.

Table 3. *p* factor values for contouring practices for each slope class (from Kirkby and Morgan [72]).

Erosion Classes	Slope Classes (%)	*p*-Value
1	1–2	0.6
2	3–8	0.5
3	9–13	0.6
4	14–16	0.7
5	17–20	0.8
6	21–25	0.9

We estimated average soil loss and produced maps for the study area for each of the different scenarios considered.

3.3. N Balance

N balance was estimated for the Foglia river and the upper river basins. It was derived from the soil system budget [73] developed by the Italian Nitrogen Network [74], to verify whether uncoupled input and output terms generate N surplus or deficit. A positive N value means that a certain nitrogen amount ("N surplus") is either retained within soils or run off to surface water and leached to groundwater. A negative N means that input terms are not sufficient to sustain crop requirements and atmospheric losses (denitrification and volatilization).

The soil system budget was calculated on an annual basis (year 2010) as the net difference between N inputs (livestock manure, synthetic fertilizers, atmospheric deposition, biological fixation, and wastewater sludge) and N outputs (crop uptake, ammonia volatilization and denitrification in soils) within the catchment's agricultural land (for details see Soana et al. [75]).

N budget calculations were performed at a spatial resolution of individual municipalities (35) using farming census data (ISTAT, National Statistics Institution, 6th Agricultural Census 2010; http://censagr.istat.it/dati.htm, accessed on 4 June 2015), then aggregated to the catchment scale employing GIS techniques.

Final data of balance was expressed in tons N yr^{-1}.

For the scenario, we calculated how the wooded buffer strips contribute to N adsorption, referred to N balance of the upper Foglia river basin. For the N adsorption provided by wooded buffer strips (3 m × 1000 m), we considered the value of 405 kg N and the economic value of 0.7 €/t for the denitrification service [76,77].

3.4. Developing Scenarios

3.4.1. Scenarios at Upper River Basin Scale

To understand how the land use change at the upper river basin scale can improve the quality of the ecological system and ecosystem services values, we suppose 4 scenarios that could be realized with the application of the AEA in the RDP of Marche Region (2014–2020). We have considered different actions that can be involved in AEA, for example, wooded buffer strips provide many ecosystem services relating to mitigating impacts of climate change and provide for water purification [78], or cover crops in vineyards can help mitigate many of the problems associated with excessive precipitation or soil erosion [79–95].

In particular, the scenarios are:

1. T0 scenario: state of the art from 2008 land use map [57] with 2% of the arable land cover as a bare surface in winter (Agriculture Regional Information System).
2. T0_bis scenario: improving from the T0 scenario, the Urban Development plan for discontinuous urban fabric, industrial units, services and public facilities, parking, roads.
3. AAA scenario: from the T0 scenario, some agri-environmental actions are developed: addition of wooded buffer strips (European Reg. n 1305/2013) [80] of 5 m along with the first drainage network and 3 m along with the secondary drainage network, grassing vineyards (measure 10.1 RDP 2014–2020), winter cover for arable land (measure 10.1 RDP 2014–2020), the addition of herbaceous vegetation (3 m) along the road neighboring arable land and along the erosion furrows (5 m), variation of P parameter about RUSLE model from 1 to 0.5 and 0.6 [72], considering that 15% of arable land (slope class < 13%) are subject to contour plowing (it is the farming practice of plowing and/or planting across a slope following its elevation contour lines. These contour lines create a water break, which reduces the formation of rills and gullies during times of heavy water run-off, which is a major cause of soil erosion).
4. AAA bis scenario: from AAA scenario, it is assumed that arable land with a slope > 20% is transformed into grassland (e.g., alfalfa). This action aims to promote land management that minimizes potential erosion. In particular, this action could permit the conversion of almost 4500 hectares of arable land in grasslands that represent a specific habitat for some species of animals and plants, increasing the biodiversity level.

3.4.2. Scenarios at Farming Scale

For the realization of scenarios at farming scale, we analyzed the local condition of agricultural practices, and we selected, based on local agronomist knowledge and on RDP measurement (2014–2020), the actions that could be undertaken by farmers in the study area, to facilitate erosion protection, water regulation and supply, and N adsorption. We selected two farms as a sample of approximately 189 (F1) and 118 ha (F2), with arable land surfaces of 157 and 89 ha, respectively.

We reject the realization of new:

- Vineyards: because of the failure of the production chain and the difficulties in promoting the product (high average age);

- Orchards: because of the high installation costs, the lack of fruit and vegetables local sector, and the inadequate training of farmers to never practiced crops;
- Olive groves: because of weather and altitude problems.

We supposed 4 scenarios for the 2 sample farms with more conservative measures from T0 to T3:

- T0: actual traditional agronomic practices;
- T1: minimum tillage practices in the 35% of arable land;
- T2: conservation agriculture techniques: no-tillage, direct seeding and working bands on arable crops, crop residues left in an agricultural field, ban on continuous cropping;
- T3: conversion of arable land to grassland, meadows permanent grass of specialized perennial crops (vines, olives, and fruit), and creation of not used buffer zones along the watersheds (5% of the farming area).

For these scenarios, we estimated the value of selected ESs and considered potential payments to farmers deriving from RDP measures adoption (€ ha^{-1}). In addition to erosion protection and water regulation and supply values, we also considered other benefits as CO_2 sequestration by soil (t ha^{-1}) by applying a carbon market price expressed as emission permit price of 31 €/t CO_2 [81], biodiversity increase (n. of earthworms) [82], fertility soil improvement (qualitative evaluation), and fuels use saving (l ha^{-1} as Kg ha^{-1} CO_2) deriving from LIFE project HELPSOIL data (http://www.lifehelpsoil.eu/en/, accessed on 10 April 2015). We did not apply the N balance at the local scale because the model was constructed for the river basin scale, and the minimum data set is municipal level.

For the water regulation and supply service, we considered the economic value as derived for the upper river basin (Supplementary Materials Table S1) but considering only the agricultural classes in retaining available water for the T1 and T2 scenarios and the grassland, buffer strips and grassing vineyards water retention value for the T3 scenario obtaining a value of 112 € ha^{-1} and 172 € ha^{-1}, respectively.

4. Results

4.1. Land Use Change

Land use transformations bring about changes in the provision of ESs. Table 4 indicates the land use change of different supposed scenarios. Data showed an increase of approximately 55 ha for industrial or commercial units, and 99 ha for urban fabric class considered as a land use unable to produce ESs. Expected scenarios provided an increase of approximately 33 ha of buffer strips (0.14%), grassing vineyards (0.1%), and a decrease of arable land with a bare surface (0.33%).

Continue grasslands increased approximately 200 ha from T0_bis to AAA scenario and strongly increased from T0 to AAA_bis scenario of approximately 4678 ha representing 28.6% of the total area (Table 4). Grassland extensively managed, as expected in the scenario, has a high capacity to deliver multiple ESs as parts of agricultural systems even if understudied [83].

4.2. Evaluation of Selected Ecosystem Services at Upper River Basin Scale

4.2.1. Water Regulation and Supply

Surplus data (mm) was derived from the water balance equation as the difference between precipitation and evapotranspiration. In the study area, the average real evapotranspiration value is approximately 566 mm/y from a minimum of 410 mm/y to a maximum of 685 mm/y. Considering evapotranspiration and climatic data, with an average precipitation value of approximately 1009 mm/y, surplus ranges from 217 mm/y to 730 mm/y with an average of 442 mm/y. Available water (m^3) was obtained using the average surplus value multiplied by the extension of each land use class.

Considering available water raster values for the extension of each land use class for the Foglia upper river basin, the study showed available water of approximately 100 Mm^3 for all scenarios without important variation.

Table 4. Land use classes extension and % for each scenario.

-	T0		T0_bis		AAA		AAA_bis	
-	Area							
-	ha	%	ha	%	ha	%	ha	%
Transitional woodland-shrub	682.0	3.0	682	3.0	681.9	3.0	681.9	3.0
Sclerophyllous vegetation	364.6	1.6	364.6	1.6	364.6	1.6	364.6	1.6
Industrial and commercial units	56.9	0.2	111.6	0.5	111.6	0.5	111.6	0.5
Sport and recreation ports	18.2	0.1	18.2	0.1	18.1	0.1	18.1	0.1
Green urban areas	352.7	1.5	333.5	1.5	330.6	1.5	330.6	1.5
Water bodies	54.6	0.2	54.6	0.2	54.6	0.2	54.6	0.2
Broad-leaved forest (Fagus)	35.6	0.2	35.6	0.2	35.6	0.2	35.6	0.2
Broad-leaved forest (Quercus)	6411.6	28.2	6410	28.2	6409	28.2	6409	28.2
Hygrophilous forest	245.7	1.1	245.1	1.1	245.1	1.1	245.1	1.1
Coniferous forest	489.8	2.2	489.6	2.2	489.3	2.1	489.3	2.1
Broad-leaved forest	0.3	0.0	0.3	0.0	0.3	0.0	0.3	0.0
Mixed forest (mainly coniferous)	113.9	0.5	113.8	0.5	113.8	0.5	113.8	0.5
Mixed forest (mainly broad leaved)	2041.5	9.0	2041	9.0	2040	9.0	2040	9.0
Construction sites	3	0.0	2.2	0.0	2.2	0.0	2.2	0.0
Water courses	48.5	0.2	48.4	0.2	48.4	0.2	48.4	0.2
Shrubby buffer strips	0	0.0	0	0.0	25.3	0.1	25.3	0.1
Arboreous buffer strips	0	0.0	0	0.0	7.8	0.0	7.5	0.0
Continue grasslands	1842.4	8.1	1838	8.1	2039	9.0	6520	28.6
Discontinue grasslands	259.1	1.1	254	1.1	252.8	1.1	252.8	1.1
Road and rail networks	578.4	2.5	569.9	2.5	569.8	2.5	569.8	2.5
Bare rocks	228	1.0	228	1.0	226.9	1.0	226.9	1.0
Arable land	8663.8	38.1	8562	37.6	8407	36.9	3926	17.2
Arable land-bare surface	75.3	0.3	72	0.3	0	0.0	0	0.0
Complex cultivation patterns	54	0.2	54	0.2	52.6	0.2	52.6	0.2
Urban fabric	124.1	0.5	223	1.0	219.6	1.0	219.6	1.0
Vineyards	18.3	0.1	18.3	0.1	0	0.0	0	0.0
Grassing vineyards	0	0.0	0	0.0	17.9	0.1	17.9	0.1
Total	22,762.3							

Considering the percentage of retained water from different land use classes (Table 1), data showed a value of approximately 37 Mm^3 of available water stored and released in the T0 and AAA scenarios and almost 40 Mm^3 in the AAA_bis scenario corresponding to an economic value for the water regulation and supply service of almost € 14 × 10^6 (Table 5). Water regulation values are less in the scenario T0_bis because of more urban surface extension with respect to the T0 scenario.

Table 5. Available water and water regulation and supply service values for each scenario.

	Scenarios			
Available Water	T0	T0_bis	AAA	AAA_bis
Mm^3	100.73	100.77	100.66	100.72
Water regulation and supply service				
Mm^3	36.91	36.18	36.77	39.85
€ × 10^6 (0.35 € m^{3-1} bulk water cost)	12.92	12.66	12.87	13.95
€ ha^{-1}	569	557	566	614

Detailed data for available water and stored and released water are described for each land use class in Supplementary Materials Table S1.

4.2.2. Erosion Protection

The RUSLE model application shows a soil loss of approximately 688,978 t/y for the T0 scenario, a slight decrease in the scenario AAA of approximately 650,000 t/y, and a hard soil loss in implementing conservative agricultural measures of AAA_bis scenario (Table 6 and Figure 3).

Table 6. Erosion protection value by erosion of each scenario.

	Scenarios			
	T0	**T0-bis**	**AAA**	**AAA_bis**
soil loss_Erosion (t y^{-1})	688,978	672,307	649,975	265,750
difference with t0 (t y^{-1})	-	−16,671	−39,002	−423,227
difference with t0 (%)	-	−2.4	−5.7	−61.4
difference with t0 (m^3 y^{-1}) (considering 1.4 gr/cm^3 soil bulk density)	-	−11,908	−27,859	−302,305
Erosion protection value (10^6 €) (41 €/m^3 Marche Region, 2010)	-	0.49	1.14	12.39
Forest area (ha)	9338	9336	9334	9334
Erosion protection value by forest (€)	784,407	784,189	784,031	784,031
Erosion protection value by applying RDP measures (€)	-	-	358,184	11,610,486
Erosion protection value by applying RDP measures (€ ha^{-1})			15.8	511

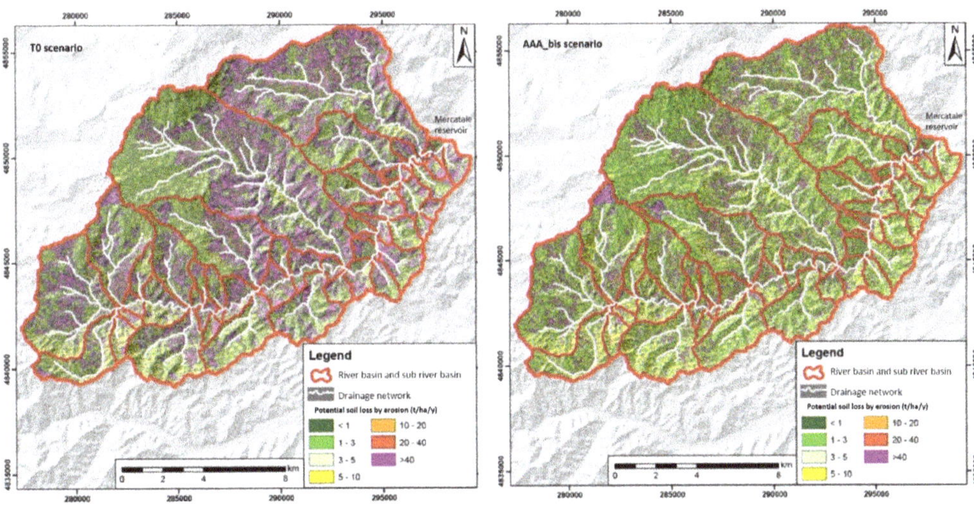

Figure 3. Potential soil loss by erosion in T0 scenario (**left**) and AAA_bis scenario (**right**).

Data showed a decrease of erosion of approximately 5.7% (39,000 t) from the T0 to AAA scenario and a very high decrease of over 60% (423,227 t) to AAA_bis scenario with the supposed transformation of arable land with a slope > 20% to grassland. From T0 to T0bis, data highlighted a slight decrease of soil loss by erosion due to a slight increase of artificial surface with a potential erosion equal to 0 in the RUSLE model.

Applying the 1.4 gr/cm^3 soil bulk density for the study area, the potential soil loss by erosion decrease is 11,908 m^3/y, 27,859 m^3/y, and 302,305 m^3/y for the T0_bis, AAA, and AAA_bis scenario, respectively.

The erosion protection value, applying the replacement cost of 41 €/m^3, is 0.49, 1.14, and 12.39 10^6 € for the T0_bis, AAA, and AAA_bis scenarios, respectively.

We considered the erosion protection value (€ ha^{-1}) provided by the forest as indicated in Morri et al. [30] for the same study area (84 € ha^{-1}) to separate the erosion protection value by applying RDP measures. The results of 84 €/ha for erosion protection by forest are similar to values reported by Li et al. [84], with a valuation developed in temperate forests.

For the AAA scenario, the erosion protection provided by RDP measures was 358×10^3 € and over 11×10^6 € for the AAA_bis scenario with values of 15.8 and 511 (€ ha^{-1}). This can be the value to put on the table to discuss the PES implication with recognizing the farmers' role to produce ESs.

4.2.3. N Balance

The application of the soil system budget as developed by the Italian Nitrogen Network shows for the Foglia river basin an N surplus of approximately 1727 Tons N yr^{-1} (Table 7).

Table 7. Soil system budget for Foglia and upper Foglia river basin.

	Soil System Budget Foglia River Basin		Upper Foglia River Basin	
Input	Tons N yr^{-1}	%	Tons N yr^{-1}	%
Livestock manure	1388	19	515	70
Synthetic fertilizers	520	7	104	14
Biological fixation	5075	71	72	10
Atmospheric deposition	172	2	45	6
Σ input	7156	100	736	100
Output	Tons N yr^{-1}	%		
Crop uptake	4918	91	292	62
NH_3 volatilization	320	6	119	25
Denitrification in soils	191	3	62	13
Σ output	5429	100	473	100
Input-output	1727		264	

The most important input is mainly due to biological fixation (71%) and livestock manure (19%), while the main output was performed by crop uptake (91%).

For the upper Foglia river basin (Figure 4), data showed a different input data: the most important N load derived from livestock manure (515 tN yr^{-1}) produced mainly by swine (45%) and cattle (35%) farming. Manure was followed in importance by synthetic fertilizers application (104 tN yr^{-1}). Biological fixation and atmospheric deposition were responsible for the remaining 16% of N input to agroecosystems. N input to agricultural lands in the upper basin municipalities ranged from 30 to 224 kg Nha^{-1}, and overall the mean input rate was 81 kg Nha^{-1}.

The total potential N output from agricultural lands was 473 tN yr^{-1} on an annual basis, with crop harvest representing the maximum sink, approximately 62% of total output. NH3 volatilization and denitrification in soils yielded annual fluxes to the atmosphere of 25% and 13% of total output, respectively. Annual N output from agricultural lands in the upper basin municipalities ranged from 27 to 92 kg Nha^{-1}, and overall, the mean output rate was 52 kg Nha^{-1}.

Upper Foglia N surplus represented 15% of the total Foglia river basin N surplus, and the higher value of N load corresponded to the area directly connected to the Mercatale reservoir that could affect water quality. Considering the buffer strips supposed in the AAA and AAA_bis scenarios (33 ha), they took up 45 t/y N, corresponding to 17% of upper river basin N surplus. Applying the value of denitrification actions, the N sequestration provided by buffer strips was €31,185.

4.3. *Evaluation of Selected Ecosystem Services at Farming Scale*

Data described in Table 8 highlight the erosion protection value of farmland 1 (F1) and farmland 2 (F2) in more conservative scenarios. It was about the range of €4450–4650 for the T1 scenario, €19,250–17,750 for the T2 scenario, and €24,900–22,700 for the T3 scenario for F1 and F2, respectively.

Table 8. Values of ecosystem services provided by different scenarios in two farms.

	Farm Extension (ha) Involved in the RDP Measurement	Payments to Farmers Deriving from RDP Measures Adoption (€)	Erosion Protection Value (€)	Erosion Decrease (%)	Water Regulation and Supply Value (€)	CO_2 Sequestration (Surface Layer) (€)	Biodiversity	Fertility Soil Improvement	Fuels Use Saving as Avoided CO_2 (€)
F1 farm									
Scenario T0	189.73	44,850						medium	
Scenario T1	55.09	8264	4451	17	6170	68,312	-	good	319
Scenario T2	157.41	39,354	19,270	74	17,630	395,257	>1.5–4 time number of earthworms	good	913
Scenario T3	179.98	50,940	24,893	95	30,956	451,922	>1.5–4 time number of earthworms	good	1043
F2 farm									
Scenario T0	118.4	27,934						medium	
Scenario T1	31.11	4667	4656	19	3484	38,576	-	good	180
Scenario T2	88.89	22,222	17,747	74	9956	223,203	>1.5–4 time number of earthworms	good	515
Scenario T3	120.16	32,414	22,726	94	20,667	301,712	>1.5–4 time number of earthworms	good	697

		F1 farm	Scenario T0	Scenario T1	Scenario T2	Scenario T3	F2 farm	Scenario T0	Scenario T1	Scenario T2	Scenario T3
Farm Extension (ha) Involved in	the RDP Measurement		189.73	55.09	157.41	179.98		118.4	31.11	88.89	120.16
Payments to Farmers Deriving from RDP Measures Adoption (€)			44.85	8264	39,354	50.94		27,934	4667	22,222	32,414
Erosion Protection Value (€)				4451	19,27	24,893			4656	17,747	22,726
Erosion Decrease (%)				17	74	95			19	74	94
Water Regulation and Supply Value (€)				6170	17.63	30,956			3484	9956	20,667
CO2 Sequestration	(Surface Layer) (€)			68,312	395,257	451,922			38,576	223,203	301,712
Biodiversity				-	>1.5–4 time number of earthworms	>1.5–4 time number of earthworms			-	>1.5–4 time number of earthworms	>1.5–4 time number of earthworms
Fertility Soil Improvement			medium	good	good	good		medium	good	good	good
Fuels Use Saving as Avoided CO_2 (€)				319	913	1043			180	515	697

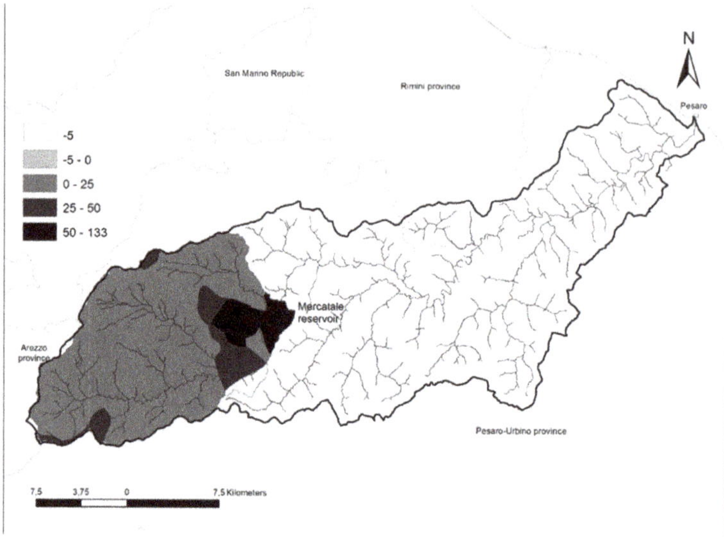

Figure 4. Spatial distribution of N surplus. Units are kg Nha^{-1}/year.

From T0 to T3 scenario, erosion decreased by 17, 74, and 95% for F1, and 19, 74, and 94% for F2, respectively, with a significant erosion decrease in the more conservative scenario at the local scale.

The water regulation and supply service, considering the value of 112 € ha^{-1} (T1 and T2 scenario) and 172 € ha^{-1} (T3), corresponded to € 3484–6170 for the T1 scenario, €9956–17,630 in T2 and €20,667–30,956 for the T3, the more conservative agricultural scenario.

Applying HELPSOIL data to the agricultural measurements provided in the different farming scenarios, we supposed carbon sequestration varies from 68,312–38,576 in T1 to 451,922–301,712 in T3 for F1 and F2, respectively. In addition, data showed a biodiversity increase in the number of earthworms between the traditional agriculture (T0) and the conservative one (T1-T2-T3), and fertility soil improvement from medium to good. Fuels use saving expressed as Kg ha^{-1} of avoided CO_2 is €319–180 in scenario T1 until €697–1043 in more conservative scenario (T3).

5. Discussion

Agricultural systems provide provisioning ESs that are essential to human well-being. Nevertheless, they also provide a range of other ESs, including regulating services and services that support provisioning.

Agricultural management practices are key to realizing the benefits associated with ESs and reducing disservices from agricultural activities applying solutions contained in RDP programs as challenges that will be magnified in the face of climate change. The European Commission aims to invest in NbS as living solutions inspired by, continuously supported by, and, using nature, designed to address various societal challenges in a resource-efficient and adaptable manner and provide simultaneously economic, social, and environmental benefits (European Commission 2015). These solutions can also contribute to the landscape and ecosystem resilience with local-based, resource-efficient, and systematic actions [20]. These solutions are useful in addressing the socio-economic challenges of the 21st century with the primary aim of maintaining and/or increasing welfare production at lower costs. The ecological restoration process—planning, site construction, monitoring, etc.,—generates economic output and employment, forming the "ecological restoration economy" [85].

Considering different RDP actions connected to regulation ecosystem services, we simulated different scenarios at the upper river basin and local scale, with solutions based on ecological functions recovery considering the economic and environmental benefits of restoration, highlighting the potential role of agriculture in developing these actions also through the RDP tools.

The ecological regulatory functions and their services have been evaluated since they are the fundamental and physiological architecture of maintenance and operation of ecosystems and basic gears for the delivery of other services and can be used to estimate the critical use thresholds compared to the other ESs [43–86].

This study analyses selected regulation ecosystem services provided by agroecosystems as erosion protection, water regulation and supply, and N balance in terms of ecological functions and services in different scenarios at the upper river basin and farming scales.

Data show the decrease of soil loss and the increase of erosion protection service provided by collecting conservative agriculture measures from actual state to a final scenario (AAA_bis or T3) both at the upper river basin and farming scales.

Considering spending for consolidation actions in the upstream Foglia river basin for 10 years (1997–2006) as a value of €6 $\times$ 10^6 (Marche Region data bank, Pesaro Urbino Protection of territory Service), the calculated value of erosion protection for 10 years (€3.6 $\times$ 10^6) corresponds to over half of spending for landslide or consolidation interventions. The value of N denitrification provided by buffer strips in the AAA and AAA_bis scenario is over €30,000.

The value of water regulation and supply provided by the most conservative scenario (AAA-bis) is almost 14 $\times$ 10^6, corresponding to half of the spending for flood defense in the period 1991–2007 for the Foglia river basin (Marche Region data bank, Pesaro Urbino Protection of territory Service).

As highlighted by Maes and Jacobs [20], the integration of the actions expressed by these scenarios represents an indication of the opportunities that farmers can implement, especially through cooperation actions (AEA), in which every action was integrated with the others so that the net effect may be evident and effective in the territory.

The synergy achieved through territorial cooperation can have a twofold effect: to improve the overall resilience of the territory by integrating the Water Directive (WFD 2000/60/EC) and the Directive 2007/60/EC on the assessment and management of flood risks, by improving the regulation ESs and on the other side, by identifying the appropriate stakeholders who can recognize the economic weight of the ecosystem regulation functions and related service.

The proposed scenarios in this study are useful for limiting erosion and increasing the water quality in the study area and, in particular, reducing problems concerning the Mercatale reservoir managed by "Consorzio di Bonifica delle Marche" for irrigation and drinking water. These measures can preserve water quality and quantity and reduce siltation on the tributary of the Mercatale reservoir, increasing its useful life. In this context, data derived from this study can be put on the discussion table with some values of ecosystem services supplied by farmers (F1 and F2) in the more conservative scenario (e.g., T3) that could be paid by the "Consorzio di Bonifica delle Marche" stakeholder as buyer of ESs, as highlighted in the PES example of Figure 5.

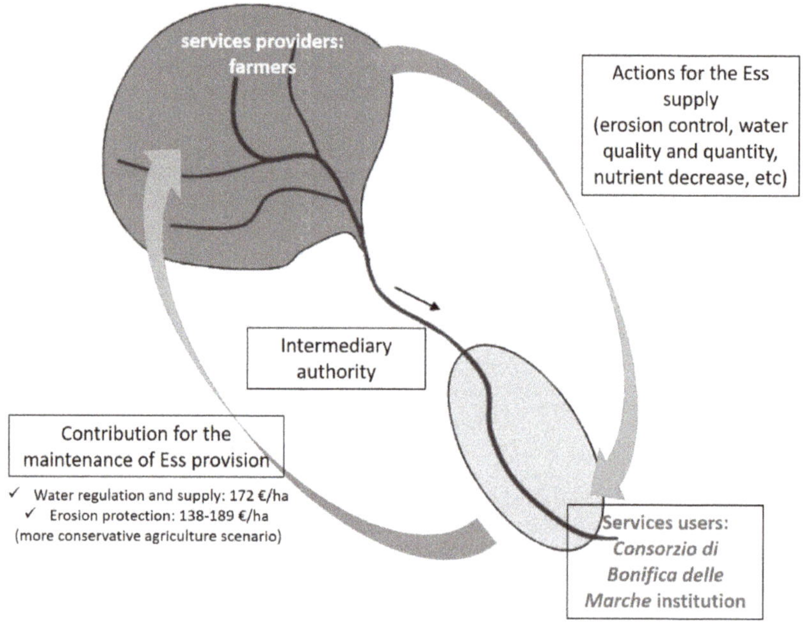

Figure 5. Hypothetic PES scheme (elaborated from Pagiola and Platais [87]) with Ecosystem Services values provided by farmers with actions as proposed in scenario T3.

At the national level, the great scientific debate on ESs and the opportunity to its implementation that generates economic output and employment, which forms the ecological restoration economy, allowed to insert in the national Law n. 221, (dated 28 December 2015) "Environmental provisions to promote green economy measures and to contain excessive use of natural resources", a specific article (no. 70) regarding the PES as a tool to compensate providers of ESs by pricing their values for users.

Furthermore, part of the value of these ecosystem services could be recognized as the cost of the natural resources that water tariffs could absorb for downstream people who use them. Ministerial Decree 39/2015 developed guidelines for defining resource and environmental costs for the different water use sectors at the national level.

These mechanisms can encourage the conservation of natural ecosystems through environmentally-friendly practices to preserve natural resources, thereby also improving the welfare in rural areas [88,89] and at the same time contribute to enriching the debate on ecosystem services provided by actions actuated by farmers.

For this reason, results of these evaluations, based on a stakeholder consultation (Consorzio di Bonifica delle Marche and regional agricultural sector), will be presented to farmers by local agricultural associations considering that farmers' awareness of their role and economic motivation play a central role in terms of developing synergic actions useful to natural capital conservation and in maintaining flows of ecosystem services [36].

Local-scale evaluation considering two different farms further highlights the suitability in terms of employment and economic output multipliers for farmers and farms, especially by agro-environment agreement that emphasizes regulating ESs management and also generating public goods that can be finalized to the PES tool identifying an ES buyer.

The ESs-based approach can be a point of convergence for policies that simultaneously address concerns for food provision and the integration policies of water regulation WFD (2000/60/EC), the Directive 2007/60/EC on the assessment and management of flood risks and biodiversity strategies.

The European Commission [90] advocates an Action Plan that better speeds up the practical implementation of nature protection directives, reversing biodiversity loss and degradation of ESs also concerning resilience to climate change and their mitigation.

For these reasons, the actions that can be implemented by farmers in the new rural development programs (RDP 2023–2027) can be considered as NbS and could positively affect the quantity, quality, and time of some ESs related to water regulation and supply, N adsorption, and erosion protection [91]. These actions can protect, sustainably manage, and restore natural or modified ecosystems, which address societal challenges (e.g., climate change, food and water security, or natural disasters) effectively and adaptively, while simultaneously providing human well-being and biodiversity benefits.

6. Conclusions

Agriculture undeniably can be a balanced tool for production and land management in harmony with the Farm to Fork (F2F) strategy, the ten-year plan developed by the European Commission to guide the transition towards a fair, healthy, and respectful food system of the environment. However, a territorial vision and planning of actions are necessary, which, if integrated, can develop emerging properties beneficial for a territory's resilience. The actions of the European agricultural strategy can be grouped within NbS, whose objectives offer solutions to the social challenges that involve working with nature as an integrated approach that could address the twin crises of climate change and biodiversity loss [92].

NbS may involve the conservation or rehabilitation of natural ecosystems and/or the enhancement or creation of natural processes in modified or artificial ecosystems, applied at both micro and macro scales [93]. Therefore, NbS focuses on protecting and restoring ecosystems, increasing functionality, and addressing societal challenges while improving human well-being and biodiversity [94].

The evaluation of ESs, therefore, represents a useful tool for developing different types of actions:

1. Build resilience scenarios to help make informed choices about the most advantageous agricultural practices and to consider these values for sustainable management and monitoring of the territory.
2. Develop synergistic actions so that the territory can offer an effective and lasting response to climate change.
3. Maintain a greater focus on ecosystem services support efforts to emphasize multi-functionality in agriculture, to manage a broader set of ecosystem services, including provisioning, but also cultural services, which rarely have a price on the market, thus highlighting the potential of the role of farmers in a very diverse landscape context.
4. Develop opportunities regarding the payment of ESs as recognition of the maintenance of functions useful to the territory and structures of collective interest, such as—in this case—the Mercatale basin for irrigation purposes.

This approach promotes considerations on the benefits provided by the activities of farmers in the conservation of natural resources and the search for synergies between food production and conservation of the resilient landscape with its ecosystem services by applying some actions of the new community agricultural policy, such as NbS, in harmony with the objectives of the Farm to Fork strategy.

Supplementary Materials: The following supporting information can be downloaded at: https://www.mdpi.com/article/10.3390/land11010057/s1. Table S1: Detailed data for available water and water stored and released for each land use classes.

Author Contributions: Conceptualization, E.M. and R.S.; methodology, E.M. and R.S.; formal analysis, E.M.; investigation, E.M. and R.S.; data curation, E.M.; writing—original draft preparation, E.M. and R.S.; writing—review and editing, E.M.; project administration, R.S.; funding acquisition, R.S. All authors have read and agreed to the published version of the manuscript.

Funding: This research was funded by Consorzio di Bonifica delle Marche institution (president Netti), the drainage consortium established in 2013 by Marche region with R.L n. 13 dealing with soil and wetlands conservation from hydrogeological instability, sustainable development of the territory and protection and enhancement of agricultural production. Agreement signed on 20/05/2014.

Informed Consent Statement: Not applicable.

Data Availability Statement: Not applicable.

Acknowledgments: We are grateful to the agronomist Filippo Biondi for his knowledge and suggestion in the choice of the farms used as a sample in this study, Michele Tromboni (Consorzio di Bonifica delle Marche) and Fabrizio Furlani (Province of Pesaro Urbino) for their involvement and support for the realization of next presentation of these data to farmers by local agricultural associations.

Conflicts of Interest: The authors declare no conflict of interest.

References

1. FAO. Global agriculture towards 2050. In *High Level Expert Forum—How to Feed the World in 2050*; FAO: Rome, Italy, 2009; Available online: http://www.fao.org/fileadmin/templates/wsfs/docs/Issues_papers/HLEF2050_Global_Agriculture.pdf (accessed on 20 November 2021).
2. European Commission. *The CAP towards 2020. Impact Assessment of Alternative Policy Options*; European Commission: Brussels, Belgium, 2011.
3. WEF. *The Global Risks Report 2020*, 15th ed.; Insight Report; WEF: Colony, Switzerland, 2020.
4. Tilman, D.; Cassman, K.G.; Matson, P.A.; Naylor, R.; Polasky, S. Agricultural sustainability and intensive production practices. *Nature* **2002**, *418*, 671–677. [CrossRef]
5. ISPRA. *Guidelines for the Hydrogeological Assessment and Its Mitigation through Measures and Interventions in Agriculture and Forestry*; Manuali e Linee Guida: Roma, Italy, 2013; p. 98. (In Italian)
6. Zhang, W.; Ricketts, T.H.; Kremen, C.; Carney, K.; Swinton, S.M. Ecosystem services and dis-services to agriculture. *Ecol. Econ.* **2007**, *64*, 253–260. [CrossRef]
7. Quinn, J.E.; Brandle, J.R.; Johnson, R.J. A farm-scale biodiversity and ecosystem services assessment tool: The healthy farm index. *Pap. Nat. Resour.* **2013**, *11*, 176–192. [CrossRef]
8. Zanten, B.T.; Verburg, P.H.; Espinosa, M.; Gomez-y-Paloma, S.; Galimberti, G.; Kantelhardt, J.; Kapfer, M.; Lefebvre, M.; Manrique, R.; Piorr, A.; et al. European agricultural landscapes, common agricultural policy and ecosystem services: A review. *Agron. Sustain. Dev.* **2012**, *34*, 309–325. [CrossRef]
9. Notte, A.L.; Marongiu, S.; Masiero, M.; Molfetta, P.; Molignoni, R.; Cesaro, L. Livestock and Ecosystem Services: An Exploratory Approach to Assess Agri-Environment-Climate Payments of RDP in Trentino. *Land* **2015**, *4*, 688–710. [CrossRef]
10. Rega, C.; Spaziante, A. Linking ecosystem services to agri-environmental schemes through SEA: A case study from Northern Italy. *Environ. Impact Assess. Rev.* **2013**, *40*, 47–53. [CrossRef]
11. Batie, S. Sustainable development: Challenges to the profession of agricultural economics. *Am. J. Agric. Econ.* **1989**, *71*, 1083–1101. [CrossRef]
12. Notte, A.L.; Scolozzi, R.; Molfetta, P.; Gubert, F.; Molignoni, R.; Franchi, R.; Pecile, A. Environment-climate payments for the rural development programs 2014–2020. The case of the autonomous province of Trento. *Ann. Di Bot.* **2014**, *4*, 91–96.
13. Page, G.; Bellotti, B. Farmers value on-farm ecosystem services as important, but what are the impediments to participation in PES schemes? *Sci. Total Environ.* **2015**, *515–516*, 12–19. [CrossRef]
14. De Clerck, F.J.; Jones, S.K.; Attwood, S.; Bossio, D.; Girvetz, E.; Chaplin-Kramer, B.; Enfors, E.; Fremier, A.K.; Gordon, L.J.; Kizito, F.; et al. Agricultural ecosystems and their services: The vanguard of sustainability? *Curr. Opin. Environ. Sustain.* **2016**, *23*, 92–99. [CrossRef]
15. Swinton, S.M.; Lupi, F.; Robertson, G.P.; Hamilton, S.K. Ecosystem services and agriculture: Cultivating agricultural ecosystems for diverse benefits. *Ecol. Econ.* **2007**, *64*, 245–252. [CrossRef]
16. Tixier, P.; Peyrard, N.; Aubertot, J.N.; Gaba, S.; Radoszycki, J.; Caron-Lormier, G.; Vinatier, F.; Mollot, G.; Sabbadin, R. Chapter Seven—Modelling Interaction Networks for Enhanced Ecosystem Services in Agroecosystems. *Adv. Ecol. Res.* **2013**, *49*, 437–480.
17. Garrod, G. Greening the CAP: How the improved design and implementation of agrienvironment schemes can enhance the delivery of environmental benefits. *J. Environ. Plan. Manag.* **2009**, *52*, 571–574. [CrossRef]
18. Poole, A.E.; DSalazar, R.B.; Macdonald, D.W. Optimizing agri-environment schemes to improve river health and conservation value. *Agric. Ecosyst. Environ.* **2013**, *181*, 157–168. [CrossRef]
19. Matthews, A. Greening agricultural payments in the EU's Common Agricultural Policy. *Bio-Based Appl. Econ.* **2013**, *2*, 1–27.
20. Maes, J.; Jacobs, S. Nature-Based Solutions for Europe's Sustainable Development. *Conserv. Lett.* **2017**, *10*, 121–124. [CrossRef]
21. Jack, B. Ecosystem Services: European Agricultural Law and Rural Development. In *Law and Agroecology—A Transdisciplinary Dialogue*; Monteduro, M., Buongiorno, P., Di Benedetto, S., Isoni, A., Eds.; Springer: Berlin/Heidelberg, Germany, 2014; pp. 127–150.
22. Whittingham, M.J. The future of agri-environment schemes: Biodiversity gains and ecosystem service delivery? *J. Appl. Ecol.* **2011**, *48*, 509–513. [CrossRef]

23. Macfadyen, S.; Cunningham, S.A.; Costamagna, A.C.; Schellhorn, N.A. Managing ecosystem services and biodiversity conservation in agricultural landscapes: Are the solutions the same? *J. Appl. Ecol.* **2012**, *49*, 690–694. [CrossRef]
24. Prager, K.G.; Reed, M.; Scott, A. Encouraging collaboration for the provision of ecosystem services at a landscape scale—Rethinking agri-environmental payments. *Land Use Policy* **2012**, *29*, 244–249. [CrossRef]
25. Dale, V.H.; Polasky, S. Measures of the effects of agricultural practices on ecosystem services. *Ecol. Econ.* **2007**, *64*, 286–296. [CrossRef]
26. De Groot, R.S.; Alkemade, R.; Braat, L.; Hein, L.; Willemen, L. Challenges in integrating the concept of ecosystem services and values in landscape planning, management and decision making. *Ecol. Complex.* **2010**, *7*, 260–272. [CrossRef]
27. TEEB. *The Economics of Ecosystems and Biodiversity Ecological and Economic Foundations*; Pushpam, K., Ed.; Earthscan: London, UK; Earthscan: Washington, DC, USA, 2010.
28. Power, A.G. Ecosystem services and agriculture: Tradeoffs and synergies. *Philos. Trans. R. Soc. B Biol. Sci.* **2010**, *365*, 2959–2971. [CrossRef]
29. Gaglio, M.; Aschonitis, V.; Gissi, E.; Castaldelli, G.; Fano, E.A. Non-Market Ecosystem Services of Agricultural Land and Priorities Towards a More Sustainable Agriculture in Italy. *J. Agric. Food Dev.* **2016**, *2*, 23–31. [CrossRef]
30. Morri, E.; Pruscini, F.; Scolozzi, R.; Santolini, R. A forest ecosystem services evaluation at the river basin scale: Supply and demand between coastal areas and upstream lands (Italy). *Ecol. Indic.* **2014**, *37*, 210–219. [CrossRef]
31. Santolini, R.; Morri, E.; Pasini, G.; Giovagnoli, G.; Morolli, C.; Salmoiraghi, G. Assessing the quality of riparian areas: The case of River Ecosystem Quality Index applied to the Marecchia river (Italy). *Int. J. River Basin Manag.* **2015**, *13*, 1–16. [CrossRef]
32. Reed, M.S.; Moxey, A.; Prager, K.; Hanley, N.; Skates, J.; Bonn, A.; Evans, C.; Glenk, K.; Thomson, K. Improving the link between payments and the provision of ecosystem services in agri-environment Schemes. *Ecosyst. Serv.* **2014**, *9*, 44–53. [CrossRef]
33. Sattler, C.; Nagel, U.J. Factors affecting farmers' acceptance of conservation measures. A case study from north-eastern Germany. *Land Use Policy* **2010**, *27*, 70–77. [CrossRef]
34. De Francesco, E.; Gatto, P.; Runge, F.; Trestini, S. Factors affecting farmers' participation in agri- environmental measures: A Northern Italian perspective. *J. Agric. Econ.* **2007**, *59*, 114–131.
35. Bertoni, D.; Cavicchioli, D.; Pretolani, R.; Olper, A. Determinants of agri-environmental measures adoption: Do institutional constraints matter? *Environ. Econ.* **2012**, *3*, 8–19.
36. Matzdorf, B.; Lorenz, J. How cost-effective are result-oriented agri-environmental measures? An empirical analysis in Germany. *Land Use Policy* **2010**, *27*, 535–544. [CrossRef]
37. Millennium Ecosystem Assessment. *Living Beyond Our Means: Natural Assets and Human Well-Being*; Island Press: Washington, DC, USA, 2005.
38. Gordon, L.J.; Finlayson, M.; Falkenmark, M. Managing water in agriculture for food production and other ecosystem services. *Agric. Water Manag.* **2010**, *97*, 512–519. [CrossRef]
39. Pretty, J.N.; Noble, A.D.; Bossio, D.; Dixon, J.; Hine, R.E.; de Vries, F.; Morison, J.I.L. Resource-conserving agriculture increases yields in developing countries. *Environ. Sci. Technol.* **2006**, *40*, 1114–1119. [CrossRef] [PubMed]
40. Baulcombe, D.; Crute, I.; Davies, B.; Dunwell, J.; Gale, M.; Jones JPretty, J.; Sutherland, W.; Toulmin, C. *Reaping the Benefits: Science and the Sustainable Intensification of Global Agriculture*; The Royal Society: London, UK, 2009.
41. Hoffmann, H.; Schomers, S.; Meyer, C.; Klas Sander, K.; Hickey, V.; Feuerbacher, A. Agriculture and Ecosystem Services. *Ref. Modul. Food Sci.* **2019**, *3*, 9–13. [CrossRef]
42. Brauman, K.A.; Daily, G.C.; Ka'eo Duarte, T.; Mooney, H.A. The Nature and Value of Ecosystem Services: An Overview Highlighting Hydrologic Services. 2Annu. *Rev. Environ. Resourc.* **2007**, *32*, 67–98. [CrossRef]
43. Santolini, R.; Morri, E.; D'Ambrogi, S. Connectivity and Ecosystem Services in the Alps. In *Alpine Nature 2030 Creating Ecological Connectivity for Generations to Come*; Federal Ministry for the Environment, Nature Conservation, Building and Nuclear Safety (BMUB) Public Relations Division: Berlin, Germany, 2016; p. 251. ISBN 978-3-00-053702-8.
44. Coates, J.B.; Tharme, R.; Herrero, M. Water-related Ecosystem Services and Food Security. In *Managing Water and Agroecosystems for Food Security*; Eline, B., Ed.; CAB International: Wallingford, UK, 2013; pp. 29–41.
45. Bartoli, M.; Racchetti, E.; Delconte, C.A.; Sacchi, E.; Soana, E.; Laini, A.; Longhi, D.; Viaroli, P. Nitrogen balance and fate in a heavily impacted watershed (Oglio River, Northern Italy): In quest of the missing sources and sinks. *Biogeosciences* **2012**, *9*, 361–373. [CrossRef]
46. Palomo, I.; Martín-López, B.; Potschin, M.; Haines-Young, R.; Montes, C. National Parks, buffer zones and surrounding lands: Mapping ecosystem service flows. *Ecosyst. Serv.* **2013**, *4*, 104–116. [CrossRef]
47. Sandhu, H.S.; Wratten, S.D.; Cullen, R.; Case, B. The future of farming: The value of ecosystem services in conventional and organic arable land. An experimental approach. *Ecol. Econ.* **2008**, *64*, 835–848. [CrossRef]
48. Romero, M.N.; Oteros-Rozas, E.; Gonzalez, J.A.; Martın-Lopez, B. Exploring the knowledge landscape of ecosystem services assessments in Mediterranean agroecosystems: Insights for future research. *Environ. Sci. Policy* **2014**, *37*, 121–133. [CrossRef]
49. Costanza, R.; D'arge, R.; De Groot, R.; Farber, S.; Grasso, M.; Hannon, B.; Limburg, K.; Naeem, S.; O'neill, R.V.; Paruelo, J.; et al. The value of the world's ecosystem services and natural capital. *Nature* **1997**, *387*, 253. [CrossRef]
50. Schirpke, U.; Marino, D.; Marucci, A.; Palmieri, M.; Scolozzi, R. Operationalising ecosystem services for effective management of protected areas: Experiences and challenges. *Ecosyst. Serv.* **2017**, *28*, 105–114. [CrossRef]

51. Scolozzi, R.; Morri, E.; Santolini, R. Delphi-based change assessment in ecosystem service values to support strategic spatial planning in Italian landscapes. *Ecol. Indic.* **2012**, *21*, 134–144. [CrossRef]
52. Scolozzi, R.; Schirpke, U.; Morri, E.; D'Amato, D.; Santolini, R. Ecosystem services-based SWOT analysis of protected areas for conservation strategies. *J. Environ. Manag.* **2014**, *146*, 543–551. [CrossRef] [PubMed]
53. Robertson, G.P.; Swinton, S.M. Reconciling agricultural productivity and environmental integrity: A grand challenge for agriculture. *Front. Ecol. Environ.* **2005**, *3*, 38–46. [CrossRef]
54. Rosa, S.D.L.; Privitera, R. Characterization of non-urbanized areas for land-use planning of agricultural and green infrastructure in urban contexts. *Landsc. Urban. Plan.* **2013**, *109*, 94–106. [CrossRef]
55. Rosa, S.D.L.; Barbarossa, L.; Privitera, R.; Martinico, F.; Greca, P.L. Agriculture and the City: A Method for Sustainable Planning of New Forms of Agriculture in Urban Contexts. *Land Use Policy* **2014**, *41*, 290–303. [CrossRef]
56. Smith, S.; Rowcroft, P.; Everard, M.; Couldrick, L.; Reed, M.; Rogers, H.; Quick, T.; Eves, C.; White, C. *Payments for Ecosystem Services: A Best Practice Guide*; Defra: London, UK, 2013.
57. Morri, E. Environmental State of Foglia River Basin: River Functionality and Ecosystem Services. Ph.D. Thesis, Urbino University, Marche Region, Italy, 2012.
58. Rusco, E.; Maréchal, B.; Tiberi, M.; Bernacconi, C.; Ciabocco, G.; Ricci, P.; Spurio, E. *Case Study—Italy Sustainable Agriculture and Soil Conservation (SoCoProject)*; JRC Technical Notes; EUR 24131 EN/9; Office for Official Publications of the European Communities: Luxembourg, 2009.
59. Bonan, G.B. Forests and climate change: Forcings, feedbacks, and the climatebenefits of forests. *Science* **2008**, *320*, 1444–1449. [CrossRef]
60. Hümann, M.; Schüler, G.; Müller, C.; Schneider, R.; Johst, M.; Caspari, T. Identification of runoff processes—The impact of different forest types and soil properties on runoff formation and floods. *J. Hydrol.* **2011**, *409*, 637–649. [CrossRef]
61. Nedkov, S.; Burkhard, B. Flood regulating ecosystem services—Mapping supply and demand in the Etropole Municipality. Bulgaria. *Ecol. Indic.* **2012**, *21*, 67–79. [CrossRef]
62. Amici, M.; Spina, R. *Average Annual and Seasonal Precipitation of Marche Region for the Period 1950–2000*; Centre for Ecology and Climate, Civil Protection Ancona, Marche Region: Ancona, Italy, 2002; p. 103. (In Italian)
63. Spina, R.; Stortini, S.; Fusari, R.; Scuterini, C.; Di Marino, M. *Climatological Characterization of the Marche Region: Average Temperature Range for the 1950–2000 Period*; Centre for Ecology and Climate, Civil Protection Marche Region: Ancona, Italy, 2002; p. 56.
64. IPPC Intergovernmental Panel on Climate Change. *WG1 Report Climate Change: The Physical Science Basis*; IPPC: Geneva, Switzerland, 2007.
65. Zhang, L.; Dawes, W.R.; Walker, G.R. The response of mean annual evapotranspiration to vegetation changes at catchment scale. *Water Resour. Res.* **2001**, *37*, 701–708. [CrossRef]
66. Allen, R.G.; Pereira, L.S.; Raes, D.; Smith, M. *Crop Evapotranspiration*; FAO Irrigation and Drainage Paper; FAO: Rome, Italy, 1998; p. 56.
67. Marche Region 2010. General Price List. Available online: http://contrattipubblici.regione.marche.it/Servizi-On-line/Servizi-on-line-di-libera-consultazione/Prezzario (accessed on 3 September 2014). (In Italian)
68. Costantini, E.A.C.; Dazzi, C. *The Soils of Italy*; Springer Science & Business Media: Berlin/Heidelberg, Germany, 2013; p. 354.
69. Renard, K.G.; Foster, G.R.; Weesies, G.A. *Predicting Soil Erosion by Water: A Guide to Conservation Planning with the Revised Universal Soil Loss Equation (RUSLE)*; USDA Agricultural Handbook: Washington, DC, USA, 1997; p. 404. Available online: http://www.ars.usda.gov/SP2UserFiles/Place/64080530/RUSLE/AH_703.pdf (accessed on 3 September 2015).
70. Diodato, N.; Ceccarelli, M.; Bellocchi, G. GIS-aided evaluation of evapotranspiration at multiple spatial and temporal climate patterns using geoindicators. *Ecol. Indic.* **2010**, *10*, 1009–1016. [CrossRef]
71. Pelton, J.; Frazier, E.; Pickilingis, E. Calculating Slope Length Factor (LS) in the Revised Universal Soil Loss Equation (RUSLE); 2014. Available online: http://gis4geomorphology.com/ (accessed on 2 January 2015).
72. Kirkby, M.J.; Morgan, R.P.C. *Soil Erosion*; Wiley: New York, NY, USA, 1980; p. 312.
73. Oenema, O.; Kros, H.; de Vries, W. Approaches and uncertainties in nutrient budgets: Implications for nutrient management and environmental policies. *Europ. J. Agron.* **2003**, *20*, 3–16. [CrossRef]
74. Bartoli, M.; Soana, E.; Laini, A.; Nizzoli, D.; Pinardi, M.; Racchetti, E.; Gardi, C.; Viaroli, P.; Acutis, M.; Salmaso, F.; et al. Cross Comparison of Nitrogen Sources, Sinks and Transport within River Basins: The Italian Nitrogen Network initiative (INN). In Proceedings of the XXIV Congresso Società Italiana di Ecologia: L'Ecologia Oggi: Responsabilità E Governance, Ferrara, Italy, 15–17 September 2014; p. 135.
75. Soana, E.; Racchetti, E.; Laini, A.; Bartoli, M.; Viaroli, P. Soil Budget, Net Export, and Potential Sinks of Nitrogen in the Lower Oglio River Watershed (Northern Italy). *Clean Soil Air Water* **2011**, *39*, 956–965. [CrossRef]
76. Soana, E.; Longhi, D.; Racchetti, E.; Pinardi, M.; Laini, A.; Bolpagni, R.; Castaldelli, G.; Bartoli, M.; Fano, E.A.; Viaroli, P. Small Services over Large Areas: The Need for New Paradigms for the Secondary Hydrographic Network. In Proceedings of the XXIII SitE Conference, Ancona, Italy, 16–18 September 2013. (In Italian)
77. Castaldelli, G.; Soana, E.; Racchetti, E.; Vincenzi, F.; Fano, A.E.; Bartoli, M. Vegetated canals mitigate nitrogen surplus in agricultural watersheds. *Agric. Ecosyst. Environ.* **2014**, *212*, 253–262. [CrossRef]
78. Cole, L.J.; Stockan, J.; Helliwell, R. Managing riparian buffer strips to optimize ecosystem services: A review. *Agric. Ecosyst. Environ.* **2020**, *296*, 106891. [CrossRef]

79. García-Díaz, A.; Bienes, R.; Sastre, B.; Novara, A.; Gristina, L.; Cerdà, A. Nitrogen losses in vineyards under different types of soil groundcover. A field runoff simulator approach in Central Spain. *Agric. Ecosyst. Environ.* **2017**, *236*, 256–267. [CrossRef]
80. European Parliament and Council. *Regulation (UE) N. 1305/2013 on Support for Rural Development by the European Agricultural Fund for Rural Development (EAFRD) and Repealing Council Regulation (EC)*; No. 1698/2005; European Parliament and Council: Strasbourg, France, 2013.
81. Tvinnereim, E.; Zelljadt, E.; Yakymenko, N.; Mazzacurati, E. *Carbon*; Point Carbon: Amsterdam, The Netherland, 2011; Available online: www.pointcarbon.com (accessed on 3 September 2015).
82. Birkás, M.; Jolánkai, M.; Gyuricza, C.; Percze, A. Tillage effects on compaction, earthworms and other soil quality indicators in Hungary. *Soil Tillage Res.* **2004**, *78*, 185–196. [CrossRef]
83. Bengtsson, J.; Bullock, J.M.; Egoh, B.; Everson, C.; Everson, T.; O'Connor, T.; O'Farrell, P.J.; Smith, H.G.; Lindborg, R. Grasslands—more important for ecosystem services than you might think. *Ecosphere* **2019**, *2*, e02582. [CrossRef]
84. Li, J.; Ren, Z.; Zhou, Z. Ecosystem services and their values: A case study in the Qinba mountains of China. *Ecol. Res.* **2006**, *21*, 597–604. [CrossRef]
85. BenDor, T.K.; Livengood, A.; Lester, W.T.; Davis, A.; Yonavjak, L. Defining and evaluating the ecological restoration economy. *Restor. Ecol.* **2015**, *23*, 209–219. [CrossRef]
86. Santolini, R.; Morri, E. *Ecological Criteria for the Introduction of Evaluation and Remuneration of Ecosystem Services (SE) in Design and Planning (Italian) in La Dimensione Europea Del Consumo Di Suolo E Le Politiche Nazionali. CRCS Rapporto*; INU Istituto Nazionale di urbanistica: Roma, Italy, 2017; pp. 149–154.
87. Pagiola, S.; Platais, G. Introduction to Payments for Environmental Services, PowerPoint presentation, ESSD Week 2005—Learning days. Available online: www.unescap.org.com (accessed on 10 September 2015).
88. Wunder, S.; Engel, S.; Pagiola, S. Taking Stock: A Comparative Analysis of Payments for Environmental Services Programs in Developed and Developing Countries. *Ecol. Econ.* **2008**, *65*, 834–852. [CrossRef]
89. Pellegrino, D.; Schirpke, U.; Marino, D. How to support the effective management of Natura 2000 sites? *J. Environ. Plan. Manag.* **2016**, *60*, 383–398. [CrossRef]
90. European Commission. *An Action Plan for Nature, Citizens and Economy*; Final Report, COM (2017) 198 final; European Commission: Brussels, Belgium, 2015.
91. European Commission. Towards an EU research and innovation policy agenda for nature-based solutions and re-naturing cities. In *Final Report of the Horizon 2020 Expert Group on Nature-Based Solutions and Re-Naturing Cities (Full Version European Commission)*; European Commission: Brussels, Belgium, 2015.
92. Seddon, N.; Chausson, A.; Berry, P.; Girardin, C.A.; Smith, A.; Turner, B. Understanding the value and limits of nature-based solutions to climate change and other global challenges. *Philos. Trans. R. Soc. B* **2020**, *375*, 20190120. [CrossRef]
93. UNESCO. Nature-based solutions for water. In *The United Nations World Water Development Report*; United Nations Educational, Scientific and Cultural Organization (UNESCO): Paris, France, 2018.
94. Cohen-Shacham, E.; Walters, G.; Janzen, C.; Maginnis, S. *Nature-Based Solutions to Address Global Societal Challenges*; IUCN: Gland, Switzerland, 2016; p. 97. Available online: https://doi.org/10.2305/IUCN.CH.2016.13.en (accessed on 20 November 2021).
95. Heuvel, J.V.; Centinari, M. Under-Vine Vegetation Mitigates the Impacts of Excessive Precipitation in Vineyards. *Front. Plant Sci.* **2021**, *12*, 713135. [CrossRef]

land

MDPI

Article

The Impact of Urbanization on Land: A Biophysical-Based Assessment of Ecosystem Services Loss Supported by Remote Sensed Indicators

Francesca Assennato [1], Daniela Smiraglia [1,*], Alice Cavalli [2], Luca Congedo [1], Chiara Giuliani [1], Nicola Riitano [1], Andrea Strollo [1] and Michele Munafò [1]

1 Italian Institute for Environmental Protection and Research (ISPRA), Via Vitaliano Brancati 48, I-00144 Rome, Italy; francesca.assennato@isprambiente.it (F.A.); luca.congedo@isprambiente.it (L.C.); chiara.giuliani@isprambiente.it (C.G.); nicola.riitano@isprambiente.it (N.R.); andrea.strollo@isprambiente.it (A.S.); michele.munafo@isprambiente.it (M.M.)

2 Department of Innovation in Biology, Agri-Food and Forestry Systems (DIBAF), University of Tuscia, Via San Camillo del Lellis SNC, I-01100 Viterbo, Italy; alice.cavalli@studenti.unitus.it

* Correspondence: daniela.smiraglia@isprambiente.it

Abstract: Urbanization and related land consumption are one of the main causes of ecosystem services loss. This is especially the case for soil-related services affecting ecosystem functions and limiting accessibility to natural resources. Satellite remote sensing and environmental databases enable in-depth analysis of urban expansion and land changes, which can be used to monitor trends in the provision of ecosystem services. This work aims to describe a multilayered approach to the assessment of biophysical loss of ecosystem services flows in Italy caused by an increase in land consumption in the period 2012–2020. The results show higher losses in wood production, carbon storage, hydrological regime regulation, and pollination in the northern regions of Italy, as well as in some southern regions, such as Campania and Apulia. Habitat quality loss is widespread throughout Italy, whereas crop production loss varies on the basis of the locations in which it occurs and the crop types involved. Loss of arable land and fodder production mainly occurs in northern regions, whereas southern regions have experienced a drop in permanent crop production. This study highlights the importance of using integrated data and methodologies for well-founded approaches, with a view to gaining a thorough understanding of ecosystem services-related processes and the changes connected therewith.

Keywords: ecosystem services flows; land monitoring; land cover; land consumption; Italy

Citation: Assennato, F.; Smiraglia, D.; Cavalli, A.; Congedo, L.; Giuliani, C.; Riitano, N.; Strollo, A.; Munafò, M. The Impact of Urbanization on Land: A Biophysical-Based Assessment of Ecosystem Services Loss Supported by Remote Sensed Indicators. *Land* **2022**, *11*, 236. https://doi.org/10.3390/land11020236

Academic Editor: Monika Kopecká

Received: 3 December 2021
Accepted: 2 February 2022
Published: 5 February 2022

1. Introduction

Compact and dispersed patterns of urban expansion [1,2] are a global phenomenon and are one of the most important factors in landscape change [3,4]. Starting from the 20th century, the increasing demand for land to be used for buildings and infrastructures led to greater urbanization of previously natural and agricultural areas [5], making urbanization a major cause of land consumption. More generally, land consumption can be considered to be the change from non-artificial land cover to artificial land cover, with a distinction having to be made between permanent consumption (due to permanent artificial land cover, such as concrete or asphalt) and non-permanent consumption (due to reversible artificial land cover or soil alteration processes, such as, soil compaction in construction sites or excavations in quarries) [6,7]. The process of urbanization, and the related increase in impervious surfaces, affects other types of land use and land cover [8], triggering a series of effects on the environment, which results in it being the main cause of land degradation [9]. The main effects consist of the loss of fertile soils [10], the adverse impact on water balance [11], the increase in surface water runoff and flood risk [12], the negative influence on local microclimates due to urban heat islands [13,14], landscape fragmentation, and the loss of biodiversity [15,16]. Such effects limit the accessibility to natural resources and the provision of ecosystem services (ES), defined as the goods and benefits that people derive from ecosystems [17,18]. ES provision is a dynamic process and can be measured, in terms of flow, as the annual potential number of services provided by ecosystems [19,20].

Recent studies have focused mainly on drivers of change and the impact thereof on ES as a relevant issue to be understood, as well as the related implications on ecosystems functions and conservation [21], also including soil functions. Soils play a crucial role in providing several so-called soil ES [22,23], in particular provisioning and regulation services [24,25], contributing for example to food and wood production, hydrological flow regulation, carbon sequestration, pollination, habitat function, and biodiversity conservation [26]. Indeed, there is, through soil functions, a direct link of soil properties to specific ES, as indicated in Adhicari et al. [27]. In recent decades, policies and strategies have highlighted the importance of ES in maintaining livelihoods, and have stressed the fact that, for the purpose of providing ES, the natural land and agricultural system depend on soil. This is also confirmed by the recent EU Soil Strategy for 2030 [28], the EU Biodiversity Strategy for 2030 [29], and the Common Agricultural Policy [30]. The increasing amount of attention that is, with a view to fostering sustainable land development, paid at the international level to ecosystems, and the need to consider how changes in ecosystems affect human well-being [17,31,32], have given rise to an assessment being made both at a global level—based on the Aichi Biodiversity Targets [33] and the 2030 Agenda [34]—and at a European level [35]. The concept of ES, and the related frameworks that have been developed in connection therewith, have been recognized as being vital for supporting territorial management policies [17,22,36]. The various initiatives aim to improve the mapping of ecosystems, become apprised of the functional interactions and the pressures to which they are subjected, as well as establish indicators that assess changes in the provision of ES, in terms of the biophysical and economic aspects thereof [37].

ES assessment is a complex process, which involves integrating different scientific fields and several databases that vary in accuracy, scale, updating, and availability. Therefore, an approach based on various thematic layers is required, so as to have an overview of ES loss that supports the resource management decision-making process. Furthermore, monitoring land use and land cover spatial distribution, especially in urban areas, is essential for providing accurate and timely information to be used as input data that allows soil-based ES provision changes to be assessed. Within this context, major advances in satellite remote sensing have improved data collection and analysis methodologies used for the purpose of detecting artificial land covers [38–41]. These include the Sentinel satellite constellations of the EU Copernicus Programme [42] which provide high-resolution (spatial, spectral, and temporal) multispectral and SAR images suitable for measuring land consumption in a detailed manner [6].

In Italy, where land degradation processes due to land consumption are increasing, especially in agricultural and peri-urban areas [43], the Italian Institute for Environmental Protection and Research (ISPRA) is in charge of a monitoring program at the national level regarding: (i) land consumption mapping, based on a Sentinel satellite images analysis that is integrated with the photointerpretation of national orthophotos; (ii) land cover mapping, based on the integration of the Copernicus Land Monitoring Services (CLMS) datasets (Corine Land Cover, High Resolution Layers, Urban Atlas, Riparian Zones, and Natura 2000) and the national land consumption map; (iii) estimate of land degradation and ES changes (biophysical and economic) caused by annual land consumption. This study is the first work that is focused on Italy and that, in trying to collect and harmonize existing data, analyzes nationwide soil-based ES loss with a detailed spatial resolution. The paper aims to describe the multilayered approach that has been adopted in assessing biophysical loss of ES flows caused by land consumption increase in the period 2012–2020, as recorded by the aforementioned national monitoring program.

After having introduced the topic of ES and ES losses due to urban expansion, we describe the procedure adopted for the purpose of assessing changes in six ES in the period under examination. We describe the input data used and the methodologies applied for the purpose of analyzing the reference data and the additional thematic layers. The section in which the results are set out shows what losses were recorded for each ES. We then discuss the results of the analysis and considering the procedure's achievements and shortcomings. Finally, we summarize the conclusions derived therefrom.

2. Materials and Methods

2.1. Study Area

Italy is located in the Mediterranean basin (southern Europe), it covers around 300,000 km^2 and is divided into 20 administrative regions. It is a peninsula characterized by two mountain ranges (the Alps and the Apennines), wide river valleys, two major islands (Sardinia and Sicily), and various other small islands. Italy's physiography and geographical position have led to it having heterogeneous features in terms of climate (ranging from a Mediterranean climate along the coastline to an alpine climate at the higher elevations) and in terms of land use/land cover, with urban areas concentrated in the plains and along the coasts, which cover 7.1% of the national territory (year 2020) (Figures 1 and 2) [44].

2.2. Reference Data

The basic cartographic maps used to detect changes in the provision of ES in Italy are the national land cover map and the national land consumption map. These maps are the common data source for all the ES analyzed in this study.

The national land cover map is based on the integration of the Copernicus Land Monitoring Services (CLMS) datasets (namely the Corine Land Cover, High Resolution Layers, Urban Atlas, Riparian Zones, and Natura 2000 datasets). The datasets show differences in the classification systems, resolutions, and updating frequencies adopted. Therefore, a harmonization procedure had to be adopted in order to make the datasets comparable. De Fioravante et al. [45] followed the EIONET EAGLE (Action Group on Land monitoring in Europe) group framework, which provides a methodology for describing land cover and land use data in a consistent manner. The CLMS data were integrated in an EAGLE compliant land cover map for the year 2012 (10 m resolution), which includes 16 classes of land cover (Figure 1).

The national land consumption map is, on the other hand, a product generated by the processing of Sentinel-1 (based on the SAR backscatter of VH polarization threshold) and Sentinel-2 (based on Normalized Difference Vegetation Index extraction and threshold value) satellite images, and photointerpretation of national orthophotos. The land consumption map is updated every year with a spatial resolution of 10 m [6,44]. National land consumption maps for the years 2012 and 2020 (Figure 2) were used for the purpose of assessing ES changes.

Land consumption is mapped both in its permanent and temporary forms following the definition given by [6,7]. By permanent consumed land, artificial surfaces composed of impervious material (e.g., concrete, asphalt or tarmacadam), both 3D (buildings) and 2D (roads), were taken into consideration (6.6% of the Italian national territory). This component is assumed to be irreversible, since extremely long amounts of time are required for restoring soil to its the natural state. Reversible consumed land is the result of compaction in construction sites, excavations in quarries, temporary coverage with artificial surfaces and other forms of soil alteration that do not imply permanent consumption, but that adversely affect the soil's natural state. However, this component is a minor (albeit dynamic) part of the total consumed land (0.5% of the Italian national territory). In light of the limited extent thereof and taking into account that for some portions of Italy's territory the distinction between permanent and temporary forms of land consumption is lacking, both components are considered as consumed land, without any exception whatsoever

(Figure 2). Figure 3 shows some examples of land consumption in urban, agricultural, and natural environments in the period under investigation.

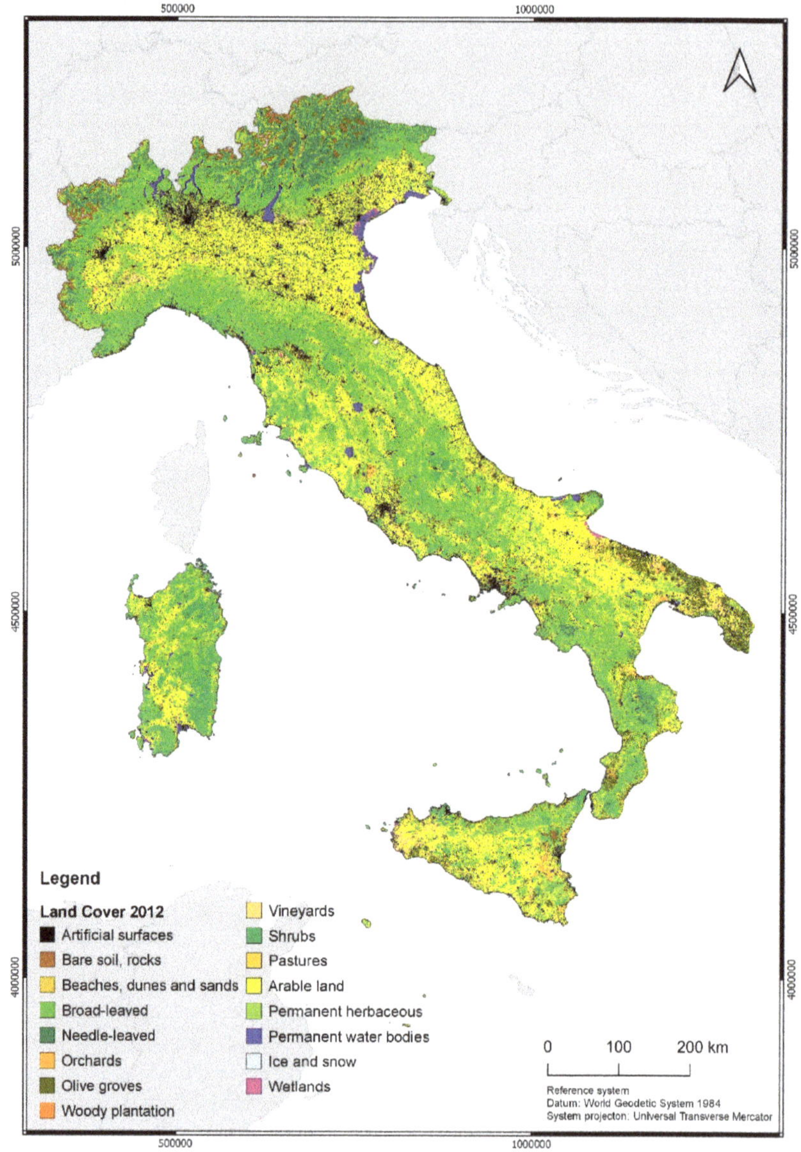

Figure 1. Land cover map for Italy (year 2012).

2.3. *Biophysical Assessment of Loss of Ecosystem Services Flows*

The biophysical evaluation of Italy's territory concerned six ES: crop production, carbon storage, wood production, habitat quality, hydrological regime regulation, and pollination. The land cover and land consumption maps were used as data input for all six ES, and other appropriate data inputs were added for each ES. Specific methodologies or already existing analysis models were adopted for each ES, taking into consideration the loss of ES flow in light of the reduction in the current yearly potential provision thereof, due to the increase in land consumption. Consumed land leads to soil-based ES being reduced to zero, since the soil no longer performs its functions. The analyses were conducted at the national and regional level and were performed using QGIS 3.20 [46] and InVEST 3.9.1 (Integrated Valuation of Ecosystem Services and Tradeoffs) [47] software.

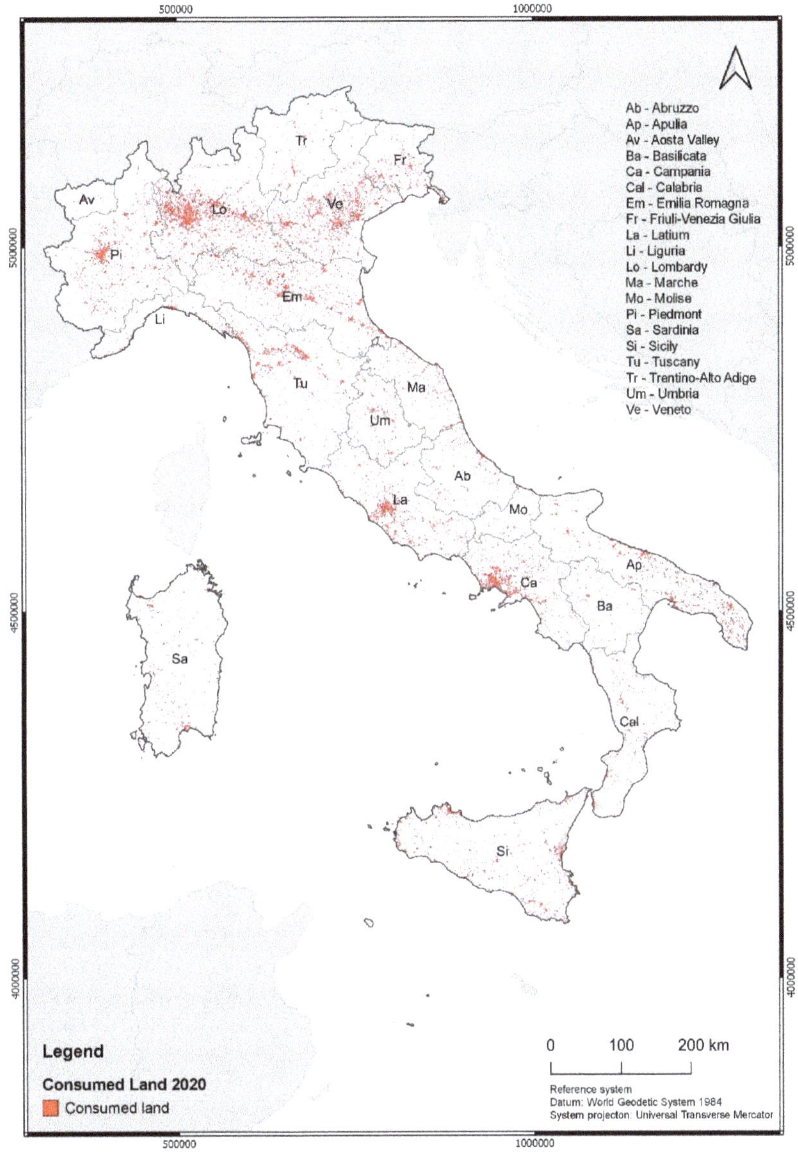

Figure 2. Land consumption map for Italy (year 2020).

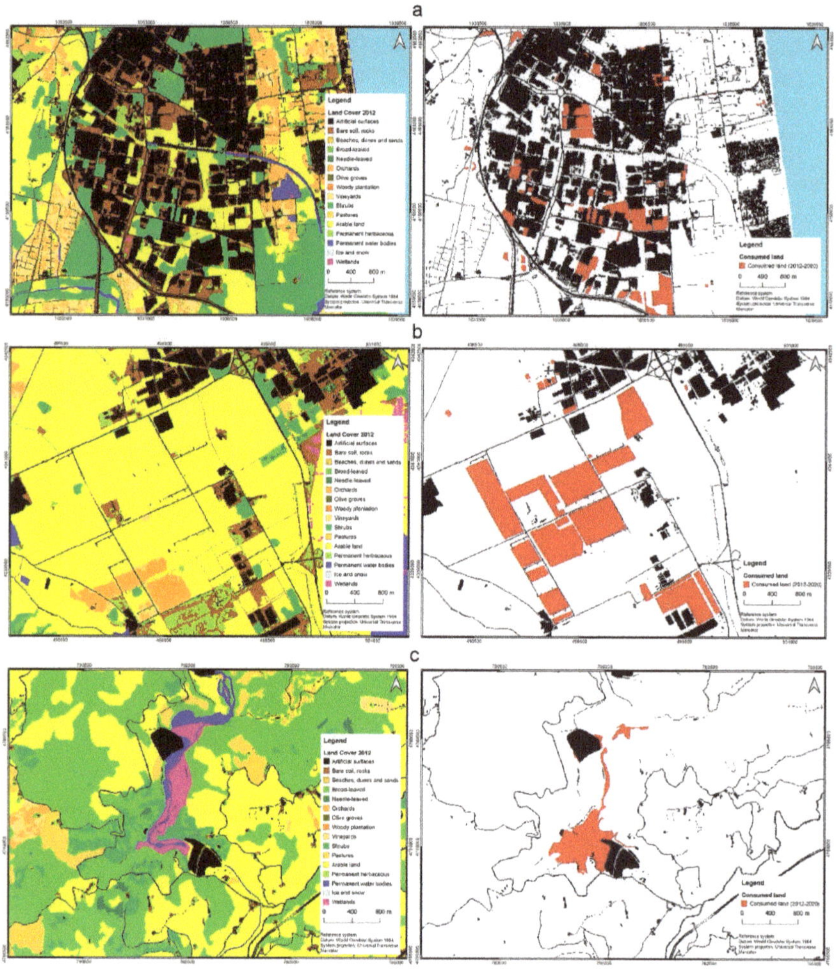

Figure 3. Examples of land consumption in urban (**a**), agricultural (**b**), and natural (**c**) environments.

2.3.1. Crop Production

Crop production is essential for food supply. Land take due to urbanization nullifies the land's capacity to cultivate and produce the resulting goods and benefits. With a view to estimating the provisioning of crop production, and the loss of flow in the period under examination due to land consumption increase, data on surface cover (hectares) and production (quintals) for each crop types were, in respect of each Italian administrative province, taken from the national census of agriculture carried out by the Italian National Institute of Statistics (ISTAT) (year 2013) [48]. The crop types were grouped into five classes (fruit trees, olive groves, vineyards, fodder, and arable land). Data on crop production were then spatialized for the purpose of obtaining the values in q/ha, and then in q/pixel compliant with the land cover and land consumption maps' spatial resolution. The amount of crop production in 2012 and the potential loss in the period from 2012 to 2020 on account of land consumption were calculated. Even though the quintal is not considered in the International System of Units, it is widely used, especially in Europe, and it is the unit of measurement with which data are officially released by ISTAT. For these reasons, the results of changes in agricultural production will be presented in quintals (1 q = 100 kg).

2.3.2. Wood Production

The production of wood raw materials is a provisioning service supplied to a large extent by natural forest and tree plantations. As is the case with crop production, the urbanization of areas covered by forests leads to wood production ES flow being completely lost. The method adopted follows the For-est model [49,50] concerning the above ground biomass, which uses the National Inventory of Forests and Forest Carbon Tanks (INFC) data (classified by administrative region and forest typology) [51] as input data. The model takes into consideration four main forest management categories: stands, coppices, plantations, and protective forests. It uses, through the Richards function [49], growing stock as the only driver for the purpose of estimating the evolution of forest carbon pools (above ground and below ground biomass, dead wood and litter, and soil organic matter) over time. The lack of detailed, homogenous and georeferenced data for all of the arboricultural systems in Italy has led to the simplification of the national methodology. The average annual index of the above ground biomass volume increase, which is calculated by the For-est model and utilizes the historical series of INFC data, is used to assess the biophysical value of the ES flow. All of the land cover map's forest classes were, therefore, considered in the same way and the same average annual index was applied to them. The change from forest area to consumed land resulted in this index being reduced to zero.

2.3.3. Carbon Storage

Carbon storage is a regulation service provided by terrestrial and marine ecosystems thanks to their ability to fix greenhouse gases [52]. This service contributes to the regulation of the climate at a global level and plays a fundamental role in the context of climate change mitigation and adaptation strategies. The model Carbon Storage and Sequestration of InVEST software was used to assess the carbon storage loss using the INFC and the National Soil Organic Carbon Map [53] as a data source. The land cover classes were associated with the INFC forest typologies [54]. Four different carbon pools, according to the Intergovernmental Panel on Climate Change classification [55], were taken into consideration for each land cover type: above ground biomass, below ground biomass, dead organic matter, and soil organic matter. Specific coefficients were used for the purpose of identifying different pools' contribution thereto [56–58].

The flow of carbon content regarding above ground and below ground biomass volume for forest types and permanent crops is expressed as a function of the growth rate taking as a reference point the growing stock data provided by the INFC for each forest type, and Caneveira et al. [59] for permanent crops. In [59], the authors proposed default carbon stock values for olive trees, vineyards, and fruit trees in which the age of plantations is known (1, 5, 10, 15, and 20 years). The average increase for these crop types is given by the difference between the value of 20 years and that of 5 years divided by the 15 years that elapse. Literature values were used for other natural and agricultural areas [60], and artificial areas are considered as zero. With regard to the carbon contained in the dead organic substance, the necromass values are deduced from those of the above ground biomass through coefficients (0.20 for evergreen plants and 0.14 for deciduous trees, source [55]), whereas those in the litter specific formulas for each species reported in the bibliography were used [56,61]. As far as the soil organic pool is concerned, the values reported in the National Soil Organic Carbon Map were used. The map contains the values of the carbon contained in the soil up to a depth of 30 cm, in raster format and with a resolution of 1 km. See [45] for a detailed description of carbon storage ES assessment.

2.3.4. Habitat Quality

This is a regulation service which, in supplying different types of habitats, contributes to human well-being by guaranteeing ecosystem functionality, biodiversity maintenance, and environmental resilience. Various anthropogenic factors have an impact on habitat quality, causing degradation and alteration of ecological processes [62]. With a view to assessing the loss of ES flow due to land consumption in the period under investigation, the Habitat Quality model of the InVEST software has been used. In this model, habitat quality is assessed in relation to different land use and land cover classes [47], on the basis of the premise that areas with higher habitat quality host a greater number of species and that habitat size and quality decrease lead to a decline in the persistence of species [63]. After having associated the land cover classes with habitat types (according to the EUNIS classification [64]), a habitat suitability value for each land cover class, for the purpose of assessing its ability to provide a habitat for biodiversity, has been assessed, ranging from 0 to 1, where 1 indicates habitat with the highest suitability for species biodiversity [65]. Three types of threats affecting habitat suitability have been taken into account: the habitat's sensitivity to being influenced by different types of threats, the impact (weight) of different threats on each habitat type, and the maximum distance at which the threats can affect the habitat. A threats impact value has been assessed, ranging from 0 (absence of threat) to 1 (presence of threat). The habitat suitability values and the threats parameters, which have been used as input data in the InVEST model, have been set through an expert-based approach [66] (Tables 1 and 2), using a questionnaire put to over 41 national experts, whose results are described in [67]. The result is a map of habitat quality in which the cell values range from 0 to 1, where 1 indicates high habitat quality. The ES was estimated ranking the range in five classes (class 1 poor habitat quality, class 5 high habitat quality) and assessing the surface variation in the period 2012–2020 for each class.

Table 1. Habitat suitability values generated by the survey (adapted from [67]).

Habitat Type	Suitability
Beaches, dune and, sands	0.74
Water bodies	0.83
Wetlands	0.96
Grasslands	0.86
Shrublands	0.81
Broadleaves forests	0.93
Conifer forests	0.82
Inland unvegetated or sparsely vegetated areas	0.55
Intensive agricultural lands	0.26
Extensive agricultural lands	0.52
Buildings and other artificial areas or impervious soils	0.09
Open urban areas	0.27

Table 2. Input values regarding the threats parameters generated by the survey: habitat sensitivity to the types of threats, impact (weight) of different threats on each habitat type, and maximum distance at which the threats can affect the habitat (adapted from [67]).

Threats / Habitat Type	Motorways; Trunks; Primary Roads	Secondary and Tertiary Roads	Residential and Service Roads	Tracks and Bridleways	Railways	Intensive Agricultural Lands	Extensive Agricultural Lands	Buildings and Other Artificial Areas or Impervious Soils
Beaches, dunes and sands	0.81	0.46	0.69	0.50	0.67	0.68	0.51	0.86
Water bodies	0.72	0.64	0.60	0.36	0.51	0.76	0.53	0.72
Wetlands	0.84	0.74	0.69	0.44	0.64	0.80	0.59	0.79
Grassland	0.80	0.71	0.63	0.42	0.60	0.75	0.52	0.72
Shrublands	0.78	0.71	0.63	0.39	0.60	0.72	0.51	0.69
Broadleaves forests	0.85	0.77	0.66	0.40	0.65	0.67	0.47	0.77
Conifers forests	0.84	0.76	0.68	0.39	0.61	0.63	0.44	0.76
Inland unvegetated or sparsely vegetated areas	0.61	0.57	0.52	0.30	0.46	0.51	0.35	0.61
Intensive agricultural lands	0.61	0.54	0.47	0.24	0.44	\	0.12	0.51
Extensive agricultural lands	0.71	0.61	0.55	0.26	0.51	0.54	\	0.62
Buildings and other artificial areas or impervious soils	\	\	\	\	\	\	\	\
Open urban areas	0.56	0.52	0.46	0.19	0.46	0.31	0.21	0.56
Weight	0.86	0.69	0.61	0.28	0.62	0.69	0.42	0.79
Distance [km]	1.5	1.0	0.9	0.3	1.6	1.6	0.6	1.7

2.3.5. Hydrological Regime Regulation

With regard to regulation services, hydrological regime regulation is an important water cycle-related ES. The evaluation thereof is based on the BigBang 1.0 (Nationwide GIS-based hydrological water budget on a regular grid) hydrological model developed by ISPRA for the purpose of evaluating the factors of hydrological balance at national level, with a GIS-based spatially distributed procedure being used that assesses the water budget components at different spatial and temporal scales [68]. Total precipitation, actual evapotranspiration, surface runoff, and groundwater recharge are evaluated on a monthly basis directly on the grid with a 1 km resolution. The procedure allows the effects produced by soil consumption increase in the period 2012–2020, in terms of increase in surface runoff, to be verified. This is done by applying the land cover and the land consumption maps, while the climatic conditions are considered as values mediated over the period in question. In this paper, the BigBang 1.0 version has been used. A BigBang 4.0 version has, however, recently been released for national accounting purposes [69] and will be tested for future improvements in ES assessment.

2.3.6. Pollination

Pollination is an important regulation service for agriculture production and ecosystems well-being [70,71] which is threatened by urbanization, agriculture intensification, and the use of pesticide and fertilizers [72,73]. The provision of this ES depends on the availability of nesting habitats and floral resources, and on the foraging distance of pollinators [74]. The InVEST Crop Pollination model was used to evaluate this ES. 50 species of pollinators (bees and bumblebees) were selected for the Italian national territory [75] (Table 3), and their characteristics were associated with the land cover classes and then analyzed, in terms of the availability of hosting pollinating species according to the types of nesting (reeds, rocks, cliffs, walls, plant stems, soil, and dead wood). The Italian national territory was divided on the basis of the altitude range (less than 800 m, from 800 to 1600 m, from 1600 to 2100 m, and over 2100 m) and the three main biogeographical regions (Mediterranean, Continental and Alpine), taking into account each of the resulting areas' different vegetation and different pollen production periods. In particular, the information used were: the type of nesting, period of activity, maximum flight distance, preferences among the varieties of nests, presence of flowers, and pollen period for the most widespread plants. The abundance of pollinators in each area were, therefore, considered as a "source" and subsequently the distribution on the agricultural surfaces to be pollinated were estimated (i.e., the potential index of abundance of pollinators that can reach an agricultural area). The result is a map in which the availability of pollinators for a potential agricultural area to be pollinated is estimated and the cell values range from 0 (low availability) to 1 (high availability). The ES flow was estimated, ranking the range in four classes (class 1 poor pollination, class 4 high pollination) and assessing the surface variation in each class in the period under investigation.

3. Results

3.1. Crop Production Loss

In 2012, agricultural production in Italy amounted to 915,705,580 quintals (q), with an average production of 45,785,279 q. Looking in detail at regional production (Figure 4), fruit trees had their highest production in Sicily and Calabria (12,861,590 q and 9,909,800 q, respectively), whereas production was below 7 q in the other regions. The region with the highest production levels for olive groves was Apulia, with 8,213,754 q, whereas the highest production level for vineyards was recorded in Apulia and Sicily (around 15,000,000 q) and in Veneto with 8,099,192 q. As far as fodder is concerned, the highest production level was recorded in Lombardy (32,761,977 q) and Emilia-Romagna (26,786,595 q), whereas it remained below 15,000,000 q in the other regions. Emilia-Romagna also recorded the highest production level for arable crop production (107,583,166), followed by Lombardy, Veneto, Piedmont, and Apulia, with around 50,000,000 q.

Table 3. Main species of selected pollinators.

Pollinators		
Andrena agilissima	*Andrena bicolor*	*Andrena carbonaria*
Andrena dorsata	*Andrena flavipes*	*Andrena morio*
Andrena minutuloides	*Andrena nigroaenea*	*Andrena nitidiuscula*
Andrena taraxaci	*Anthidium manicatum*	*Anthophora dispar*
Anthophora plumipes	*Bombus hortorum*	*Bombus humilis*
Bombus lapidaries	*Bombus lucorum*	*Bombus pascuorum*
Bombus pratorum	*Bombus ruderatus*	*Bombus terrestris*
Ceratina cucurbitina	*Colletes succinctus*	*Dasypoda altercator*
Eucera longicornis	*Eucera nigrescens*	*Halictus scabiosae*
Halictus sexcinctus	*Halictus maculatus*	*Heriades truncorum*
Hoplitis adunca	*Hoplitis anthocopoides*	*Hylaeus angustatus*
Hylaeus communis	*Hylaeus clypearis*	*Lasioglossum calceatum*
Lasioglossum leucozonium	*Lasioglossum nitidulum*	*Lasioglossum pauxillum*
Lasioglossum villosulum	*Megachile parietina*	*Megachile rotundata*
Osmia bicornis	*Osmia caerulescens*	*Osmia cornuta*
Osmia leaiana	*Panurgus calcaratus*	*Stelis nasuta*
Tetraloniella salicariae	*Xylocopa violacea*	

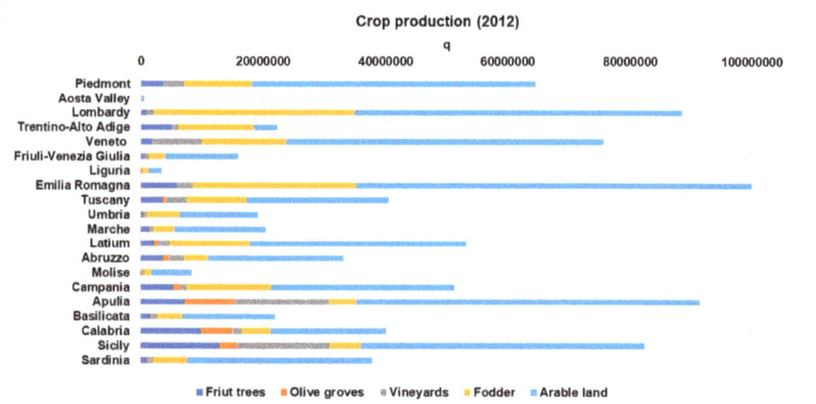

Figure 4. Estimation of crop production (quintals) for five agricultural classes in 2012 at the regional level.

In Italy, a loss of approximately 4,154,559 q of agricultural products due to land consumption in the period 2012–2020 was estimated (0.45% of percentage loss). The greatest drop occurred in the class of arable land (2,533,940 q, 0.43%), followed by fodder, fruit trees, vineyards, and olive groves, with a loss of approximately 974,403 (0.56%), 307,691 (0.45%), 247,670 (0.41%), and 90,853 q (0.42%) in the products generated therefrom, respectively (Table 4). At the regional level (Figure 5, Table 4), Emilia-Romagna is the region that has the highest variation in arable land production (376,972 q, 0.35%), followed by Veneto (372,463 q, 0.72%), and Lombardy (270,171 q, 0.50%). In the other regions, the loss is between 50,000 and 100,000 q and it is less than 50,000 q in only five regions, with the lowest value being recorded in the Aosta Valley, where the loss was approximately 3,536 q, which corresponded to a loss of 0.90%. As far as the fodder category is concerned, the greatest loss occurred in Lombardy (approximately 250,000 q, 0.76%), followed by Campania (130,000 q, 0.64%), Veneto (131,000 q, 0.75%), Trentino-Alto Adige, Emilia-Romagna, and Latium (79,000, 87,000 and 86,500 q, respectively), corresponding to a 0.60% loss in the regional fodder production in Latium and Trentino-Alto Adige, and 0.35% in Emilia-Romagna. In the other regions, the loss was less than 20,000 quintals, with the exception of Piedmont (around 42,000 q). The lost production of olive groves caused by land consumption was greater in Apulia, with a drop of 50,000 q (0.60%), whereas it was less than 10,000 q in the other regions. As far as the fruit trees category is concerned, Calabria and Sicily lost almost 40,000 (0.41%) and 75,000 q (0.58%), Trentino-Alto Adige 39,000 q (0.76%), and Campania and Apulia 30,000 q (0.65% and 0.49%, respectively). On the other hand, the loss was lower (below 10,000 q) in the other regions. Finally, the variation in vineyards due to soil consumption in the period under examination led to a reduction of about 10,000 q in all of the regions, with the exception of Apulia (58,200 q, 0.38%), Sicily (60,160 q, 0.40%), and Veneto (56,000 q, 0.69%).

Table 4. Percentage loss of crop production due to land consumption for five agricultural classes in the period 2012–2020 at the regional level.

Regions	Crop Typologies					Total
	Fruit Trees	Olive Groves	Vineyards	Fodder	Arable Land	
Piedmont	0.19	0.00	0.17	0.38	0.32	0.31
Aosta Valley	0.67	\	1.67	0.79	0.90	0.88
Lombardy	0.30	0.75	0.29	0.76	0.50	0.59
Trentino-Alto Adige	0.76	0.00	0.90	0.64	0.66	0.68
Veneto	0.95	0.68	0.69	0.94	0.72	0.76
Friuli-Venezia Giulia	0.35	0.00	0.45	0.71	0.50	0.53
Liguria	0.38	0.21	0.52	0.21	0.41	0.34
Emilia Romagna	0.23	0.07	0.13	0.32	0.35	0.34
Tuscany	0.23	0.18	0.17	0.26	0.28	0.26
Umbria	0.19	0.23	0.27	0.20	0.40	0.34
Marche	0.15	0.36	0.43	0.53	0.43	0.43
Latium	0.31	0.25	0.57	0.65	0.57	0.58
Abruzzo	0.40	0.54	0.51	0.29	0.57	0.51
Molise	0.09	0.47	0.32	0.22	0.22	0.23
Campania	0.65	0.32	0.60	0.93	0.64	0.71
Apulia	0.49	0.60	0.38	0.47	0.35	0.39
Basilicata	0.26	0.29	0.33	0.35	0.32	0.32
Calabria	0.41	0.22	0.42	0.33	0.29	0.32
Sicily	0.58	0.44	0.40	0.39	0.38	0.42
Sardinia	0.19	0.13	0.21	0.19	0.26	0.24
Italy	0.45	0.42	0.41	0.56	0.43	0.45

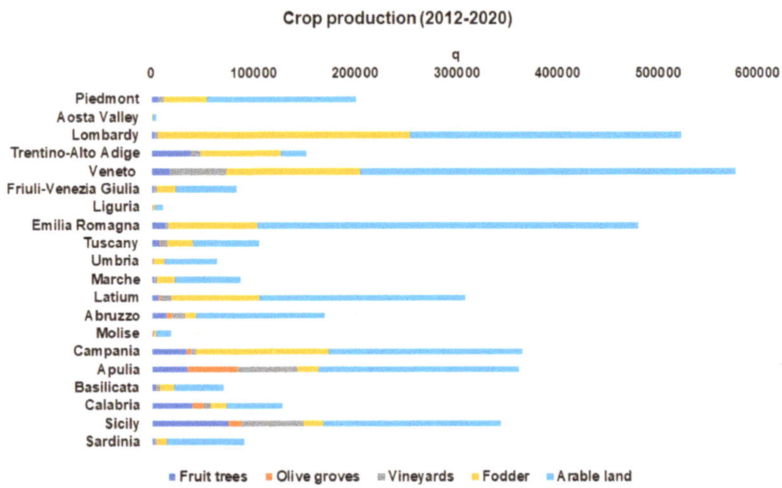

Figure 5. Estimation of crop production loss (quintals) due to land consumption for five agricultural classes in the period 2012–2020 at the regional level.

3.2. Wood Production Loss

At the national level, a total loss of 595,063 m^3 of wood production was estimated. At the regional level, Piedmont is the region with the highest loss (63,520 m^3), followed by Trentino-Alto Adige, Veneto, Lombardy, and Campania with a loss of around 50,000 m^3. In the other regions, the loss of flow is between 10,000 and 50,000 m^3, apart from Molise and the Aosta Valley, which have the lowest values (5,628 and 6.963 m^3 respectively) (Figure 6).

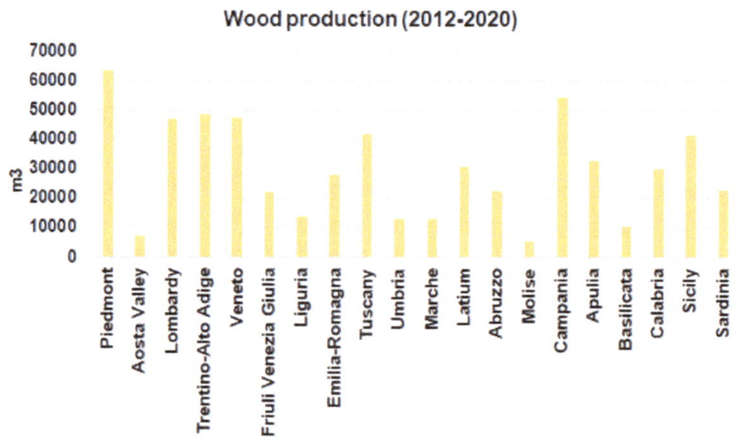

Figure 6. Estimation of wood production loss (m^3) due to land consumption in the period 2012–2020 at the regional level.

3.3. Carbon Storage Loss

In Italy, a loss of about 17,000 tons of stored carbon due to land consumption was estimated between 2012 and 2020. At the regional level, the greatest loss occurred in Piedmont, Trentino-Alto Adige, and Veneto (1902, 1731, and 1686 tons, respectively), with a considerable contribution also being provided by the Lombardy, Campania, and Sicily regions (between 1200 and 1500 tons). Values below 250 tons were estimated for the Marche, Umbria, Basilicata, Molise, and Aosta Valley regions (Figure 7).

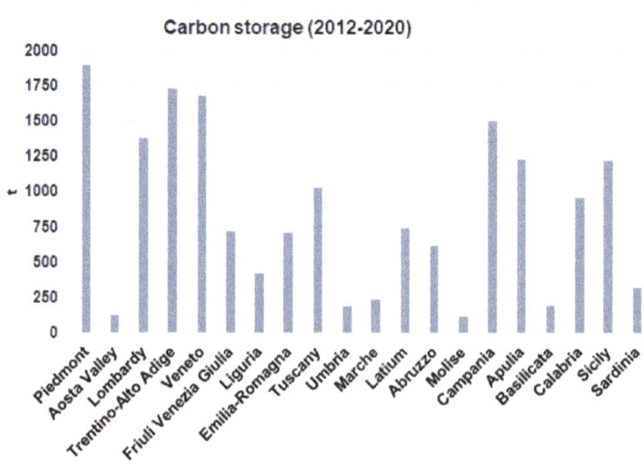

Figure 7. Tons of carbon storage lost due to land consumption in the period 2012–2020 at the regional level.

3.4. Habitat Quality Loss

Figure 8 shows the surface variations of ES flow at the regional level for each class of habitat quality. Class 1 (poor quality) records an increase of above 1% for all of the regions (between 1% and 3%), with the highest percentages being recorded for the Trentino-Alto Adige, and Abruzzo regions. Classes 2 and 3 record a slight decrease, with only Veneto and Apulia reaching 1% for class 2 and class 3, respectively. Class 4 has not recorded considerable changes, with the exception of Trentino-Alto Adige and the Aosta Valley, where increases of 2% and 1.4%, respectively, have been recorded. Class 5 has recorded a slight decrease, with the greatest changes being recorded for Sicily, Trentino-Alto Adige, and Campania (0.5–0.6%).

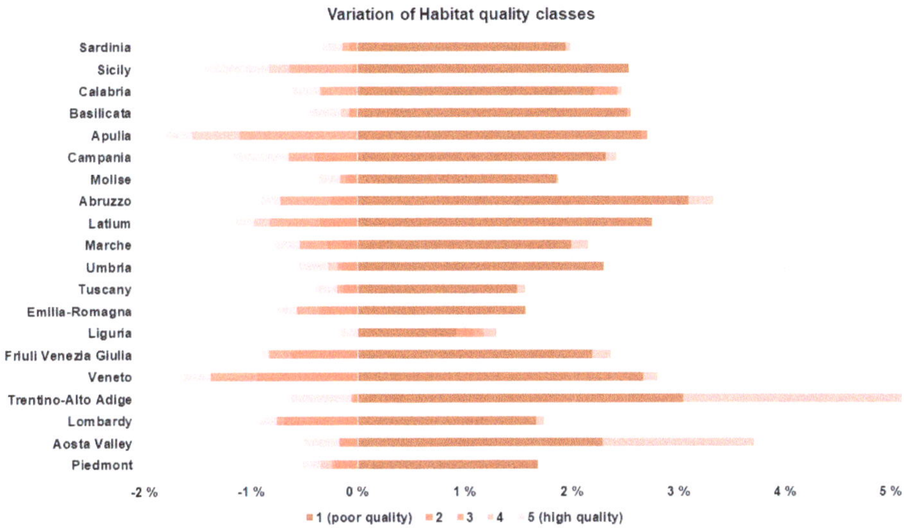

Figure 8. Percentage surface variation of habitat quality classes due to land consumption in the period 2012–2020 at the regional level.

3.5. Hydrological Regime Regulation Loss

An estimate was made, in the period 2012–2020, of the impact that the increase in land consumption had on surface runoff (which represents the hydrological regime regulation service). Figure 9 shows the variations in hm^3 in this service caused by the consumption of land in Italy. At the national level, an increase of about 250 hm^3 of surface runoff was estimated. The biggest increases were recorded in the Veneto and Lombardy regions (50 and 44 hm^3, respectively), whereas the smallest increases (less than 10 hm^3) were recorded in the Umbria, Basilicata, Liguria, Molise, and Aosta Valley regions.

3.6. Pollination Loss

Figure 10 shows the surface variations of ES flow at the regional level for each class of pollination. Class 1 (poor pollination) records a slight increase, with the highest value being recorded for Apulia (0.7%). Class 2 records an increase in almost all regions, with the highest percentages to be found in Trentino-Alto Adige and Campania (0.9%). Classes 3 and 4 record a decrease in all regions and the greater variations have been estimated in Veneto, Friuli-Venezia Giulia, Lombardy, Campania, and Marche (between −1% and −3%) for class 3, and the Aosta Valley and Trentino-Alto Adige (−3.4% and −1.6%, respectively) for class 4.

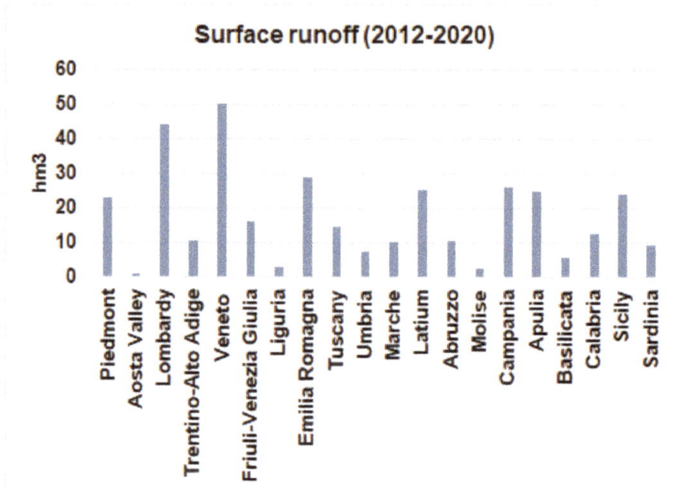

Figure 9. Surface runoff variations due to land consumption in the period 2012–2020 at the regional level.

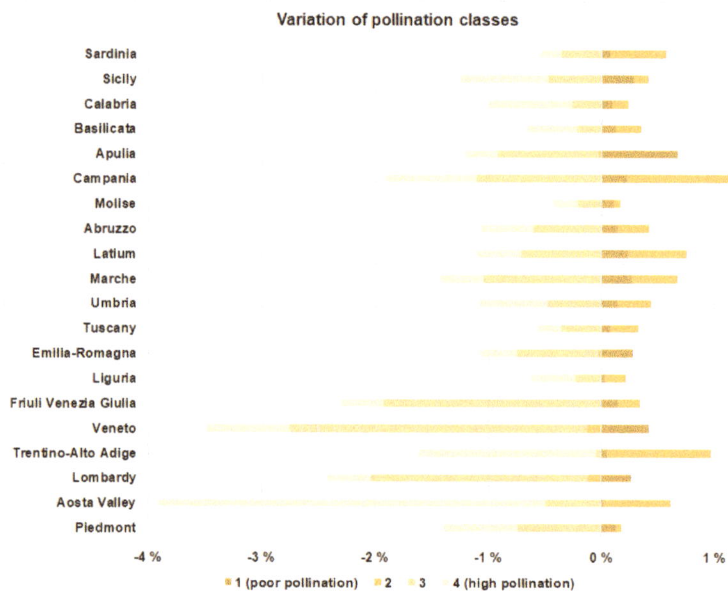

Figure 10. Percentage surface variation of pollination classes due to land consumption in the period 2012–2020 at the regional level.

4. Discussions

This study has assessed, from a biophysical point of view and with a multilayered approach, the changes occurring in the provision of ES flows as a result of land consumption in Italy This has made it possible to integrate and harmonize the large amount of heterogeneous data available in Italy so as to better interpret the ongoing land dynamics and how these affect the availability of goods and services offered by ecosystems. The results have explained how transformations have occurred in Italy over the selected time interval, which have mainly been driven by urbanization. The loss of crop production varies in Italy, depending on the type of crops involved. Loss of arable land and fodder production mainly affects the northern regions, whereas southern regions show a decrease in permanent crop production. Even if the percentage variations in crop production are low, they are nevertheless important because, in many cases, agricultural areas are based on important biological and cultural practices that are difficult to replace [76], and they also play a role in carbon storage and pollination assessment. Wood production and carbon storage losses affect the northern regions more, as well as the Campania region. The northern regions also show an increase in water runoff, with potential consequences for the hydrological regime regulation service. As far as habitat quality is concerned, the increase in the lowest quality class is widespread throughout Italy. The northern regions and the Campania region have recorded a drop in pollination ES availability. Since urban expansion is a process affecting more the lowlands, the higher losses in ES flows have mainly been recorded in the northern regions which include the Po plain, and the plains of Campania and Apulia in the south. These findings are in line with the demographic dynamics model of wealthier countries proposed since the late 1980s [77]. Indeed, whereas compact urban growth has been frequently associated with population increase and economic development, which is typical of most of the European countries after World War II [5], the advanced economies have been characterized in the last decades by the outward expansion of low-density settlements [78,79], often occurring in fertile agricultural areas [9,80]. Other research studies dealing with the impact of urbanization on ES have confirmed losses of stored carbon, agricultural production, habitat quality, and flood mitigation in developing urban areas [25,81,82], also leading to land reclamation and deforestation in rural areas [82].

The accurate localization of these processes has benefited from the detailed information provided by the input data. The high resolution of Sentinel satellites images and CLMS datasets and their high frequency availability enable an in-depth analysis of land consumption changes consistent with the EU requests and useful for the purpose of monitoring trends in ES provisions and flows to be conducted. The outcomes show that the proposed approach is effective in giving an overall assessment of the ES by using spatially explicit data and models. Indeed, these products of the monitoring program are already gathered in the national Environmental Data Yearbook [83], which is the official collection of nationwide data and information concerning the environment in Italy. Furthermore, they are used for national environmental monitoring policies, such as the National Capital Accounting [84], and could be a point of reference for the implementation of the various environmental policies, first of all the EU Soil Strategy for 2030 [28], in which the important role played by ES in maintaining ecosystems and human well-being is highlighted.

However, there is still room for improvement, especially with regard to models and input data. The need for more detailed information has arisen. For example, the inappropriate use of conventional agriculture technology and improper forest management practices can have negative short-term effects on crop and wood production, but can also have adverse effects on other ES in the long term. As a result thereof, a national evaluation of biophysical loss of ES must be integrated with more detailed information, for example on agricultural practices, forest management practices, and water balance at river basin level. A further example to be found in this study concerns the wood production analysis. An up-to-date database that localized arboriculture plants, forest types, age, provision, and forms of management would be needed for the purpose of developing a methodology able to overcome the simplification and achieve better results.

Furthermore, one of the main shortcomings arising therefrom concerns the data updating frequency. While the national land cover and land consumption maps are generated by analyzing datasets, satellite images, and remote-sensed indicators obtained from the Copernicus program, which guarantee a frequent data updating that is compliant with the requirements of the national monitoring program, other data concerning thematic issues required for conducting a biophysical assessment of the ES selected in this study are provided less promptly and are updated less regularly. These differences confirm the need, with a view to having more accurate and reliable results, for improved national spatial data that enhance spatial and temporal dynamics evaluations in the provision of goods and benefits. A data release agreement that is more suitable for a monitoring plan should be considered, on account of the replicability of the methodology that can only be guaranteed by satellite and spatial data being available on a regular basis. This is an important issue to take into consideration to make the analysis of soil-based ES a consolidated procedure for the purpose of increasing the effectiveness of sustainable spatial planning. ES spatial distribution and evolution over time is, indeed, able to gather and aggregate complex information [85] that can be used by decision-makers as a tool for land sustainability evaluation [86] that encourages development with less ES degradation [87].

5. Conclusions

This study is the result of a first attempt to present an assessment of nationwide biophysical loss ES flows caused by land consumption as an indicator of urbanization in Italy at a detailed spatial resolution, dealing with the problem of the lack of accurate and available information for the whole study area, and trying to collect and harmonize existing data from different scientific fields. Indeed, the concept of ES, and the related frameworks developed at the international level, have been recognized as fundamental for supporting territorial management policies, and the importance of biophysical quantifications has been emphasized, with a view to supporting economic evaluations and natural capital management. The multitemporal analysis set out herein may serve as a tool for stakeholders and policymakers that allows them to understand the impact of human activities, and in particular urbanization, on land cover pattern, the rates of variation, and the causes that give rise to changes in the provision of ES. Furthermore, remote sensing and GIS technologies make it possible to analyze, assess, and map ES at different spatial and temporal scales for the purpose of understanding how the provision thereof varies according to territorial dynamics and different underlying socioeconomic circumstances, and how the synergies and trade-offs between the various ES in certain landscapes occur.

The evaluation of soil-based ES is becoming increasingly important in the context of land planning, since urban planning predominantly has an impact on soil resources. The six ES investigated herein are strictly connected with soil properties and functions, and modifications of this resource on account of urban expansion can affect the ecosystems' ability to provide goods and benefits. Therefore, soil-based ES are a valuable reference-point and provide useful indications to administrations for the purpose of sustainable development policies and land planning, both at the national and local scale, even though they have to date not yet found a stable and codified position in the land use planning process.

Moreover, our study underlines the importance of using multilayered thematic data and the detailed localization of rapid changes that support the investigation of how urbanization affects ES and provides suggestions for how urban planning can be conducted with a view to ES protection, sustainable urban development, and human well-being. For this reason, this paper has pointed out the importance of using integrated data and methodologies, as well as more consistent and detailed data inputs. This allows well-founded approaches to be adopted, with a view to gaining both a thorough understanding of ES-related processes and the changes therein that are taking place, as well as acquiring a reference framework for the purpose of also gaining further insights that evaluate ES at different scales, from national to local levels. Further developments and improvements in the methodology are foreseen, especially as far as the search for updated and accurate thematic data and the use of new analysis models is concerned. In this respect, the Horizon 2020 project (SERENA: Soil Ecosystem seRvices and soil threats modElling aNd mApping) is currently conducting an in-depth analysis of this approach at the European level, whereas the LIFE preparatory project (NewLife4Drylands: Remote sensing oriented nature based solutions towards a NEW LIFE FOR DRYLANDS) is doing so at the local level. The results generated therefrom may serve as a basis for tackling multiscale landscape management and planning issues.

Author Contributions: Conceptualization, F.A. and D.S.; Methodology, F.A., A.S. and M.M.; Formal analysis, A.C., L.C., C.G., N.R. and A.S.; Writing—original draft preparation, D.S.; Writing—review and editing, F.A., D.S., C.G., L.C., C.G., N.R., A.S. and M.M.; Supervision, F.A. and M.M. All authors have read and agreed to the published version of the manuscript.

Funding: This manuscript is not funded by a specific project grant.

Data Availability Statement: The data supporting the findings of this study are openly available at https://groupware.sinanet.isprambiente.it/uso-copertura-e-consumo-di-suolo/library/consumo-di-suolo.

Acknowledgments: The authors want to acknowledge the European Union's Horizon 2020 research and Innovation programme that under the ongoing project SERENA (grant agreement No. 862695) will support further development of the methodology presented in this study at the European level. Moreover, the authors want to acknowledge the ongoing project NewLife4Drylands (LIFE20 PRE/IT/000007) that gives the opportunity to conduct an in-depth analysis of this approach at the local level.

Conflicts of Interest: The authors declare no conflict of interest.

References

1. Bruegmann, R. *Sprawl: A Compact History*; University of Chicago Press: Chicago, IL, USA, 2005.
2. Schneider, A.; Woodcock, C.E. Compact, dispersed, fragmented, extensive? A comparison of urban growth in 25 global cities using remotely sensed data, pattern metrics and census information. *Urban Stud.* **2008**, *45*, 659–692. [CrossRef]
3. Elvidge, C.D.; Tuttle, B.T.; Sutton, P.C.; Baugh, K.E.; Howard, A.T.; Milesi, C.; Bhaduri, B.; Nemani, R. Global Distribution and Density of Constructed Impervious Surfaces. *Sensors* **2007**, *7*, 1962–1979. [CrossRef] [PubMed]
4. Stathakis, D.; Perakis, K.; Savin, I. Efficient segmentation of urban areas by the VIBI. *Int. J. Remote Sens.* **2012**, *33*, 6361–6377. [CrossRef]
5. Smiraglia, D.; Salvati, L.; Egidi, G.; Salvia, R.; Giménez-Morera, A.; Halbac-Cotoara-Zamfir, R. Toward a New Urban Cycle? A Closer Look to Sprawl, Demographic Transitions and the Environment in Europe. *Land* **2021**, *10*, 127. [CrossRef]
6. Strollo, A.; Smiraglia, D.; Bruno, R.; Assennato, F.; Congedo, L.; De Fioravante, P.; Giuliani, C.; Marinosci, I.; Riitano, N.; Munafò, M. Land consumption in Italy. *J. Maps* **2020**, *16*, 113–123. [CrossRef]
7. Riitano, N.; Dichicco, P.; De Fioravante, P.; Cavalli, A.; Falanga, V.; Giuliani, C.; Mariani, L.; Strollo, A.; Munafò, M. Land consumption in italian coastal area. *Environ. Eng. Manag. J.* **2020**, *19*, 1857–1868. [CrossRef]
8. Grimm, N.B.; Faeth, S.H.; Golubiewski, N.E.; Redman, C.L.; Wu, J.; Bai, X.; Briggs, J.M. Global Change and the Ecology of Cities. *Science* **2008**, *319*, 756–760. [CrossRef] [PubMed]
9. EC (European Commission). Guidelines on Best Practice to Limit, Mitigate or Compensate Soil Sealing. Available online: https://ec.europa.eu/environment/soil/pdf/guidelines/pub/soil_en.pdf (accessed on 17 October 2021).
10. Ceccarelli, T.; Bajocco, S.; Perini, L.; Salvati, L. Urbanisation and Land Take of High Quality Agricultural Soils–Exploring Long-term Land Use Changes and Land Capability in Northern Italy. *Int. J. Environ. Res.* **2014**, *8*, 181–192. [CrossRef]
11. Haase, D. Effects of urbanisation on the water balance–A long-term trajectory. *Environ. Impact Assess. Rev.* **2009**, *29*, 211–219. [CrossRef]
12. Yang, B.; Li, M.-H. Ecological engineering in a new town development: Drainage design in The Woodlands, Texas. *Ecol. Eng.* **2010**, *36*, 1639–1650. [CrossRef]
13. Shiguang, M.; Fei, C.; Qingchun, L.; Shuiyong, F. Impacts of Urban Processes and Urbanization on Summer Precipitation: A Case Study of Heavy Rainfall in Beijing on 1 Aug 2006. *J. Appl. Meteorol. Climatol.* **2011**, *50*, 806–825. [CrossRef]
14. Zhou, X.; Wang, Y.-C. Dynamics of Land Surface Temperature in Response to Land-Use/Cover Change. *Geogr. Res.* **2011**, *49*, 23–36. [CrossRef]
15. Collinge, S.K. Ecological consequences of habitat fragmentation: Implications for landscape architecture and planning. *Landsc. Urban Plan.* **1996**, *36*, 59–77. [CrossRef]
16. Bennett, A.F.D.; Saunders, A. Habitat fragmentation and landscape change. In *Conservation Biology for All*; Sodhi, N.S., Ehrlich, P.R., Eds.; Oxford University Press: Oxford, UK, 2010. [CrossRef]
17. MEA (Millennium Ecosystem Assessment). *Ecosystems and Human Wellbeing: Synthesis*; World Resource Institute: Washington, DC, USA, 2005.
18. Haines-Young, R.H.; Potschin-Young, M.B. Common International Classification of Ecosystem Services (CICES) V5.1 and Guidance on the Application of the Revised Structure. Available online: www.cices.eu (accessed on 30 November 2021).
19. Schröter, M.; Barton, D.N.; Remme, R.P.; Hein, L. Accounting for capacity and flow of ecosystem services: A conceptual model and a case study for Telemark, Norway. *Ecol. Indic.* **2014**, *36*, 539–551. [CrossRef]
20. Hein, L.; Bagstad, K.; Edens, B.; Obst, C.; de Jong, R.; Lesschen, J.P. Defining Ecosystem Assets for Natural Capital Accounting. *PLoS ONE* **2016**, *11*, e0164460. [CrossRef] [PubMed]
21. Pereira, P. Ecosystem services in a changing environment. *Sci. Total Environ.* **2020**, *702*, 135008. [CrossRef] [PubMed]
22. Ghaley, B.B.; Porter, J.R.; Sandhu, H.S. Soil-based ecosystem services: A synthesis of nutrient cycling and carbon sequestration assessment methods. *Int. J. Biodivers. Sci. Ecosyst. Serv. Manag.* **2014**, *10*, 177–186. [CrossRef]

23. Baveye, P.C.; Baveye, J.; Gowdy, J. Soil "Ecosystem" Services and Natural Capital: Critical Appraisal of Research on Uncertain Ground. *Front. Environ. Sci.* **2016**, *4*, 41. [CrossRef]
24. Haines-Young, R.H.; Potschin-Young, M.B. The links between biodiversity, ecosystem services and human well-being. In *Ecosystem Ecology: A New Synthesis*; Raffaelli, D.G., Frid, C.L.J., Eds.; Cambridge University Press: Cambridge, UK, 2010; pp. 110–139. [CrossRef]
25. Lauf, S.; Haase, D.; Kleinschmit, B. Linkages between ecosystem services provisioning, urban growth and shrinkage—A modeling approach assessing ecosystem service trade-offs. *Ecol. Indic.* **2014**, *42*, 73–94. [CrossRef]
26. Dominati, E.; Patterson, M.; Mackay, A. A framework for classifying and quantifying the natural capital and ecosystem services of soils. *Ecol. Econ.* **2010**, *69*, 1858–1868. [CrossRef]
27. Adhikari, K.; Hartemink, A.E. Linking soils to ecosystem services—A global review. *Geoderma* **2016**, *262*, 101–111. [CrossRef]
28. European Commission. EU Soil Strategy for 2030 Reaping the Benefits of Healthy Soils for People, Food, Nature and Climate. Available online: https://eur-lex.europa.eu/legal-content/EN/TXT/?uri=CELEX%3A52021DC0699 (accessed on 18 January 2022).
29. European Commission. Biodiversity Strategy for 2030. Available online: https://ec.europa.eu/environment/strategy/biodiversity-strategy-2030_en (accessed on 18 January 2022).
30. European Commission. Key Policy Objectives of the New CAP. Available online: https://ec.europa.eu/info/food-farming-fisheries/key-policies/common-agricultural-policy/new-cap-2023-27/key-policy-objectives-new-cap_en (accessed on 21 January 2022).
31. The Economics of Ecosystems and Biodiversity. Mainstreaming the Economics of Nature: A Synthesis of the Approach, Conclusions and Recommendations of TEEB. Available online: http://www.teebweb.org/publication/mainstreaming-the-economics-of-nature-a-synthesis-of-the-approach-conclusions-and-recommendations-of-teeb/ (accessed on 30 November 2021).
32. Natural Capital Coalition. Natural Capital Protocol. Available online: http://naturalcapitalcoalition.org/protocol/ (accessed on 30 November 2021).
33. Convention on Biological Diversity. Aichi Biodiversity Targets. Available online: https://www.cbd.int/sp/targets/ (accessed on 30 November 2021).
34. United Nations. The Sustainable Development Agenda. Available online: https://www.un.org/sustainabledevelopment/development-agenda/ (accessed on 30 November 2021).
35. La Notte, A.; Vallecillo Rodriguez, S.; Polce, C.; Zulian, G.; Maes, J. *Implementing an EU System of Accounting for Ecosystems and Their Services. Initial Proposals for the Implementation of Ecosystem Services Accounts*; Publications Office of the European Union: Luxembourg, 2017. [CrossRef]
36. De Groot, R.S.; Wilson, M.; Boumans, R. A typology for the description, classification and valuation of ecosystem functions, goods and services. *Ecol. Econ.* **2002**, *41*, 393–408. [CrossRef]
37. Maes, J.; Egoh, B.; Willemen, L.; Liquete, C.; Vihervaara, P.; Schagner, J.P.; Grizzetti, B.; Drakou, E.G.; La Notte, A.; Bouraoui, F.; et al. *Mapping and Assessment of Ecosystems and their Services. An Analytical Framework for Ecosystem Assessments under Action 5 of the EU Biodiversity Strategy to 2020*; Publications Office of the European Union: Luxembourg, 2013. Available online: http://ec.europa.eu/environment/nature/knowledge/ecosystem_assessment/pdf/MAESWorkingPaper2013.pdf (accessed on 28 November 2021).
38. García, P.; Pérez, E. Mapping of soil sealing by vegetation indexes and built-up index: A case study in Madrid (Spain). *Geoderma* **2016**, *268*, 100–107. [CrossRef]
39. Smiraglia, D.; Rinaldo, S.; Ceccarelli, T.; Bajocco, S.; Salvati, L.; Ricotta, C.; Perini, L. A cost-effective approach for improving the quality of soil sealing change detection from landsat imagery. *Eur. J. Remote Sens.* **2014**, *47*, 805–819. [CrossRef]
40. Criado, M.; Santos-Francés, F.; Martínez-Graña, A.; Sánchez, Y.; Merchán, L. Multitemporal Analysis of Soil Sealing and Land Use Changes Linked to Urban Expansion of Salamanca (Spain) Using Landsat Images and Soil Carbon Management as a Mitigating Tool for Climate Change. *Remote Sens.* **2020**, *12*, 1131. [CrossRef]
41. Luti, T.; Segoni, S.; Catani, F.; Munafò, M.; Casagli, N. Integration of Remotely Sensed Soil Sealing Data in Landslide Susceptibility Mapping. *Remote Sens.* **2020**, *12*, 1486. [CrossRef]
42. ESA Sentinel Online. Available online: https://sentinel.esa.int/web/sentinel/home (accessed on 30 November 2021).
43. Assennato, F.; Di Leginio, M.; d'Antona, M.; Marinosci, I.; Congedo, L.; Riitano, N.; Luise, A.; Munafò, M. Land degradation assessment for sustainable soil management. *Ital. J. Agron.* **2020**, *15*, 299–305. [CrossRef]
44. ISPRA. *Consumo di Suolo, Dinamiche Territoriali e Servizi Ecosistemici*; ISPRA: Rome, Italy, 2021.
45. De Fioravante, P.; Strollo, S.; Assennato, F.; Marinosci, I.; Congedo, L.; Munafò, M. High resolution land cover integrating Copernicus products: A 2012–2020 map of Italy. *Land* **2022**, *11*, 35. [CrossRef]
46. QGIS.org. Available online: http://www.qgis.org (accessed on 30 November 2021).
47. Sharp, R.; Douglass, J.; Wolny, S.; Arkema, K.; Bernhardt, J.; Bierbower, W.; Chaumont, N.; Denu, D.; Fisher, D.; Glowinski, K.; et al. InVEST 3.9.1. *User's Guide*. Available online: https://invest-userguide.readthedocs.io/en/latest/# (accessed on 30 November 2021).
48. ISTAT–Censimenti dell'Agricoltura. Available online: https://www.istat.it/it/censimenti-permanenti/censimenti-precedenti/agricoltura (accessed on 30 November 2021).
49. Federici, S.; Vitullo, M.; Tulipano, S.; De Lauretis, R.; Seufert, G. An approach to estimate carbon stocks change in forest carbon pools under the UNFCCC: The Italian case. *iForest* **2008**, *1*, 86–95. [CrossRef]
50. ISPRA. *Italian Greenhouse Gas Inventory 1990–2016. National Inventory Report 2018*; ISPRA: Rome, Italy, 2018.

51. National Inventory of Forests and Forest Carbon Tanks. Available online: https://www.inventarioforestale.org/ (accessed on 30 November 2021).
52. Hutyra, L.; Yoon, B.; Alberti, M. Terrestrial carbon stocks across a gradient of urbanization: A study of the Seattle, WA region. *Glob. Change Biol.* **2011**, *17*, 783–797. [CrossRef]
53. National Soil Organic Carbon Map. Available online: https://scienzadelsuolo.org/carta_italiana_carbonio_organico.php (accessed on 30 November 2021).
54. Marchetti, M.; Sallustio, L.; Ottaviano, M.; Barbati, A.; Corona, P.; Tognetti, R.; Zavattero, L.; Capotorti, G. Carbon sequestration by forests in the National Parks of Italy. *Plant Biosyst.* **2012**, *146*, 1001–1011. [CrossRef]
55. Penman, J.; Gytarsky, M.; Hiraishi, T.; Krug, T.; Kruger, D.; Pipatti, R.; Buendia, L.; Miwa, K.; Ngara, T.; Tanabe, K. Good Practice Guidance for Land Use, Land-Use Change and Forestry. Available online: https://www.ipcc.ch/publication/good-practice-guidance-for-land-use-land-use-change-and-forestry/ (accessed on 14 January 2022).
56. Vitullo, M.; De Lauretis, R.; Federici, S. La contabilità del carbonio contenuto nelle foreste italiane. *Silvae* **2007**, *9*, 91–104.
57. ISPRA. *Italian Greenhouse Gas Inventory 1990–2012. National Inventory Report 2014*; ISPRA: Rome, Italy, 2014.
58. Di Cosmo, L.; Gasparini, P.; Tabacchi, G. A national-scale, stand-level model to predict total above-ground tree biomass from growing stock volume. *For. Ecol. Manag.* **2016**, *361*, 269–276. [CrossRef]
59. Canaveira, P.; Manso, S.; Pellis, G.; Perugini, L.; De Angelis, P.; Neves, R.; Papale, D.; Paulino, J.; Pereira, T.; Pina, A.; et al. Biomass Data on Cropland and Grassland in the Mediterranean Region, Final Report for Action A4 of Project MediNet. Available online: https://www.lifemedinet.com/documents (accessed on 30 November 2021).
60. Sallustio, L.; Quatrini, V.; Geneletti, D.; Corona, P.; Marchetti, M. Assessing land take by urban development and its impact on carbon storage: Findings from two case studies in Italy. *Environ. Impact Assess. Rev.* **2015**, *54*, 80–90. [CrossRef]
61. ISPRA. *Italian Greenhouse Gas Inventory 1990–2015. National Inventory Report 2017*; ISPRA: Rome, Italy, 2017.
62. Romano, B.; Zullo, F. Land urbanization in Central Italy: 50 years of evolution. *J. Land Use Sci.* **2014**, *9*, 143–164. [CrossRef]
63. Terrado, M.; Sabater, S.; Chaplin-Kramer, B.; Mandle, L.; Ziv, G.; Acuña, V. Model development for the assessment of terrestrial and aquatic habitat quality in conservation planning. *Sci. Total Environ.* **2016**, *540*, 63–70. [CrossRef]
64. European Environment Agency. EUNIS Babitat Type Hierarchical View. Available online: http://eunis.eea.europa.eu/habitats-code-browser.jsp (accessed on 14 January 2022).
65. Leh, M.D.K.; Matlock, M.D.; Cummings, E.C.; Nalley, L.L. Quantifying and mapping multiple ecosystem services change in West Africa. *Agric. Ecosyst. Environ.* **2013**, *165*, 6–18. [CrossRef]
66. Kuhnert, P.M.; Martin, T.G.; Griffiths, S.P. A guide to eliciting and using expert knowledge in Bayesian ecological models. *Ecol. Lett.* **2010**, *13*, 900–914. [CrossRef]
67. Sallustio, L.; De Toni, A.; Strollo, A.; Di Febbraro, M.; Gissi, E.; Casella, L.; Geneletti, D.; Munafò, M.; Vizzarri, M.; Marchetti, M. Assessing habitat quality in relation to the spatial distribution of protected areas in Italy. *J. Environ. Manag.* **2017**, *201*, 129–137. [CrossRef]
68. Braca, G.; Bussettini, M.; Ducci, D.; Lastoria, B.; Mariani, S. Evaluation of national and regional groundwater resources under climate change scenarios using a GIS-based water budget procedure. *Rend. Lincei Sci. Fis. Nat.* **2019**, *30*, 109–123. [CrossRef]
69. ISPRA. *Il Bilancio Idrologico Gis BAsed a Scala Nazionale su Griglia regolare–BIGBANG: Metodologia e Stime. Rapporto Sulla Disponibilità Naturale Della Risorsa Idrica*; ISPRA: Rome, Italy, 2021.
70. Zhang, W.; Ricketts, T.H.; Kremen, C.; Carney, K.; Swinton, S.M. Ecosystem services and disservices to agriculture. *Ecol. Econ.* **2007**, *64*, 253–260. [CrossRef]
71. Klein, A.-M.; Vaissière, B.E.; Cane, J.H.; Steffan-Dewenter, I.; Cunningham, S.A.; Kremen, C.; Tscharntke, T. Importance of pollinators in changing landscapes for world crops. *Proc. Royal Soc. B* **2006**, *274*, 303–313. [CrossRef] [PubMed]
72. International Union for Conservation of Nature. Nearly One in 10 Wild Bee Species Face Extinction in Europe While the Status of More than Half Remains Unknown. Available online: http://www.iucn.org/?19073/Nearly-one-in-ten-wild-bee-species-face-extinction-in-Europe-while-the-status-of-morethan-half-remains-unknown (accessed on 30 November 2021).
73. Xiao, Y.; Li, X.; Cao, Y.; Dong, M. The diverse effects of habitat fragmentation on plant-pollinator interaction. *Plant Ecol.* **2016**, *217*, 857–868. [CrossRef]
74. Nogué, S.; Long, P.R.; Eycott, A.E.; de Nascimento, L.; Fernández-Palacios, J.M.; Petrokofsky, G.; Vandvik, V.; Willis, K.J. Pollination service delivery for European crops: Challenges and opportunities. *Ecol. Econ.* **2016**, *128*, 1–7. [CrossRef]
75. Ricciardelli D'Albore, G.; Intoppa, F. *Fiori e Api. La Flora Visitata Dalle Api e Dagli Apoidei in Europa*; Calderini Edagricole: Bologna, Italy, 2000.
76. Biasi, R.; Brunori, E.; Smiraglia, D.; Salvati, L. Linking traditional tree-crop landscapes and agro-biodiversity in Central Italy using a database of typi-cal and traditional products: A multiple risk assessment through a data mining analysis. *Biodivers Conserv* **2015**, *24*, 3009–3031. [CrossRef]
77. Lesthaeghe, R. The unfolding story of the second demographic transition. *Popul. Dev. Rev.* **2010**, *36*, 211–251. [CrossRef]
78. Salvati, L.; Serra, P. Estimating Rapidity of Change in Complex Urban Systems: A Multidimensional, Local-Scale Approach. *Geogr. Anal.* **2016**, *48*, 132–156. [CrossRef]
79. Galster, G.; Hanson, R.; Ratcliffe, M.R.; Wolman, H.; Coleman, S.; Freihage, J. Wrestling Sprawl to the Ground: Defining and Measuring an Elusive Concept. Hous. *Policy Debate* **2001**, *12*, 681–717. [CrossRef]
80. Bhat, P.A.; Shafiq, M.U.; Mir, A.A.; Ahmed, P. Urban sprawl and its impact on land use/land cover dynamics of Dehradun City, India. *Int. J. Sustain. Built Environ.* **2017**, *6*, 513–521. [CrossRef]

81. Eigenbrod, F.; Bell, V.A.; Davies, H.N.; Heinemeyer, A.; Armsworth, P.R.; Gaston, K.J. The impact of projected increases in urbanization on ecosystem services. *Proc. R. Soc. B* **2011**, *278*, 3201–3208. [CrossRef]
82. Lyu, R.; Zhang, J.; Xu, M.; Li, J. Impacts of urbanization on ecosystem services and their temporal relations: A case study in Northern Ningxia, China. *Land Use Policy* **2018**, *77*, 163–173. [CrossRef]
83. Environmental Data Yearbook. Available online: https://annuario.isprambiente.it/content/environmental-data-yearbook-2019 (accessed on 15 January 2022).
84. Rapporto sullo Stato del Capitale Naturale. Available online: https://www.mite.gov.it/pagina/il-rapporto-sullo-stato-del-capitale-naturale-italia. (accessed on 15 January 2022).
85. Burkhard, B.; Kroll, F.; Nedkov, S.; Müller, F. Mapping ecosystem service supply, demand and budgets. *Ecol. Indic.* **2012**, *21*, 17–29. [CrossRef]
86. Swetnam, R.D.; Fisher, B.; Mbilinyi, B.P.; Munishi, P.K.; Willcock, S.; Ricketts, T.; Mwakalila, S.; Balmford, A.; Burgess, N.D.; Marshall, A.R.; et al. Mapping socio-economic scenarios of land cover change: A GIS method to enable ecosystem service modelling. *J Environ. Manag.* **2011**, *92*, 563–574. [CrossRef] [PubMed]
87. Peng, J.; Tian, L.; Liu, Y.; Zhao, M.; Hu, Y.; Wu, J. Ecosystem services response to urbanization in metropolitan areas: Thresholds identification. *Sci. Tot. Environ.* **2017**, *607–608*, 706–714. [CrossRef]

land

MDPI

Article

Integrated Approaches to Ecosystem Services: Linking Culture, Circular Economy and Environment through the Re-Use of Open Spaces and Buildings in Europe

Liana Ricci

School of Architecture, Planning and Environmental Policy University College, D04 V1W8 Dublin, Ireland; liana.ricci@ucd.ie

Abstract: Green and blue infrastructure, nature-based solutions, and cultural and built heritage play a key role in enhancing ecosystem services provision and shaping urban quality and communities' wellbeing calling for an integrated approach to ecosystem services in urban policy and planning and decision-making. On the other side, under-used spaces and buildings have social, cultural, economic, as well as ecological functions and benefits, which are essential to sustainable urban development. The EU has been developing and implementing policies for an integrated approach to urban development and sustainable land use through the implementation of the Urban Agenda for the EU and fourteen associated Partnerships. Thus, it engaged a broad range of institutions and stakeholders across Europe in promoting local projects and sharing best practices on sustainable land use and nature-based solutions, the circular economy, and cultural heritage. This paper reviews the experiences of cities involved in the Partnerships of the Urban Agenda for the EU by illustrating how they related to different modes of ecosystem governance and associated challenges, discussing how three case studies integrate different dimensions of ecosystem services and regeneration in under-used areas and what type of knowledge as well as regulation and governance modes they have developed for supporting innovation in land use planning and management for urban ecosystem services. The results show that appropriate alternative regulations and policies are little explored and that cities adopt an integrated approach, combining cultural, environmental, economic, and social dimensions in their interventions, directly or indirectly enhancing the benefits of built and natural heritage and urban ecosystems in under-used areas. However, some issues, such as nature-based solutions and climate change, are still partially integrated into the projects while priority is given to the cultural, aesthetic, and economic dimensions.

Keywords: ecosystems governance; knowledge; regulations; Urban Agenda for the EU; under-used spaces; urban regeneration

Citation: Ricci, L. Integrated Approaches to Ecosystem Services: Linking Culture, Circular Economy and Environment through the Re-Use of Open Spaces and Buildings in Europe. *Land* **2022**, *11*, 1161. https://doi.org/10.3390/land11081161

Academic Editors: Luca Congedo, Francesca Assennato and Michele Munafò

Received: 5 July 2022
Accepted: 22 July 2022
Published: 26 July 2022

1. Introduction

The pressing climate change challenges and ongoing socio-economic and health crises have increased the need for inclusive and sustainable public and green spaces in cities as a means to enhance ecosystem service provision via spatial planning [1–3] and new modes of governance [4,5].

Green and blue infrastructure, nature-based solutions (NBS), and built and cultural heritage play a key role in shaping urban quality and community wellbeing. On the other side, under-used spaces and buildings have social, cultural, economic, as well as ecological functions and benefits, which are essential to sustainable urban development. With the COVID-19 pandemic, the way of living in urban spaces has changed, the demand for accessible green and blue public spaces is rising, and the quality of local environments and social inclusion has gained more relevance.

In the EU and globally, there is a large amount of vacant and underused open spaces and buildings that are considered both a challenge and an opportunity to address the issues mentioned above. The stock of industrial or military heritage sites in the EU has the potential to contribute to re-activate and enhance the multifunctional and multi-dimensional role of green spaces, ecosystems, and buildings, providing a broad range of ecological, economic, and social benefits to urban communities [6–8]. In addition to their environmental functions, underused spaces and buildings have physical and intangible cultural and natural heritage with recreational, aesthetic, therapeutic, and social and cultural interaction values having relevant impacts on residents' health and wellbeing, and on access to services, particularly for most vulnerable people. They can provide places for community engagement and participation in social and cultural events, and for social inclusion and interaction [9], contributing to reduce and prevent social, economic, environmental, and territorial inequalities (i.e., cultural ecosystem services).

At the policy level, greater emphasis on the multiple environmental and cultural dimensions has been posed in the Pact of Amsterdam and the Urban Agenda for the EU and later in the new Leipzig Charter, New European Bauhaus and EU Green Deal.

In connection with these policy frameworks and with the Decision V/6 of the Fifth Ordinary Meeting of the Conference of the Parties to the Convention on Biological Diversity [10], urban policies and planning are called on to increasingly adopt an integrated approach to ecosystem services, to find appropriate knowledge, tools, regulations and modes of governance and institutions to effectively integrate the different dimensions of ecosystem services, particularly in the vacant and under-used spaces.

The paper aims to improve the understanding of integrated approaches and governance of ecosystem services by reviewing experiences of cities involved in the Partnerships of the Urban Agenda for the EU, with the aim of combining sustainable land use, the circular economy, and cultural, built, and natural heritage in under-used spaces.

The following sections review the key literature on the need for an integrated approach to ecosystem services, then illustrate the key governance challenges addressed in the Partnerships of the Urban Agenda of the EU.

After presenting the method and the analytical framework based on different modes of ecosystem governance and related challenges, the paper illustrates the case studies selected under the relevant EU Partnerships and discusses insights regarding an integrated approach to ecosystem services and governance modes to enhance the understanding of innovative land use planning for improving under-used areas while addressing land take and urban sprawl challenges.

The paper found that the cases studies show how public institutions can work with stakeholders and citizens to simultaneously address environmental sustainability, social inclusion, and circular economy challenges.

The experiences from Poland and Spain also highlight the need for additional development and implementation of new forms of governance and regulations able to integrate the different dimensions of ecosystem services and support the re-use of underused or derelict spaces and buildings where ecosystems intertwine with old urban structures and infrastructure (e.g., industrial sites). The paper opens the door to future research on the "regulation of commons" as the institutions involved in urban policy and planning are struggling to find appropriate processes and forms to hand over the management of under-used public spaces and services to communities and social groups that are already engaged or interested in management of ecosystems and other services and spaces in under-used areas.

1.1. Integrated Approaches to Ecosystem Services

The debate on the ecosystem service developed since the 1970s has produced several approaches to assessing the benefits to society of nature for supporting decision-making. However, only in the 1990s did the integration of ecosystem services into policies and plans emerge with an increasing number of researchers from diverse backgrounds and disciplines advocating the adoption of an ecosystem services perspective for better decision-making [6].

This trend continued and intensified after the Millennium Ecosystem Assessment 2005 [11,12], which introduced a framework for analyzing socio-ecological systems, defining ecosystem services as:

> "Ecosystem services are the benefits people obtain from ecosystems. These include provisioning services such as food, water, timber, and fiber; regulating services that affect climate, floods, disease, wastes, and water quality; cultural services that provide recreational, aesthetic, and spiritual benefits; and supporting services such as soil formation, photosynthesis, and nutrient cycling. [...] The human species, while buffered against environmental changes by culture and technology, is fundamentally dependent on the flow of ecosystem services." [11]

Although this definition and the ecosystem categories are widely applied and have had a broad influence on the policy and scientific debate [13], there are diverging approaches questioning the appropriateness of these categories and how they may be applied in spatial planning [3] and policies. For instance, the "services cascade" approach distinguishes between ecological structures and processes created and the benefits that people eventually obtain. This is crucial to certain types of ecosystem services, such as cultural services, that may exist but not being accessible or their usefulness may be perceived differently in different contexts. Hence, for understanding the relevance and value of 'ecosystem service' it is essential knowing the spatial, social, and cultural context and values in addition to the structure and dynamics of ecological systems [14]. Ecosystem services are produced in cultural landscapes over which cities develop [15].

The cultural dimension of ecosystem services is recognized in several policy documents, such as the European Landscape Convention, Millennium Ecosystem Assessment, the Convention on Biological Diversity, and the Economics of Ecosystems and Biodiversity, and it provides insights into relationships between natural and cultural environments, to assess and quantify their value, and to integrate it into planning, conservation, and management [16]. At the same time, the concept of cultural ecosystem services raises awareness on the range of services that ecosystems provide to humans and promotes the assessment of ecosystems and landscape services as key knowledge for sustainable development and governance, participatory and environmental decision-making, environmental democracy, and environmental justice [17].

However, research on cultural ecosystem services and integrated approaches to ecosystem services remains relatively underdeveloped and there is a need for a deeper understanding of terminology, categorization and attribution of cultural benefits [18], of how culture and nature co-produce a "sense of place" [19], of how to integrate human values and cultural practices into ecosystem management [20], as well as analytical tools, policy instruments, and governance modes to operationalise cultural and integrated ecosystem service concepts and approaches.

Although the legacies associated with past uses of urban vacant spaces and buildings are known to affect ecosystems, there is a dearth of research investigating the ecological functions of vacant urban land [21], as well as a need to understand the social dimension and socio-cultural values of ecosystem services together with their biophysical and economic aspects [22–24], and to adopt a holistic approach covering a broader range of ecosystem services, widening the assessment methodologies incorporating participatory knowledge production and research, ecosystem-based planning, and governance and regulations.

Together with an integrated approach for ecosystem services participation, co-design, co-production, and co-management of space have been increasingly considered essential to foster inclusion and develop sustainable planning and management solutions and strategies. Public authorities play a key role in finding appropriate tools and governance models, in engaging citizens and relevant non-profit, private sector, and other relevant actors in developing and implementing plans and projects for the re-use or derelict spaces and buildings while enhancing ecosystem benefits.

1.2. Ecosystem Governance Challenges in the Urban Agenda for the EU

Within the extensive debate on the role of nature-based solutions and integrated approaches to ecosystem services, there is still a dearth of knowledge on how cultural, economic, and environmental dimensions have been integrated and operationalized in European cities and beyond.

In the last two centuries, most urban spaces in Europe have evolved substantially [25]. Although with differences between regions and socio-economic phases, in recent decades, many cities accumulated a large stock of vacant or under-used spaces and buildings (e.g., industrial sites and infrastructures, construction sites, and large health facilities such as mental hospitals) due to the economic and social transformation processes and policy changes [26–28]. The large amount of vacant and under-used open spaces and buildings represents both a challenge and an opportunity for developing integrated approaches to ecosystem service, for combining sustainable land use, circular economy, and cultural revival and social inclusion. These opportunities have been in some cases exploited through projects and experimentations led by citizens, community groups, private sectors, universities, public authorities or a partnership of these actors.

However, a comprehensive review of plans and projects exploring the potential of an integrated approach to ecosystem services is lacking. Similarly, while a few recent efforts have been made to illustrate what an ecosystem services planning approach might entail [29], an appraisal of the implications of the existing experiences adopting a holistic approach to ecosystem services for urban policy and planning is missing. This paper contributes to address these lacunae by reviewing initiatives and projects from relevant actions of three of the 14 Partnerships of the Urban Agenda for the EU: the Culture and Cultural Heritage Partnership and the strictly connected Sustainable Use of Land and Nature-based Solutions and Circular Economy Partnerships. The Partnerships of the Urban Agenda for the EU have been introduced in the EU programming period 2014–2020 to engage cities, Member States, the European Commission, NGOs and other stakeholders to work together on the development and implementation of Actions to address cities' challenges while contributing to the green and digital transitions objectives among others. The review addresses the types of solutions, plans, regulations, and knowledge innovations introduced for the sustainable use and management of under-used spaces within these actions.

Addressing the above mentioned gap is essential to understand how to operationalise the ecosystem services concepts and to integrating their cultural, socio-economic and environmental dimension and benefits, as well as to understand the role of governance [30] and urban planning in the supply of and access to ecosystem services [4,31,32] and in supporting their sustainable management.

The Urban Agenda for the EU aims at involving urban authorities in achieving the three pillars of EU policy-making and implementation: better regulation, better knowledge, and better funding. For the purpose of this article, are considered the first two pillars as they are connected to challenges arising from two possible scenarios as observed by the partners of the Urban Agenda.

A first scenario, as illustrated also by the Coordinator of the Culture and Cultural Heritage Partnership of the Urban Agenda for the EU between 2014 and 2021, is when there is a group of active citizens or a community group with a strong interest in management and that recognises the value of an under-used space or building, and is possibly already using it informally. In this case, the local authority acknowledges the interest and commitment and through a participatory process can assign the management of the space or building to citizens, community groups, NGOs or other no-profit organisations. In the second scenario, the local authority is not aware of the location and conditions of under-used spaces of buildings and there is a need to acquire knowledge about them, in order to identify appropriate management strategies and interventions [33,34]. Thus, the challenges are to identify both appropriate processes and regulatory instruments for spaces that are informally managed, defining urban regulation for "urban commons" and tools to map the under-used spaces, as well as building participatory decision-making processes to re-use and regenerate the spaces.

The local authority can collect, transformation and design ideas and options for under-used spaces, but in most cases doesn't have the possibility to delegate or assign the management of spaces or buildings to citizens, NGOs, associations or community groups, due to the public procurement rules and lack of alternative tools or regulation. However, there are some cases where a regulation of "common goods", as place of value and interest or identity for a community, has been created, defining innovative urban regulations, such as "civic pats", "collaboration agreements", or "citizenship agreements", for the development or reuse of public buildings and areas[1].

These cases follow a trend that is supported by scholars arguing that a well-structured participatory process, open to all the relevant stakeholders and with clear objectives of providing public services and space for common interest, can be a valid alternative to predominant and market-based approaches, such as procurement [35], and can lead to the identification of an appropriate "common good" (or urban commons) regulation for the assignment of management or service provision to a diverse range of suitable actors.

Although, there is an extensive debate on urban commons and ecosystems commons, this paper rather than engaging with this literature focus on illustrating experiences from selected case studies to understand what insights they provide on alternative regulations, governance, and knowledge production for an integrated approach to ecosystem services in under-used spaces and buildings.

2. Materials and Methods

The paper uses a qualitative case study methodology with a descriptive aim. It analyses three projects implemented under the selected Partnerships of the Urban Agenda for the EU in Poland and Spain, with the aim to understand:

a. the extent to which projects under the selected Partnerships integrate cultural, social, and environmental dimensions of ecosystem services and adopt a holistic approach
b. what knowledge and regulations have been experimented with, developed, and applied in the projects to foster the integration of the different dimensions and types of eco-system services, and
c. what insights do they provide for informed policies and planning to enhance the benefits of ecosystems.

The projects have been selected together with the Partnership coordinator and action leader and with the contractors supporting the European Commission Directorate-General for Regional and Urban Policy in the implementation of the Urban Agenda for the EU[2]. The data regarding the project's objectives, initiatives, and interventions have been collected through desk review of action plans, of handbooks and reports delivered under the relevant Partnerships and actions, of project documents, and through interviews with a Partnership coordinator and action leader of the Partnerships on Culture and Cultural heritage, and on Sustainable Land-use and Nature-based Solutions and Circular Economy of the Urban Agenda for the EU.

To address the three points above the case studies are analysed using four governance modes as categories adapted from Primmer et al., 2015 and Winkler et al., 2021 [4,5,30]: technical and knowledge governance; collaborative and participatory governance; transformational governance and regulatory governance. These categories are linked to four corresponding challenges: the knowledge challenge, engagement and empowerment challenge, and policy and regulation challenges (Table 1).

Table 1. Governance modes adapted from Primmer et al., 2015 and Winkler et al., 2021 and linked to ecosystem services integration challenges.

Governance Mode	Ecosystem Service Integration Challenge
Technical and knowledge governance	Knowledge challenge
Collaborative and participatory governance	Engagement and empowerment challenge
Transformational governance	Policy challenge
Hierarchical/Regulatory governance	Regulation challenge (commons)

This analytical framework aims at providing an insight into different strategies and initiatives for integrated ecosystem services in under-used used spaces and buildings. It allows to investigate, first, how cultural, environmental, and socio-economic dimensions are integrated in the analysed projects and what form of governance was adopted; second, how the projects and interventions addressed the knowledge and participatory challenges, particularly in terms of community engagement and empowerment, and the regulation challenge in terms of the regulation of commons.

The following session presents the results from the review of the projects of selected Partnerships and related actions of the Urban Agenda for the EU: the Partnerships on Sustainable Use of Land and Nature-based Solutions and that on Circular Economy, and the Culture and Cultural Heritage Partnership.

3. Integrating Sustainable Land Use, Circular Economy and Cultural and Natural Heritage

The interview with the Leader of Actions 3 and 9[3] of the Urban Agenda Partnerships on Sustainable Use of Land and Nature-based Solutions and the Circular Economy provided insights on projects and initiatives at the interfaces between the identification and management of under-used land and the re-use of public spaces and building under the circular economy framework (Catalonia). The "Handbook Sustainable & Circular Re-Use of Spaces & Buildings" [36] prepared under the two complementary actions of the Partnerships mentioned above highlighted that the re-use and revitalisation of under-used urban areas can be fostered through appropriate laws or policy, or through specific programmes targeting the re-use and management of spaces and buildings. Cities and urban authorities engaged in the development and implementation of solutions for the circular reuse of space and buildings and the transition towards circular economy, and in exploring methods for mapping and collecting best practices in the management and activation under-used land and spaces.

The activation of under-used land and the land recycling [37], including the redevelopment of previously developed land, ecological upgrading of green urban areas, and re-naturalisation of land, is closely linked to the sustainable re-use of brownfields and former industrial sites. This circular re-use of land in urban or peri-urban areas aims at avoiding land take and preventing urban sprawl by providing a competitive alternative to the development of greenfields.

For activating under-used spaces to exploit their potential for reducing land take and soil sealing, it is essential that public and private actors and stakeholders establish a collaborative partnership and that relevant actors (from both public and private sectors) have adequate data on under-used land and on how it could be designed, both for temporary uses, and long-term and permanent land uses. This knowledge is strictly linked to cultural values and local culture and built, natural, and cultural heritage as triggers for the identification of sustainable and inclusive solutions and for community engagement in design and implementation phases.

In this perspective, the interview with the Coordinator of the Partnership on Culture and Cultural Heritage, provided experiences on the revitalisation, management and activation of citizens and social groups through cultural heritage, including cultural services in under-used areas (Murcia), with a focus on Actions 2 and 3[4] of the Partnership.

The following section, integrating inputs form the Action Leader of the Sustainable Use of Land and Nature-based Solutions and the Circular Economy, and from the Coordinator of the Partnership on Culture and Cultural Heritage, with data from EU reports and projects' documents review, illustrates the implementation of strategies and solutions developed in Poland with the National Model for urban revitalisation and in Catalonia and Murcia in Spain.

3.1. Integrated Regeneration Programme in Poland: National Model for Urban Revitalisation

Since 2015 Poland have been adopting a new approach to urban revitalization defined as an upgrading process of derelict and under-used areas which integrates actions for the local community, spaces, and economy and activation of local stakeholders in the revitalisation program. An Act on revitalisation was adopted in 2015 and stated that municipalities should identify their derelicts and under-used areas and buildings and special revitalisation zones. Under this policy framework the Ministry of Investment and Economic Development launched the Model for urban revitalisation (regulation challenge/hierarchical-regulatory governance) subsidising projects for more than 40 mln PLN (about 9 mln EUR) to support cities in implementing innovative ideas and share knowledge on revitalization processes. The programme included a call for the proposal of models for revitalisation activities in one of the following areas of intervention: social policy, participation in public life, housing, shaping of space, environmental protection, protection of heritage, stimulation of the economy, urban mobility, and financing of revitalisation. The Ministry received about 250 applications from different municipalities with city status and selected the 23 best projects[5]. Three of the selected cities required large-scale revitalisation interventions and were provided with grants and expert support by the Ministry in implementing pilot projects on participation in public life, housing policy, shaping of space, and revitalization management.

One of the selected projects, "New Downtown of the City of Ełk formula for revitalization", focused on the participation, activation, and creation of new public spaces as three key pillars of revitalisation. The city of Elk built a model to engage residents both in the revitalization programme and related projects (participation), a Model School of Social Animator and Educators (activation) and developed the concept of Social Revitalization Centre, a programme for managing housing both addressing revitalization challenges and creating inclusive public spaces (new public spaces). The Model School of Social Animator and Educators implemented small projects in tenements' backyards such as the organisation of a parade of "Enlightened" with residents and local theatre to illustrate and promote the results of the project. The event was organised by Elk Association Active foot, Human Foundation and the Forum Institute Social Animators LEX, a consortium of NGOs. Since August 2017, the city has implemented Street activities (*streetworkerów*) and 10 mini-projects together with the residents for a value 50,000 PLN (about 10,000 Euro).

Under the framework of the project the local authority called for initiatives for increasing residents' engagement in activities for improving their living environment; for leisure, sports, culture and education, health care services and integration of seniors, children and youths social groups; for improving knowledge and awareness on environmental issues, such as green transportation, pollution and sustainable use of green areas and on the management of urban space and engaging community in urban planning decision-making (engagement and empowerment challenge/collaborative and participatory governance; regulation challenge/hierarchical-regulatory governance). In addition, corporate social responsibility (CSR) activities have been performed for building and improving relationships between business actors and local community and local authorities.

The city of Lodz with the project "On the Trail of Textile Architecture. Revitalisation of *Księży Młyn*" was selected for one of the three large scale projects under the urban revitalisation programme. The project consists in the revitalisation of the 6.5-hectare area of Priest's Mill (*Księży Młyn*) a historical settlement and conservation of 25 city-owned multi-family workmen's residences built in the 1870s and 1880s by the owner of a textile factory, as well as the conversion of 15% of the residential buildings into spaces for retail, social, and cultural activities such as dynamic artist workshops.

The project also created an original model of mediator with mediation, conflict and management competences applied to eight priority revitalization projects and for engaging residents of 700 housing units. The team of mediators first supported residents' in finding solutions to address their everyday life problems, then expanded its scope to socio-cultural animation activities for engaging and activating residents and a variety of stakeholders (e.g., administrators of infrastructure networks, artists, police), as well as widening the areas of intervention of the revitalisation's projects. Mediators also manage the involvement of employees in activities, such as cross-financing, cooperation with entrepreneurs, communities' participation in local initiatives, and the creation of a network between institutions (engagement and empowerment challenge/collaborative and participatory governance).

Together with the mediator supporting residents in change and relocation, consultation and participation activities relied on local community organisers for connecting, activating, and working with residents and institutions to create an integrated neighbourhood, and on an area manager for representing residents and institutions, and monitoring and taking part to all meetings and activities relevant to the future of local residents, workforce and organisations and enterprises. The project allowed the historic tenement houses to regain their residential function with contemporary living standards and preserve the open green public space designed in the 19th century and the original urban fabric of the area. Most of the apartments are city-owned and include communal flats, providing shelter to those who cannot afford to rent an apartment on the open market and also create social mix. The continuous and direct interaction with residents helps to prioritise those who need a house most and contributes to enhancing ownership and building trust in local authorities.

The project turned under-used areas, considered unsafe and derelict, into one of the most interesting spaces in the city due to their environment, atmosphere, and historical characteristics. It contributed to the circular re-use of vacant properties, introducing new functions and flexible uses within under-used buildings and spaces originally designed for other activities, and creating job opportunities and service delivery with a positive social impact.

The transformation is also considered an example of sustainable urban action as it preserved the urban fabric and the historical features creating "Monuments of History" always involving the citizens in the process. It illustrates the key role and potential of local residents and their knowledge and relationship with the physical and cultural heritage of the area, in the planning and implementation of revitalization activities, overall including: building renovation and modernisation, open space management; spatial transformation of residential building into spaces for productive activities; establishment of a Social Integration Club, Residents Club, social economy entities, artist workshops, programming social, artistic and cultural projects. The activities are integrated with Lodz's major initiatives and plans for the restoration of green areas and the creation of new parks (e.g., the Lososiowa street in Lodz was established in 2014, an old industrial area).

The concept of the Blue-Green Network was adopted to guide integrated planning and management of green and blue areas in Lodz, making the city's blue-green infrastructure a key pillar of Lodz's urban resilience. Ecohydrological measures were introduced as a part of the small-scale water retention programme for Łódź and the Blue-Green Network project, including new stormwater management for enhancing water retention potential of the city's waterways and a pilot project for the renaturalisation of the Sokołówka River[6] [38]. The city of Lodz also aimed at integrating biodiversity and culture in planning and managing the urban green spaces of under-used areas, the concept of biocultural diversity as reflected in the Green Circle of Tradition and Culture (GCTC) emphasises that green spaces are shaped by cultural processes as they result from historical cultural and land use practices. The way in which land was managed by different social groups and for different purposes in the past is reflected in the green areas the city has today [38].

In this perspective, green spaces and historical palaces and factories with their historical parks and rivers are examples of biocultural diversity, being used by different groups of people. Within the GCTC, the sites *Księży Młyn* (Priest's Mill), with its under-used and well-preserved factory, buildings, workers' residences, river and lake, and historical parks, is emblematic of the integration of natural and cultural heritage reflecting historical land-use [39].

3.2. Integrating Sustainable Land Use and Circular Economy in Catalonia, Spain

INCASÓL, the Catalan Land Institute of the Government of Catalonia, implemented several initiatives and programmes for the regeneration of vacant industrial sites, revival and protection of cultural, built and natural heritage, and green infrastructure. They included the support administrations in the design and implementation of an Urban Redevelopment Programme; management and implementation of large regeneration projects of vacant or partially vacant industrial sites (Industrial Colonies), or tailored interventions on specific sites mostly owned and managed by the Ministry of Planning and Sustainability.

3.2.1. Catalonia Urban Redevelopment Programme

The Catalan Land Institute, Incasòl (*Institut Català del Sòl*) has supported a long-standing Urban Redevelopment Program to provide support for particularly urban development issues to administrations at different levels (regulation challenge/hierarchical-regulatory governance). The programme identifies municipalities with social problems and urban problems and, e.g., helps municipalities with high-density issues in redeveloping public spaces and buildings that are fully or partially owned by the public, transforming them into new sustainable and accessible areas. The transformation may include the relocation of some buildings or residents, or demolition activities to provide appropriate public and green spaces. All the projects take into account environmental sustainability (e.g., in terms of building material and energy efficiency) and aesthetic standards (e.g., improving street views and cityscape) and aim to provide social mix to avoid creating exclusion and segregation between neighbourhoods. The programme has been working in 19 neighbourhoods and the realisation of projects takes 4–6 years. The programme combines the regeneration of public open green spaces with the realisation of new affordable residential areas. It includes new affordable housing for rent, usually built on municipal land targeting both residents with no access to social housing or struggling in accessing the house marked.

3.2.2. Industrial Colonies: Regeneration of Colonia Sedó

Incasòl also implemented regeneration projects in two under-used or abandoned industrial settlements, the former textile factories of Colònia Güell and Colonia Sedó. They have been regenerated mixing different uses, including industrial use, housing, and green infrastructure. Colonia Sedó, covers an area of 427.46 ha and is located in the Esparraguera municipality 35 km away from Barcelona. It was built in the mid-19th century on the banks of the Llobregat River integrating housing, facilities and services (shops, schools, a church and social centre), and various types of infrastructure for the workers of the textile factory. The residential part of Colonia Sedó was acquired by Incasòl (i.e., Government of Catalonia) in 2003 and the Colony is now a 44% public owned and 56% private site owned by several small and medium-sized companies. To address the obsolescence, specialization, and segregation issues of the Colony site, Incasòl has worked together with the City Council to develop a strategy (regulation challenge/hierarchical-regulatory governance) for the rehabilitation of the residential complex and of the old Fonda's building, the old inn used to provide meals and accommodation to residents. Colonia Sedó was also part of the EUROPAN project[7] on "Reinventing Rurality and Productive Heritage" with the aim to exploit its significant historical-heritage value and the role of the natural environment and resources, such as water, to integrate and connect the industrial urban fabric with the cultural heritage and the sustainability and circular economy dimension. Environmental and heritage-related objectives have been combined through initiatives for renewable energies and the sustainable use of water resources and old hydraulic infrastructure, for the integration of urban agriculture into the open spaces in residential areas, for enhancing the connectivity between the existing urban fabric and surrounding areas of natural interest (e.g., protected natural areas of the Natural Area Plan and under the Habitats and Birds Directive Sites Montserrat-Roques Blanques-riu Llobregat), and for integrating conservative actions of historic heritage-listed assets (e.g., old aqueduct listed in the Inventory of Catalan Architectural Heritage) with the realisation of new buildings and uses [40].

3.3. Integrating Cultural and Natural Heritage Management and Revitalisation in Murcia, Spain

The Municipality of Murcia, in Spain, led Actions 2 and 3 of the Culture and Cultural Heritage Partnership of the Urban Agenda for the EU, aiming at the revitalisation, management, and activation of culture and cultural sectors, public spaces, and cultural heritage, including cultural ecosystem services in under-used areas. The actions refer to the two possible scenarios and related challenges described in Section 1.2. They include challenges of identifying both analytical tools and participatory decision-making processes, and regulatory instruments for under-used spaces and buildings that are informally managed.

Murcia has been a really active administration on Culture and Heritage, Murcia City Council implemented the Urban DNA initiative for the economic and social revitalisations of three neighbourhoods: El Carmen, Sta. Eulalia, La Paz. The initiative is part of the City Strategy 2020 aimed at the revitalisation and regeneration of neighbourhoods based on the specific physical, social, and cultural peculiarities and identity of each neighbourhood.

Urban DNA includes physical interventions and social innovation actions to improve the urban environment and social relations and design new services within the neighbourhoods, while introducing circular re-use of space. A broad range of actors, including almost all the departments of the Municipality of Murcia (urban development, participation, public works, decentralization, institutional relations, urban mobility, urban security, environment, health, economy, culture, development of European projects), universities, private sector, local associations, community groups, and residents, were involved in a five-step process, representing a governance model for the reuse of spaces and buildings.

Phase 1 was aimed at understanding the contexts and focused on multiple spatial and socio-analyses of the neighbourhoods, resulting in a series of urban atlases (knowledge challenge/technical and knowledge governance). Phases 2 and 3 of co-design consisted of participatory needs assessment and community activation respectively (engagement and empowerment challenge/collaborative and participatory governance). Residents and different social groups and stakeholders first engaged in various group works, analyses, and debates to integrate the outputs of Phase 1 with inputs, ideas or project proposals developed in phase 2, and then, in a second stage of co-design, provided with technical support of the local authority in the implementation of the proposed social, cultural, urban interventions (Phase 3). In Phase 4, citizens and community groups and relevant public bodies worked together to realise the interventions defined in the previous phases (e.g., planting trees or improvement of public spaces and urban infrastructure). In Phase 5, the Municipality of Murcia assigns the management of the project and of the space to a designated association and neighbourhood group, while the urban authority remains responsible for exceptional maintenance of the interventions (regulation challenge/hierarchical-regulatory governance).

Following this governance model, projects and interventions integrated socio-economic, cultural, and environmental dimensions of ecosystem services. In El Carmen neighbourhood, some of the interventions included the realisation of new garden, Painter Pedro Flores, a green space of 2500 m^2, with improved accessibility and children's innovative playground, landscaping and tree plantation, and irrigation infrastructure.

4. Results

From the review of the three cases, it emerges that all the cities adopt an integrated approach combining cultural, environmental, and economic and social dimensions of their intervention, directly or indirectly enhancing the benefits of built, cultural, and natural heritage, as well as the circular re-use of spaces, buildings, and urban ecosystems in under-used areas.

Lodz and Elk in Poland are also examples of the integration of affordable housing, regeneration, as well as the revitalization and re-use of open public space. The Catalonia redevelopment programme and factory regeneration integrates social dimensions with the environmental dimensions, linking access to housing with access to cultural and environmental services (e.g., demolition in dense urban areas for creation of green public spaces). However, some issues, such as nature-based solutions and climate change, are still moderately integrated in the projects, while priority is given to the cultural, aesthetic, and economic dimensions of the interventions.

All the cases studies widely address the engagement and empowerment challenge by proposing different modes of collaborative and participatory governance. The case of Murcia shows an adaptive collaborative governance approach focused on collective learning and bottom-up participation, originating from actors who affect or are affected by the conditions of ecosystems and under-used spaces and services. This approach also translated into a "learning by doing" process with changes and improvements made during the project implementation (e.g., issues in the coordination between companies and municipal services led to introducing changes intensifying coordination efforts).

In all the cases studies, the knowledge challenge is also addressed, collecting and operationalizing knowledge on the project site/s and community aspirations for the transformation of the space (technical and knowledge governance). However, only Murcia and Lodz define a specific implementation process for integrating technical and community knowledge to strengthen the collaboration of public authorities, citizens, community groups, the business sector, and other relevant stakeholders in decision-making (Murcia), and create dedicated roles (mediator in Lodz) to facilitate and manage the interventions. In Murcia, the citizens' engagement, and collaboration between local administration and residents, were crucial for the design and implementation of the interventions. The Urban DNA initiatives provided opportunities to raise awareness and enhance community knowledge and planning capacity on the circular reuse of public spaces and buildings. The direct engagement and co-design with citizens, community groups, research and education institutions, as well as the private sector, was directed to help to attribute value to local, place-based, social, cultural, economic and environmental features and generate opportunities for economic development, social innovation, and environmental sustainability.

Although the analysed projects introduce procedural and organisational changes, they do not develop specific new regulatory mechanisms or regulation of commons. In the case of Catalonia, it emerges how the state actors and formal institutions play a major role in project design and decision-making. There are hierarchical, vertical interactions and collaboration (hierarchical/regulatory governance) and technical support, involving Incasól and the Ministry of Planning and Sustainability, and the city administrations in the development of regeneration and re-use strategies.

The case studies provide limited evidence of transformational governance mode, activities, or strategies for catalysing the behavioral change of relevant actors and policy changes are not integrated in the projects and programmes, limiting the scope of the ecosystem integration approaches.

5. Discussion and Conclusions

The results show that there is a strong integration of cultural, environmental and socio-economic dimensions, with an emphasis on the cultural value, environment and ecosystems, and land recycling. However, the focus on the role ecosystem services is still partially explored while more attention is given to the urban regeneration and brownfields redevelopment rather than re-naturalisation.

The project interventions address the knowledge and participatory challenges, providing different solutions and strategies, however the regulation and policy dimensions are mainly limited to the collaboration among different public institutions and levels of governance. Alternative tools and regulations, such as common good regulations, are little explored and implemented in the Partnerships. Future research, both case studies and theoretical reviews, concerning alternative processes and forms to assign the management of common spaces and services to communities and social groups that are already engaged or interested in management of ecosystems in under-used spaces, would provide valuable insights for innovation regulations and urban and land-use planning.

The projects illustrate the key role of the public authorities and institutions in addressing the sustainable challenges where the market has no interest to intervene. However, this may generate constraints for a bottom-up approach to effectively engage and empower segregated social groups and citizens, underpinning gentrification.

Concerning the relevance of the results for urban and land-use planning and management, it should be considered that the review of the case studies is limited to three experiences from the Partnerships of the Urban Agenda for the EU, and that most of the projects focused on the re-use of public spaces and building. A systematic review of additional experiences within and beyond the EU Partnerships, also including the exploration of the potential of the re-use of private spaces and buildings, would be very relevant.

In addition, further research, including quantitative and detailed data on funds and resources of the analysed projects and programmes, on the dimension, land use, and land cover of the areas of interventions, on size and composition of the governance networks and structures, and on expected or measured social and economic impacts and targets, would provide significant insights concerning innovative urban planning, useful to operationalise an integrated approach to ecosystem services in Europe.

Funding: This research received no external funding.

Informed Consent Statement: Not applicable.

Acknowledgments: The data and information included in this paper are partially based on research undertaken as follow-up of a collaboration with Ecorys Brussels NV, on the drafting of the article 10th, part of a series of articles based on the 14 Partnerships of the Urban Agenda for the EU "The Green City Dimension Reconquering public spaces by interlinking design, inclusion, and sustainability". The collaboration integrated information provided by Cristina Clotet Ollé Action Leader of Actions SL03 and CE09 from the Catalan Land Institute INCASÒL, and by Sandra Gizdulich from the Italian Agency for the Territorial Cohesion.

Conflicts of Interest: The authors declare no conflict of interest.

Notes

1. Some examples of this innovative urban regulations are in the Emilia Romagna Region in Italy where, in line with regional Law, the Municipality of Reggio Emilia is experimenting "citizenship agreements" and co-design actions for the reuse of urban assets, spaces and buildings; and in the city of Naples, Turin and Salerno, where the local authorities are developing local regulations to introduce "collaboration pacts" as a tool to assign the management of spaces or buildings to citizens or community groups following a specific procedure to include temporary uses of the "common goods" and require that management objectives are in line with the public interest and spaces and activities are open to the public.
2. Framework Contract 'Support to the implementation of the Urban Agenda for the EU through the provision of management, expertise, and administrative support to the Partnerships', signed between the European Commission (Directorate-General for Regional and Urban Policy) and Ecorys.
3. The interview with the Action Leader of Actions 3 "Identifying and managing under-used land of the Sustainable use of land and nature-based solutions" and Action 9 "Manage the re-use of buildings and spaces in a circular economy under the complementary", of the Urban Agenda Partnerships on Sustainable Use of Land and Nature-based Solutions and Circular Economy, in addition to the projects in the selected case studies, illustrated the complementarity and the collaboration within the two actions.

4 Action 2 is "Partnership focus on Street Invasion, Atomisation and Cultural Reactivation", and Action 3 is "Cultural Hubs for Innovation, Modernisation and Enhancement".

5 The 23 City selected in the programme are Chorzów, Dabrowa Górnicza, Dobiegniew, Ełk, Grajewo, Hrubieszów, Konin, Leszno, Lublin, Milicz, Opole Lubelskie, Rybnik, Słupsk, Stalowa Wola, Starachowice, Szczecin, Warszawa, Włocławek, Wrocław, Żyrardów, Bytom, Łódź, Wałbrzych.

6 The Pilot project is part of the EU SWITCH programme (Sustainable Water Management Improves Tommorow's Cities' Health) https://cordis.europa.eu/project/id/18530 (accessed on 4 November 2021).

7 Information about the EUROPAN Project and Colonia Sedó are accessible at: https://www.europan-europe.eu/en/session/europan-16/site/esparreguera-colonia-sedo-es#specific-documents (accessed on 4 November 2021).

References

1. Geneletti, D.; Adem Esmail, B.; Cortinovis, C.; Arany, I.; Balzan, M.; van Beukering, P.; Bicking, S.; Borges, P.A.; Borisova, B.; Broekx, S.; et al. Ecosystem services mapping and assessment for policy- and decision-making: Lessons learned from a comparative analysis of European case studies. *One Ecosyst.* **2020**, *5*, e53111. [CrossRef]
2. Gómez-Baggethun, E.; Barton, D.N. Classifying and valuing ecosystem services for urban planning. *Ecol. Econ.* **2013**, *86*, 235–245. [CrossRef]
3. Lennon, M.; Scott, M. Delivering ecosystems services via spatial planning: Reviewing the possibilities and implications of a green infrastructure approach. *Town Plan. Rev.* **2014**, *85*, 563–587. [CrossRef]
4. Winkler, K.J.; Rodrigues, J.G.; Albrecht, E.; Crockett, E.T.H. Governance of ecosystem services: A review of empirical literature. *Ecosyst. People* **2021**, *17*, 306–319. [CrossRef]
5. Keune, H.; Bauler, T.; Wittmer, H. Ecosystem services governance: Managing complexity? In *Ecosystem Services: Global Issues, Local Practices*; Jacobs, S., Dendoncker, N., Keune, H., Eds.; Elsevier: Amsterdam, The Netherlands, 2014; pp. 135–155.
6. Costanza, R.; d'Arge, R.; de Groot, R.S.; Farber, S.; Grasso, M.; Hannon, B.; Limburg, K.; Naeem, S.; O'Neill, R.V.; Paruelo, J.; et al. The value of the world's ecosystem services and natural capital. *Nature* **1997**, *387*, 253–260. [CrossRef]
7. Daily, G.C.; Soderquist, T.; Aniyar, S.; Arrow, K.; Dasgupta, P.; Ehrlich, P.R.; Folke, C.; Jansson, A.M.; Jansson, B.O.; Kautsky, N.; et al. The value of nature and the nature of value. *Science* **2000**, *289*, 395–396. [CrossRef] [PubMed]
8. De Groot, R.S. *Functions of Nature: Evaluation of Nature in Environmental Planning, Management and Decision Making*; Wolters-Noordhoff: Groningen, The Netherlands, 1992.
9. Hølleland, H.; Skrede, J.; Holmgaard, S.B. Cultural Heritage and Ecosystem Services: A Literature Review. *Conserv. Manag. Archaeol. Sites* **2017**, *19*, 210–237. [CrossRef]
10. COP5 (Fifth Conferences of Parties) of Convention of Biological Diversity (2000) COP 5 Decision V/6: Ecosystem Approach. Available online: http://www.cbd.int/decision/cop/?id=7148 (accessed on 4 July 2022).
11. Millennium Ecosystem Assessment. *Ecosystems and Human Well-Being: Policy Responses*; Island Press: Washington, DC, USA, 2005.
12. Reid, W.V.; Mooney, H.A.; Cropper, A.; Capistrano, D.; Carpenter, S.R.; Chopra, K.; Dasgupta, P.; Dietz, T.; Duraiappah, A.K.; Hassan, R.; et al. *Ecosystems and human well-being—Synthesis: A Report of the Millennium Ecosystem Assessment*; Island Press: Washington, DC, USA, 2005. Available online: https://edepot.wur.nl/45159 (accessed on 4 July 2022).
13. Carpenter, S.R.; Mooney, H.A.; Agard, J.; Capistrano, D.; Defries, R.S.; Díaz, S.; Dietz, T.; Duraiappah, A.K.; Oteng-Yeboah, A.; Pereira, H.M.; et al. Science for managing ecosystem services: Beyond the Millennium Ecosystem Assessment. *Proc. Natl. Acad. Sci. USA* **2009**, *106*, 1305–1312. [CrossRef] [PubMed]
14. Haines-Young, R.H.; Potschin, M.B. The links between biodiversity, ecosystem services and human wellbeing. In *Ecosystem Ecology: A New Synthesis*; Raffaelli, D.G., Frid, C.L.J., Eds.; Cambridge University Press: Cambridge, UK, 2010.
15. James, P.; Tzoulas, K.; Adams, M.D.; Barber, A.; Box, J.; Breuste, J.; Elmqvist, T.; Frith, M.; Gordon, C.; Greening, K.L.; et al. towards an integrated understanding of green space in the European built environment. *Urban For. Urban Green.* **2009**, *8*, 65–75. [CrossRef]
16. Vasiljevic, N.; Gavrilovic, S. Cultural Ecosystem Services. In *Life on Land. Encyclopedia of the UN Sustainable Development Goals*; Leal Filho, W., Azul, A., Brandli, L., Özuyar, P., Wall, T., Eds.; Springer: Cham, Switzerland, 2019. [CrossRef]
17. *European Landscape Convention*; European Treaty Series, No. 176; Council of Europe: Firenze, Italy, 2000.
18. Haines-Young, R. Report of Results of a Survey to Assess the Use of CICES, 2016; Support to EEA Tasks under the EU MAES Process. 2016. Available online: https://cices.eu/content/uploads/sites/8/2016/07/Report-on-Survey-Results_19072016_Upload.pdf (accessed on 4 July 2022).
19. Ryfield, F.; Cabana, D.; Brannigan, J.; Crowe, T. Conceptualizing 'sense of place' in cultural ecosystem services: A framework for interdisciplinary research. *Ecosyst. Serv.* **2019**, *36*, 100907. [CrossRef]
20. Wang, S.; Fu, B.; Wei, Y.; Lyle, C. Ecosystem services management: An integrated approach. *Curr. Opin. Environ. Sustain.* **2013**, *5*, 11–15. [CrossRef]
21. Nassauer, J.I.; Raskin, J. Urban vacancy and land use legacies: A frontier for urban ecological research, design, and planning. *Landsc. Urban Plan.* **2014**, *125*, 245–253. [CrossRef]
22. Bryan, B.A.; Raymond, C.M.; Crossman, N.D.; Macdonald, D.H. Targeting the management of ecosystem services based on social values: Where, what, and how? *Landsc. Urban Plan.* **2010**, *97*, 111–122. [CrossRef]

23. Raymond, C.M.; Bryan, B.A.; MacDonald, D.H.; Cast, A.; Strathearn, S.; Grandgirard, A.; Kalivas, T. Mapping community values for natural capital and ecosystem services. *Ecol. Econ.* **2009**, *68*, 1301–1315. [CrossRef]
24. De Vreese, R.; Leys, M.; Dendoncker, N.; Van Herzele, A.; Fontaine, C.M. Images of nature as a boundary object in social and integrated ecosystem services assessments. Reflections from a Belgian case study. *Ecosyst. Serv.* **2016**, *22*, 269–279. [CrossRef]
25. Hall, P.; Hay, D. *Growth Centres in the European Urban System*; Heinemann Educational Books: London, UK, 1980.
26. Turok, I.; Mykhnenko, V. The trajectories of European cities, 1960–2005. *Cities* **2007**, *24*, 165–182. [CrossRef]
27. Wolff, M.; Wiechmann, T. Urban growth and decline: Europe's shrinking cities in a comparative perspective 1990–2010. *Eur. Urban Reg. Stud.* **2018**, *25*, 122–139. [CrossRef]
28. European Commission. The State of European Cities 2016: Cities Leading the Way to a Better Future. European Commission (Directorate-General for Regional and Urban Policy) and UN Habitat. 2017. Available online: https://ec.europa.eu/regional_policy/sources/policy/themes/cities-report/state_eu_cities2016_en.pdf (accessed on 4 July 2022).
29. Scott, M.; Lennon, M.; Haase, D.; Kazmierczak, A.; Clabby, G.; Beatley, T. Nature-based solutions for the contemporary city/Re-naturing the city/Reflections on urban landscapes, ecosystems services and nature-based solutions in cities/Multifunctional green infrastructure and climate change adaptation: Brownfield greening as an adaptation strategy for vulnerable communities?/Delivering green infrastructure through planning: Insights from practice in Fingal, Ireland/Planning for biophilic cities: From theory to practice. *Plan. Theory Pract.* **2016**, *17*, 267–300.
30. Primmer, E.; Jokinen, P.; Blicharska, M.; Barton, D.N.; Bugter, R.; Potschin, M. Governance of ecosystem services: A framework for empirical analysis. *Ecosyst. Serv.* **2015**, *16*, 158–166. [CrossRef]
31. Jax, K.; Furman, E.; Saarikoski, H.; Barton, D.N.; Delbaere, B.; Dick, J.; Duke, G.; Görg, C.; Gómez-Baggethun, E.; Harrison, P.A.; et al. Handling a messy world: Lessons learned when trying to make the ecosystem services concept operational. *Ecosyst. Serv.* **2018**, *29 Pt C*, 415–427. [CrossRef]
32. Potschin-Young, M.; Haines-Young, R.; Görg, C.; Heink, U.; Jax, K.; Schleyer, C. Understanding the role of conceptual frameworks: Reading the ecosystem service cascade. *Ecosyst. Serv.* **2018**, *29*, 428–440. [CrossRef]
33. Urban Agenda for the EU Circular Economy Action Plan. Partnership on Circular Economy. 2018. Available online: https://ec.europa.eu/futurium/en/system/files/ged/ua_ce_action_plan_30.11.2018_final.pdf (accessed on 20 October 2021).
34. Urban Agenda for the EU Sustainable Use of Land and Nature-Based Solutions Partnership Action Plan. Partnership on Sustainable Land Use. 2018. Available online: https://ec.europa.eu/futurium/en/system/files/ged/sul-nbs_finalactionplan_2018.pdf (accessed on 20 October 2021).
35. Foster, S.R.; Iaione, C. The City as a Commons. *Yale Law Policy Rev.* **2016**, *34*, 281–349. [CrossRef]
36. Handbook—Sustainable and Circular re-use of spaces and buildings. Partnership on Circular Economy and Sustainable Land Use. 2020. Available online: https://ec.europa.eu/futurium/en/circular-economy/handbook-sustainable-and-circular-re-use-spaces-and-buildings.html (accessed on 4 July 2022).
37. European Environment Agency; Louwagie, G. *Land Recycling in Europe: Approaches to Measuring Extent and Impacts*; EU Publications Office: Luxembourg, 2016.
38. Szpakowska-Loranc, E.; Matusik, A. Łódź—Towards a resilient city. *Cities* **2020**, *107*, 102936. [CrossRef]
39. Davies, C.; Hansen, R.; Rall, E.; Pauleit, S.; Lafortezza, R.; Bellis, Y.; Santos, A.; Tosics, I. Green Infrastructure Planning and Implementation—The Status of European Green Space Planning and Implementation Based on an Analysis of Selected European City-Regions; 2015. Report 5.1. Available online: https://ign.ku.dk/english/green-surge/rapporter/D5_1_Green_Infrastructure_Planning_and_Implementation1.pdf (accessed on 4 July 2022).
40. Vilanova, A.; Moya, S. Plan Director De La Colonia Sedó en Esparreguera. Documento General (I y II). 2005. Serveis Tècnics Esparreguera. Available online: https://drive.google.com/drive/folders/1ZJ5DKE736Vpifsq3i7KSSFwkVtvyE0d9 (accessed on 20 October 2021).

land

Article

Spatiotemporal Patterns and Drivers of the Carbon Budget in the Yangtze River Delta Region, China

Qi Fu [1,2,3,†], Mengfan Gao [1,†], Yue Wang [1], Tinghui Wang [1], Xu Bi [4,*] and Jinhua Chen [1,2,3,*]

1 School of Politics and Public Administration, Soochow University, Suzhou 215123, China; fuqi@suda.edu.cn (Q.F.); 20194202061@stu.suda.edu.cn (M.G.); 20204202067@stu.suda.edu.cn (Y.W.); 20204002013@stu.suda.edu.cn (T.W.)
2 The Institute of Regional Governance, Soochow University, Suzhou 215123, China
3 Research Institute of Metropolitan Development of China, Soochow University, Suzhou 215123, China
4 College of Resources and Environment, Shanxi University of Finance and Economics, Taiyuan 030006, China
* Correspondence: bixu@mail.bnu.edu.cn (X.B.); jhchen@suda.edu.cn (J.C.)
† These authors contributed equally to this work.

Abstract: Improving our understanding of the patterns and drivers of regional carbon budgets is critical to mitigating climate change regionally and globally. Different from previous research, our study attempts to reveal the comprehensive impact of climate change and human activities factors on the carbon budget. Based on the Carnegie–Ames–Stanford approach (CASA) model, the IPCC inventory method, the ordinary least squares (OLS) regression model, the Geodetector model, and the geographically weighted regression (GWR) method, we investigated the spatiotemporal patterns of the carbon budget in the Yangtze River Delta (YRD) region from 2000 to 2015 and analyzed the effects of climate change and human activities on the carbon budget. The results showed that the carbon budget in the YRD region changed from 271.33 million tons in 2000 to −1193.76 million tons in 2015. During this period, the changes in the carbon budget per unit area in the four provinces all showed a decreasing trend, among which Shanghai decreased the most, followed by Jiangsu, Zhejiang and Anhui. In terms of spatial pattern, the carbon budget of the YRD region has a "core-edge" structural feature. The closer it is to Shanghai, the core area, the more severe the carbon budget deficit; the farther from it, the greater the carbon budget surplus. Overall, we found that human activities have a greater impact on the carbon budget than climate change. The top three drivers were, in order, changes in population density, GDP per capita, and unused land, with q values of 0.3317, 0.1202, and 0.0998, respectively. Locally, the impact of the drivers on the carbon budget shows obvious spatial heterogeneity. In particular, the population density was negatively correlated with carbon budget changes in the entire study area, and the coefficients of GDP per capita and unused land were negative in most counties. Based on the results, we put forward suggestions for restricting population flow among the core area and the peripheral area, promoting industrial innovation in the core area and ecological protection in the peripheral area, as well as implementing three-dimensional space development in the core area and controlling the expansion of construction land in the peripheral area. Our study can provide a scientific basis for low-carbon development in the YRD region. The methodology and findings of this study can provide references for similar studies in other urbanized regions around the world.

Keywords: carbon budget; spatiotemporal patterns; drivers; Yangtze River Delta

Citation: Fu, Q.; Gao, M.; Wang, Y.; Wang, T.; Bi, X.; Chen, J. Spatiotemporal Patterns and Drivers of the Carbon Budget in the Yangtze River Delta Region, China. *Land* **2022**, *11*, 1230. https://doi.org/10.3390/land11081230

Academic Editors: Luca Congedo, Francesca Assennato and Michele Munafò

Received: 6 July 2022
Accepted: 1 August 2022
Published: 3 August 2022

1. Introduction

Since the industrial revolution, the global economy has developed rapidly, and various energy sources have been widely used. The resulting emissions of greenhouse gases (including CO_2, CH_4, and N_2O), especially CO_2, have led to the gradual acceleration of the rise in the global average temperature over the past 200 years [1–3]. The need to mitigate climate change has become a global consensus [4]. Terrestrial ecosystems have a strong carbon sequestration function, and terrestrial vegetation can convert and store atmospheric CO_2 into organic matter through photosynthesis, playing a key role in mitigating global warming [5]. Many studies have demonstrated that carbon sequestration in terrestrial ecosystems can offset substantial anthropogenic carbon emissions [6–8]. Therefore, it is critical to uncover the relationship between carbon emissions and carbon sequestration and the drivers behind them.

The carbon budget of terrestrial ecosystems refers to the difference between carbon sequestration and carbon emissions [9]. Conducting carbon budget research can help clarify the regional carbon emission reduction pressure and carbon sink potential [10–12]. Preliminary research on regional carbon budgets mainly focused on natural ecosystems, such as forests, grasslands, farmlands, and wetlands [13–16]. In recent years, with the deepening of global change research and the proposal of a low-carbon economy, carbon budget research covering natural and human social and economic activities has begun to attract academic attention [9,17]. The assessment methods of the carbon budget mainly include field surveys, empirical models, remote-sensing models, IPCC inventory methods, etc. [18]. Field surveys can be used to obtain measured data on vegetation and soil carbon densities or the net carbon exchange between ecosystems and the atmosphere. This method is suitable for studies at the scale of sample sites and ecosystems [19]. At larger spatial scales, statistical models are usually established to evaluate carbon budgets based on empirical relationships, and such studies are typified by Houghton's bookkeeping model [20]. In recent decades, with the advancement of remote sensing and GIS technology, remote-sensing models (such as the CASA model) have become an effective technical means to assess carbon sequestration [21]. The IPCC inventory method is widely used in carbon emission accounting due to its simple calculation and high practicability [22]. Combining remote-sensing models with the IPCC inventory approach can greatly improve the efficiency and accuracy of carbon budget assessments [23]. To date, researchers have carried out many carbon budget studies based on these methods in different regions of the world [21,24,25]. However, how to apply scientific research results to guide carbon management in practice is still a challenge [26].

Before developing carbon management strategies, policymakers should understand what underlying factors affect carbon budgets and where and when these impacts occur [27,28]. Climate change and human activities are considered the two main driving factors of carbon budget dynamics [21,29–32]. Climatic change affects the ecosystem carbon budget mainly by changing vegetation phenology, photosynthesis, respiration, soil moisture and evapotranspiration [10,33]. Human activities, especially the impacts of land use/land cover change (LUCC) on carbon budgets, are important factors leading to the current increase in atmospheric CO_2 concentrations [34,35]. Studies by Houghton et al. showed that from 1850 to 1990, global LUCC led to the emission of 124 Pg C into the atmosphere, of which 108 Pg C came from the reduction in forest ecosystem area [20]. In addition, some researchers have focused on other indicators of human activity, such as the impact of population growth and GDP growth on the carbon budget [9,36,37]. However, most of the current driving analysis studies only focus on one of climate change [38,39] or human activities [40,41], and comprehensive analysis of carbon budget changes caused by multiple drivers is rare.

With the rapid development of the national economy, China has become a major CO_2 emitter [42]. To mitigate global warming, China has made unremitting efforts to increase forest and grass areas and reduce energy consumption in recent years [43]. The Chinese government has pledged to peak carbon emissions by 2030 and be carbon-neutral by 2060 [44]. Located in East China, the Yangtze River Delta (YRD) region is the largest urban agglomeration and is one of the regions with the fastest urbanization [45]. In the past few decades, energy has played an important role in promoting the rapid development of the YRD region, but it has also caused a high level of carbon emissions [46]. The unbalanced regional development in this area has become increasingly prominent, and it is urgent to formulate differentiated regional development policies. To date, some researchers have carried out research on the carbon budget at the national scale in China [47]. Other researchers have conducted many studies on either carbon sequestration or carbon emissions in the YRD region [22,45,48–50]. However, studies on the spatiotemporal patterns of the carbon budget and its drivers in this region have not yet been performed.

Different from previous research, our study tried to reveal the comprehensive impact of human activities and climate change factors on the carbon budget and to narrow the gaps in carbon budget research in the YRD region. We estimated the carbon budgets of 308 county-level administrative units in the YRD region from 2000 to 2015 and explored the drivers of the spatiotemporal patterns of the carbon budget. The purpose of our study was to find scientific carbon management pathways in the YRD region. Given this challenge, three questions were addressed: (1) Is there a certain characteristic in the spatiotemporal pattern of the carbon budget in the YRD region? (2) Which drivers have a greater impact on the carbon budget and where do these impacts occur? (3) How should we carry out carbon management to reduce the imbalance of regional carbon budget? Our results provide a scientific basis for the low-carbon and sustainable development in the YRD region, and our methodology provides a reference for other rapidly urbanizing regions in the world.

2. Materials and Methods

2.1. Study Area

The YRD region includes Shanghai, Jiangsu, Zhejiang, and Anhui provinces (Figure 1), comprising 41 cities and 308 county-level administrative units (including counties and districts, which are uniformly expressed as counties in our research). The study area covers an area of 358,000 square kilometers, accounting for approximately 3.7% of China's total area. The YRD region is located in the lower reaches of the Yangtze River (114°54′–123°10′ E and 27°02′–35°08′ N). The region has abundant rainfall, with an annual precipitation of 704–2000 mm and an annual average temperature of 12.2–18.9 °C. The water system is well-developed and the vegetation is mainly subtropical evergreen broad-leaved forest.

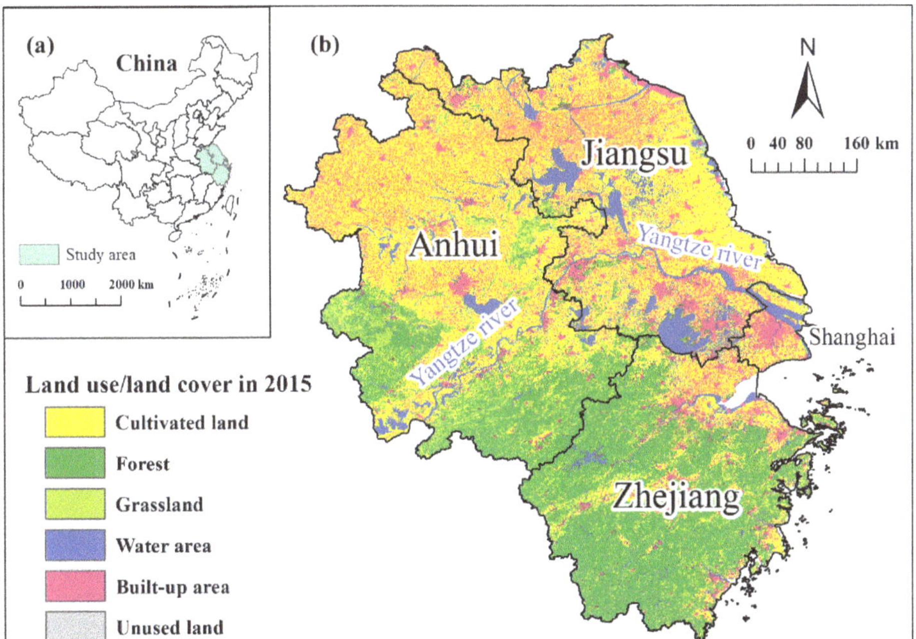

Figure 1. (**a**) The location of the YRD region in China; (**b**) the land use/land cover (LULC) pattern of the YRD region in 2015.

The YRD is one of the most densely populated areas in China, and it has the fastest urbanization process in the entire country [45]. During 2000 and 2015, the process of urban expansion led to a significant decline in ecosystem carbon sequestration [51]. However, carbon emissions in the YRD region continued to increase during this period. Researchers confirmed that the coupling coordination between carbon sequestration and carbon emission in this region was in an imbalanced state [52]. As of 2015, carbon sequestration had offset less than 5% of carbon emissions in eight core cities in the YRD region [53].

2.2. Data Sources

The data required to assess the carbon sequestration, carbon emissions, and drivers are mainly meteorological data, land use data, normalized difference vegetation index (NDVI), population density data, gross domestic product (GDP) per capita data, administrative boundary data, nighttime light data, and energy consumption data (Table 1).

Table 1. Data used in this study.

Data Name	Data Type	Year	Source
County administrative boundaries in the YRD region	Vector data	2015	Resource environment data cloud platform (http://www.resdc.cn/Default.aspx accessed on 30 May 2021)
Monthly total precipitation	Site data	2000, 2015	China Meteorological Data Network (http://data.cma.cn/ accessed on 5 June 2021)
Average temperature	Site data	2000, 2015	China Meteorological Data Network (http://data.cma.cn/ accessed on 5 June 2021)
Solar radiation	Site data	2000, 2015	China Meteorological Data Network (http://data.cma.cn/ accessed on 15 June 2021)
Normalized difference vegetation index	1 km × 1 km Raster data	2000, 2015	Resource environment data cloud platform (http://www.resdc.cn/Default.aspx accessed on 2 July 2021)
Land use/land cover	30 m × 30 m Raster data	2000, 2015	Resource environment data cloud platform (http://www.resdc.cn/Default.aspx accessed on 5 July 2021)
Population density	1 km × 1 km Raster data	2000, 2015	Resource environment data cloud platform (http://www.resdc.cn/Default.aspx accessed on 5 July 2021)
GDP per capita	1 km × 1 km Raster data	2000, 2015	Resource environment data cloud platform (http://www.resdc.cn/Default.aspx accessed on 5 July 2021)
Energy emission data	Text data	2000, 2015	Statistical yearbook of CNKI (https://data.cnki.net/Yearbook accessed on 6 August 2021)
Night lights data	1 km × 1 km Raster data	2000, 2015	Harvard University Database (https://dataverse.harvard.edu/dataset.xhtml?persistentId=doi:10.7910/DVN/YGIVCD accessed on 8 October 2021)

2.3. Methods

The analytical workflow for this study consisted of three key steps. First, we used the CASA model, the IPCC inventory method and the PSO-BP neural network model to evaluate the carbon sequestration and carbon emissions in the Yangtze River Delta in 2000 and 2015, and then calculated the carbon budget. Second, we investigated the spatiotemporal pattern of the carbon budget based on statistical analysis and GIS spatial analysis. Third, we used the OLS model to eliminate the variables with multicollinearity from multiple potential driving factors and analyzed the driving forces from the global and local perspectives based on the Geodetector model and the GWR model, respectively. The flow chart of this study is shown in Figure 2.

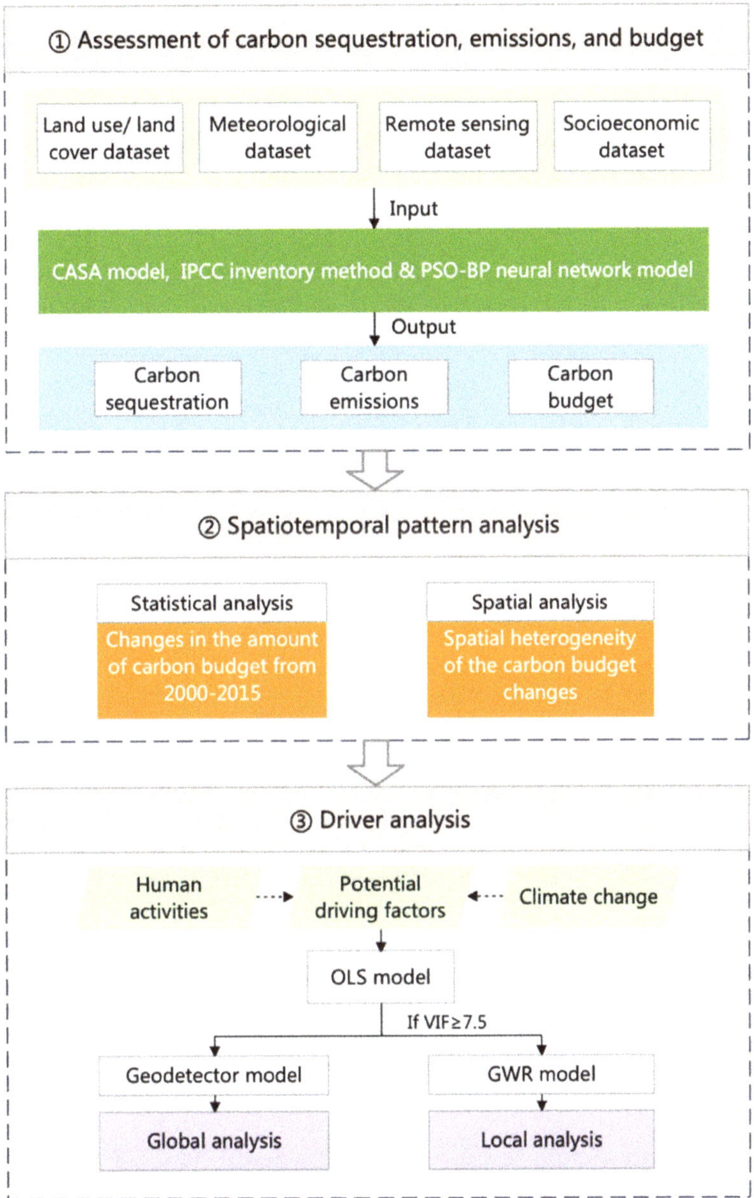

Figure 2. The flow chart of this study.

2.3.1. Assessment of Carbon Sequestration, Emissions, and Budget

Carbon Sequestration

The CASA model was used to estimate the *NPP* in the YRD region [54], and then, the amount of carbon sequestration in vegetation was obtained. The *NPP* estimated in the model can be determined from the photosynthetically active radiation absorbed by the plants $(APAR)$ and the actual photosynthetic efficiency (ε):

$$NPP(x,t) = APAR(x,t) \times \varepsilon(x,t) \tag{1}$$

In Equation (1), $APAR(x,t)$ is the photosynthetically active radiation absorbed by pixel x in month t (MJ·m^{-2}), and $\varepsilon(x,t)$ is the actual light energy utilization rate of pixel x in month t (MJ·m^{-2}). For the specific calculation, we referred to the research of Yang et al. [55].

According to vegetation photosynthesis, every 1 g of dry matter produced by the ecosystem can absorb 1.62 g of CO_2, and the carbon content of dry matter accounts for approximately 45% of the total NPP [56]. Therefore, the equation for calculating carbon sequestration (CS) is as follows:

$$CS = (NPP \div 0.45) \times 1.62 \tag{2}$$

Carbon Emissions

In previous studies, the total CO_2 emission data for China were usually calculated by referring to the method of calculating the pollutant emission coefficient described in the national Greenhouse Gas Inventory Guidelines published by the IPCC in 2006. This method has high practicality and applicability, and it has been widely used to calculate energy and fuel emissions [57–59]. In this paper, the method of Han et al. [22] was used. The specific formula are as follows:

$$AE = \sum_{m,n,t} (EC_{m,n,t} \times NCV_{m,n} \times EF_{m,n}) \tag{3}$$

where m, n, and t are the industry, fuel type, and year of investigation, respectively; AE is the carbon emissions (in metric tons), and EC is the energy consumption (in metric tons). NCV is the net calorific value (MWh/t) and EF is the CO_2 emission factor (t CO_2/MWh). The details of NCV and EF refers to the study of Han et al. [22]. Due to the lack of energy consumption data for some counties in the YRD region, in our study, a particle swarm optimization–back propagation (PSO-BP) neural network model was used to simulate the carbon emissions of counties with missing data using the MATLAB platform with reference to Chen et al. [60]. To prevent overfitting, the MATLAB platform uses the method of dividing the data into three parts: training, validation, and testing. Only the training data were used in the training step, and the other two datasets that were not used in the training were used for testing. The collected carbon emission data for the counties were used as training data and testing data, and the counties with missing data were predicted and simulated. The training sample size was 70% of the output layer data in the different years. We achieved good simulation results (see Supplementary Figure S1). The PSO-BP neural network codes used in this study are shown in the Supplementary File (Code).

Carbon Budget

Zhao et al. [61] defined the regional carbon budget as the comparative relationship and balance between carbon sequestration and carbon emissions caused by all natural and man-made activities in a certain region in a specific period of time. Our study calculated the carbon budget with reference to Li et al. [62], and the formula is as follows:

$$CB = CS - AE \tag{4}$$

In Equation (4), CB is the amount of carbon budget, CS is the amount of carbon sequestration, and AE is the amount of carbon emissions. If the value of CB is positive, it means that the carbon budget is in surplus; if the value of CB is negative, it means that the carbon budget is in deficit.

2.3.2. County-Level Data Analysis

Using the county level as the study unit is conducive to a more specific analysis of the temporal and spatial characteristics of regional carbon budgets and the formulation of more targeted emission reduction measures. Based on the Zonal Statistics tool of ArcGIS 10.8, our study counted and analyzed the values of the carbon budget and drivers at the county level. Then, we further divided the county-level statistical values into six categories to better demonstrate their evolution characteristics.

2.3.3. Driver Analysis

Referring to the research by Sun et al. [63] and after considering the natural status, development characteristics, and data availability of the YRD region, at the county scale, we selected 11 indicators as the potential drivers of carbon budget changes. The drivers included climate change (average temperature, average annual precipitation, and annual solar radiation per unit area) and human activities (proportion of cultivated land, forestland, grassland, water area, built-up area, unused land; population density; and per capita GDP). Changes in all potential drivers from 2000 to 2015 were first calculated and then regressed with changes in carbon budget per unit area.

After selecting the potential drivers, we first performed collinearity diagnosis based on the ordinary least squares (OLS) regression model and eliminated potential drivers with a variance inflation factor (VIF) greater than 7.5. Then, a Geodetector and geographically weighted regression (GWR) model were used to carry out global and local driving factor analysis. Geodetector can be used to detect spatial heterogeneity and find out the driving mechanism behind it [64]. The global impact of drivers on the carbon budget changes was analyzed by the factor detection tool in Geodetector. The GWR model is a local regression model based on the OLS model, which can reflect the degree of influence of different geographical variables on the region [65].

We used the OLS and GWR tools in ArcGIS 10.8 for the modeling. The Geodetector can be accessed at the website www.geodetector.cn (accessed on 10 November 2021).

3. Results

3.1. Changes in Carbon Sequestration and Emissions

From 2000 to 2015, both carbon sequestration and carbon emissions in the YRD region showed an increasing trend, where carbon sequestration increased by 4.18%, while carbon emissions increased by 215.86% (Table 2). Among the four provinces, Anhui and Jiangsu both increased carbon sequestration to some extent, while the other two showed a decreasing trend. Shanghai, in particular, saw a 14.51% reduction in carbon sequestration. In terms of carbon emissions, Jiangsu, Zhejiang, Shanghai and Anhui were ranked according to the increase in the total amount. According to the increase in carbon emissions per unit area, the rankings were Shanghai (433.78 t/ha), Jiangsu (64.20 t/ha), Zhejiang (32.16 t/ha) and Anhui (14.50 t/ha).

Table 2. Quantitative changes in carbon sequestration and emissions during 2000–2015.

		Carbon Sequestration (Mt)			Carbon Emission (Mt)		
	Area (ha)	**2000**	**2015**	**2000–2015**	**2000**	**2015**	**2000–2015**
YRD region	3.59×10^7	968.82	1009.36	40.54	697.49	2203.12	1505.63
Anhui	1.40×10^7	368.86	394.82	25.95	161.22	364.38	203.16
Jiangsu	1.07×10^7	225.53	243.19	17.67	286.39	974.57	688.18
Zhejiang	1.06×10^7	359.41	358.51	−0.90	145.96	485.21	339.25
Shanghai	0.63×10^6	15.02	12.84	−2.18	103.92	378.96	275.04

3.2. *Spatiotemporal Dynamics of the Carbon Budgets*

3.2.1. Changes in the Amount of Carbon Budget from 2000 to 2015

In 2000, the total carbon budgets of Anhui, Zhejiang and the entire study area were in surplus, while Jiangsu and Shanghai were in deficit (Figure 3a). Specifically, the ranking of carbon budgets was as follows: Zhejiang, Anhui, Jiangsu and Shanghai, which were 213.45, 207.65, −60.86 and −88.90 million tons, respectively. In 2015, only Anhui's carbon budget was in surplus, and the other three provinces were in deficit. Jiangsu had the largest carbon budget deficit gaps (−731.37 million tons), followed by Shanghai (−366.12 million tons) and Zhejiang (−126.70 million tons). In terms of changes in carbon budgets from 2000 to 2015, all four provinces showed a decreasing trend. The order of carbon budget reduction was Jiangsu, Zhejiang, Shanghai and Anhui. From 2000 to 2015, the carbon budget of the entire YRD region decreased by 1465.09 million tons.

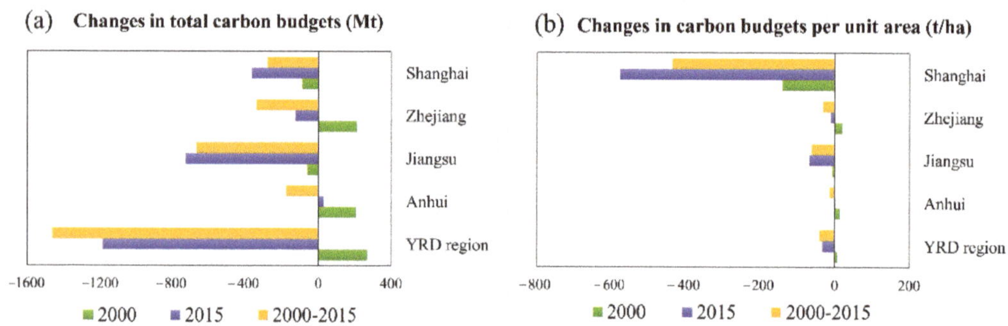

Figure 3. Quantitative changes in carbon budgets during 2000–2015.

In 2000, the carbon budget per unit area was negative in Shanghai, Jiangsu, and the YRD region and positive in Anhui and Zhejiang (Figure 3b). In 2015, only Anhui's carbon budget per unit area was positive, and the other three provinces were all negative, especially Shanghai, which reached −577.43 t/ha. From 2000 to 2015, the changes in the carbon budget per unit area in the four provinces all showed a decreasing trend, among which Shanghai decreased the most (−437.21 t/ha), followed by Jiangsu (−62.55 t/ha), Zhejiang (−32.24 t/ha) and Anhui (−12.65 t/ha).

3.2.2. Spatial Heterogeneity of the Carbon Budget Changes

We investigated the changes in the total carbon budget (Figure 4) and in the carbon budget per unit area (Figure 5) of each county in the YRD region from 2000 to 2015. In 2000, the carbon budget of most counties in the study area was in surplus, while the counties in deficit were mainly distributed in southern Jiangsu, northern Zhejiang and Shanghai (Figure 4a; Figure 5a). In 2015, more counties went from surplus to deficit in the carbon budget, and deficit gaps were increasing in many areas (Figure 4b; Figure 5b). From the results of carbon budget changes, most counties showed a decreasing trend from 2000 to 2015, and the areas with a large decrease were located in Shanghai, southern Jiangsu and northern Zhejiang; only a few counties showed an increasing trend, which were scattered in northern Jiangsu, southern Zhejiang and northern Anhui (Figure 4c; Figure 5c).

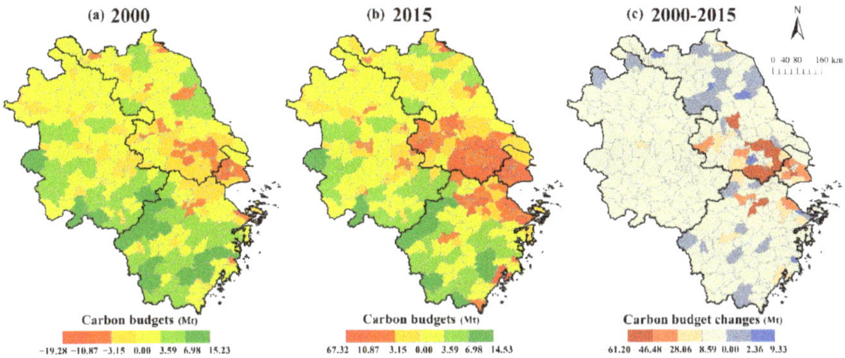

Figure 4. Spatial distribution and changes in the carbon budget from 2000 to 2015 in the YRD region.

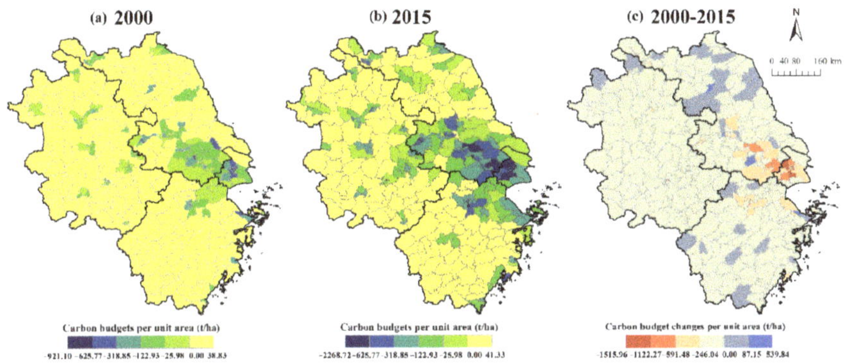

Figure 5. Spatial distribution and changes in the carbon budget per unit area from 2000 to 2015 in the YRD region.

3.3. Driver Analysis

By using the carbon budget changes in the YRD region as the dependent variable and taking climate change (average temperature, average annual precipitation, and annual solar radiation per unit area) and human activities (proportion of cultivated land, forestland, grassland, water area, built-up area, unused land; population density; and per capita GDP) as the independent variables, the OLS model was first constructed. According to the OLS results, the variance inflation coefficient (VIF) values of the proportion of cultivated land and built-up area exceeded 7.5. To avoid multicollinearity, these two factors were excluded, and the remaining nine factors were selected. The GWR tool in ArcGIS provides standard error coefficients that measure the reliability of each coefficient estimate. These estimates are more confident when the value of the standard error coefficient is relatively small. Spatial distributions of standard error coefficients for each driver can be found in the Supplementary File (Figure S2).

3.3.1. Global Analysis

Our study analyzed the impact of each driver on the carbon budget changes based on the factor detection module in Geodetector. The magnitude of the q value represents the influence of each driver. The results showed that changes in population density had the greatest impact on the carbon budget from 2000 to 2015, followed by changes in GDP per capita, the proportion of unused land, and average temperature per unit area, with q values of 0.3317, 0.1202, 0.0998, and 0.0928, respectively (Figure 6). Population, GDP, and unused land changes are factors in human activities, and temperature changes are factors in climate change. The q values of the remaining drivers were not high, but they could still reflect the differences in the impact of different factors on the carbon budget changes. Changes in the proportion of grassland and annual average precipitation had less effect on the carbon budget.

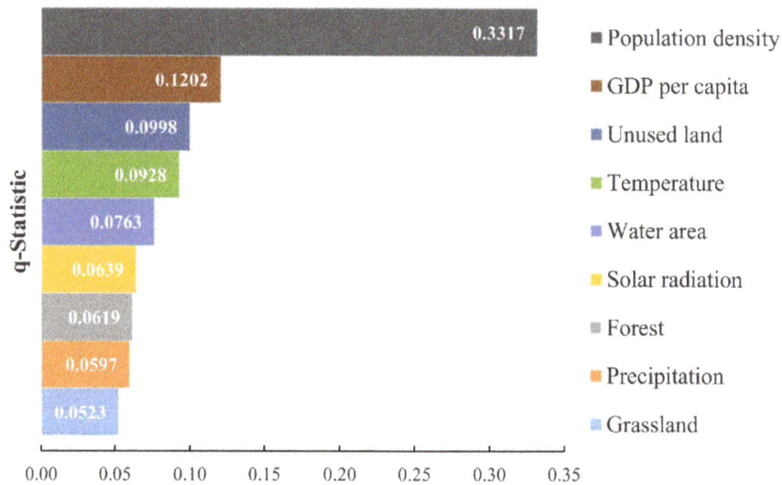

Figure 6. Impact of various drivers on carbon budget changes based on Geodetector analysis. All factors passed the significance test with p values less than 0.01. The land use type refers to the proportion of the total area, and the climate change factor refers to the quantity value per unit area.

3.3.2. Local Analysis

The results of GWR analysis showed that there is obvious spatial heterogeneity in the impact of various driving factors on carbon budget changes (Figure 7). Temperature and carbon budget changes were negatively correlated in southern Anhui and southern Jiangsu (around the Yangtze River) and were positively correlated in other regions. The areas with positive precipitation coefficients were mainly located in the east and south of Jiangsu, and the other areas in the YRD region had negative values. The areas with negative solar radiation coefficients were mainly located along the eastern coast of the YRD region, while most of the western regions were positive. The forest coefficient showed the characteristics of a circle. The coefficient value centered on Shanghai was positive and the largest, and the farther away from the center, the smaller the coefficient value which gradually became negative. The influence of grassland on carbon budget changes also had the characteristics of a circle, but it had become centered in eastern Zhejiang, and the evolution trend in coefficient values was opposite to that of forest. The spatial distribution of water area coefficients was similar to that of solar radiation, with negative values mainly distributed in the eastern coastal zone of the study area and positive values in other regions. The coefficients of unused land were high in the southeast and low in the northwest. Except for some counties in the western part of Anhui, where the coefficients were negative, all other regions were positive. Population density was negatively correlated with carbon budget changes in the entire study area, and the minimum coefficient values were located in the western fringe counties of Anhui and the junctions of Jiangsu, Zhejiang and Shanghai. The per capita GDP coefficients were positive in a few counties in eastern Zhejiang and negative in other areas; the minimum values were located in the middle of the YRD region.

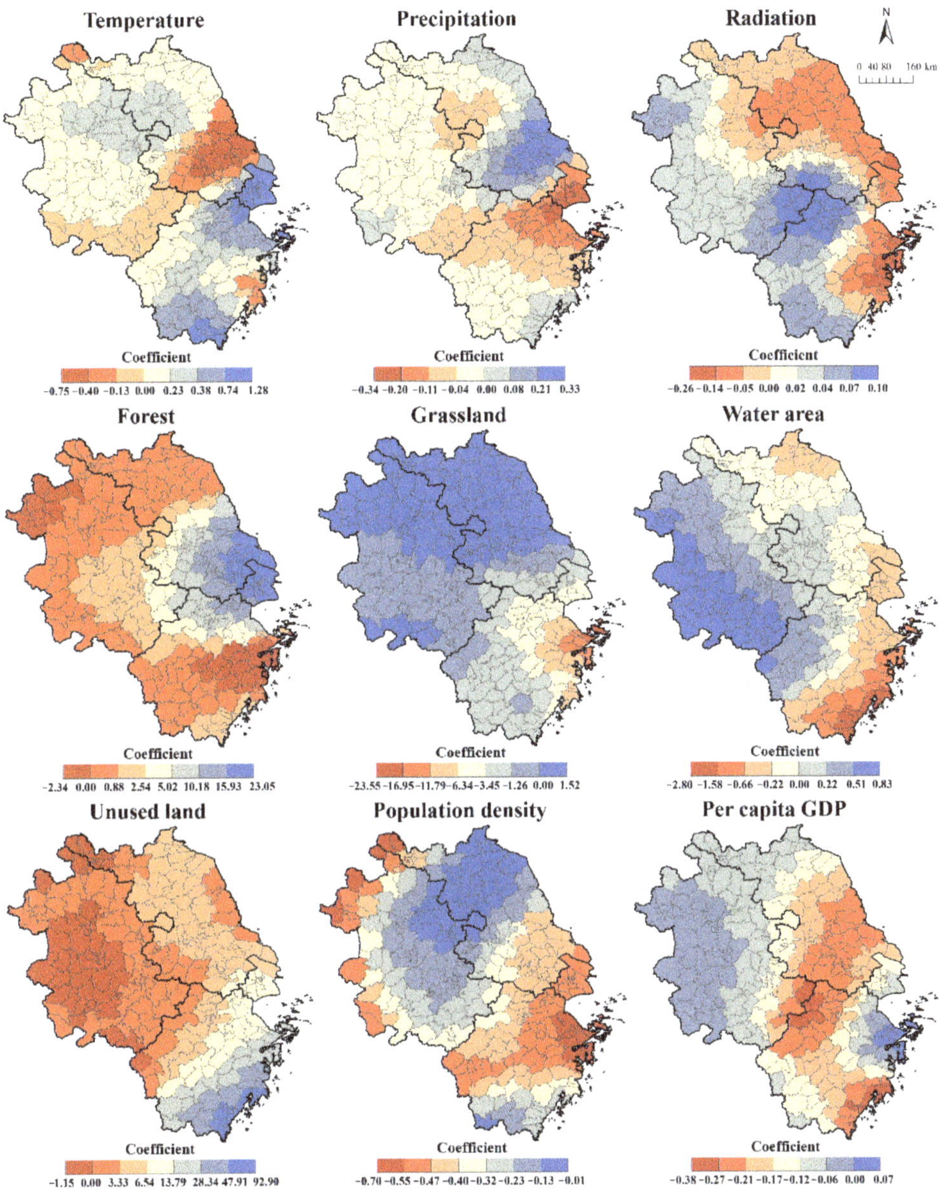

Figure 7. Spatial distribution of the GWR coefficients between carbon budget changes and each driver in the YRD region. A positive coefficient indicates a positive correlation between the driver and the carbon budget changes, and a negative coefficient indicates a negative correlation.

4. Discussion

In this paper, we combined the remote-sensing model and the IPCC inventory method to explore the spatiotemporal pattern changes in the carbon budget in the YRD region of China and analyzed the drivers leading to the changes. Different from previous studies, this study focused on county-scale analysis and analyzed the comprehensive effects of multiple drivers on carbon budgets. The issues we tried to discuss were threefold: (1) to determine the characteristics of the spatiotemporal pattern of the carbon budget in the YRD region, (2) to analyze the driving effect of different factors on carbon budget changes, and (3) to discuss the implications, limitations and future perspectives of this study.

4.1. Spatiotemporal Patterns of Carbon Budgets

Our study found that the increase in carbon sequestration in the YRD region from 2000 to 2015 was much smaller than the increase in carbon emissions, and the carbon budget deficit areas were mainly located in Shanghai, southern Jiangsu, and northern Zhejiang, which is consistent with the findings of Tao et al. [53]. The rapid urbanization process, the increase in impervious surface area, and the large energy consumption in these areas were the main reasons for this phenomenon. As the core city of the YRD region, Shanghai is the growth pole of regional development. Many studies have shown that the socioeconomic development of the YRD region has a "core-edge" structure [66–68]. We found that the spatial pattern characteristics of carbon budgets in the YRD region are as follows: with Shanghai as the center, it gradually changes from a carbon budget deficit to a surplus as the distance increases, and the closer to Shanghai, the greater the deficit. This indicated that the spatial pattern of the carbon budget in the YRD region also has a "core-edge" structural feature. Although few studies have focused on the carbon budget in the YRD region, the results of some carbon sequestration or carbon emission studies conducted in this region can also support our speculation. For example, the research conducted by Sun et al. in the YRD region showed that the ecosystem carbon storage gradually increased with the distance from Shanghai [69]; Han et al. showed that the closer to Shanghai, the greater the value of net carbon emissions (the difference between carbon sources and carbon sinks in each pixel) [22]; Liu et al. demonstrated that the spatial network of the carbon emission efficiency in the YRD region had a core–edge structure [70]. Similarly, the carbon emissions of the Pearl River Delta, [71,72], Los Angeles, Moscow, Tokyo, New Delhi, and Sao Paulo also have this spatial pattern [73]. According to the first law of geography, the correlation between objects is related to the distance. Generally, the closer the distance is, the greater the correlation between objects, and vice versa. The carbon budget is also applicable to this law; that is, adjacent areas have similar pattern characteristics due to their similar dominant functional positioning. Shanghai and its surrounding areas have a high degree of geographical connection, and the dominant function is economic development. The large amount of energy consumption and the limited area of natural ecosystems make the region's carbon budget deficit serious.

4.2. Drivers of Carbon Budget Changes

The factors influencing the carbon budget of terrestrial ecosystems vary in different regions [74]. In this study, changes in population density, per capita GDP, and unused land area accounted for the top three impacts on the carbon budget in the YRD region (Figure 6). This shows that compared with climate change factors, human activities have a greater impact on the carbon budget in the YRD region. Moreover, we found that the effects of different drivers on the spatiotemporal pattern of the carbon budget were different (Figure 7). The results indicated that changes in population density have a much larger impact on the carbon budget than other drivers and were negatively correlated across all regions. This is consistent with Tan et al. [37], who concluded that population growth has a positive effect on China's carbon emissions. For the core area, an increase in population means an increase in food consumption, housing needs and transportation needs, resulting in more energy consumption and carbon emissions [75]. For the peripheral area, the primary industry practitioners continue to flow to the secondary and tertiary industries of the core area or local [76]. The reduction in the labor force in the primary industry means an enhancement of mechanization; the development of the secondary industry means an increase in energy consumption. These will all increase carbon emissions. Our study showed that an increase in per capita GDP in the YRD region leads to a reduction in the carbon budget. On the one hand, industrial development needs to obtain a large quantity of raw materials, energy and resources from the region for the production process, resulting in carbon emissions; on the other hand, the construction of factories occupies forestland, grassland and wetlands, resulting in reduced carbon sequestration [77]. The increase in population and GDP in the YRD region is also accompanied by the development of much unused land into construction land, such as roads and high-rise buildings. These urban infrastructures and high-rise buildings will increase the demand for energy-intensive raw materials (such as steel, cement, etc.), thereby increasing carbon emissions [78]. In addition, a large increase in impervious surfaces comes at the cost of encroaching on urban public space, which will weaken the carbon sequestration function of natural ecosystems.

The relationship between water area and carbon budget is negatively correlated in the east of the study area and positively correlated in the west, which is related to the imbalance of economic development in the east and west of the YRD region. The eastern coastal areas have a high degree of urbanization, and the construction and development of cities pay more attention to the quality of life of residents. In recent years, many blue infrastructures, such as lakes and wetlands, have been added to cities in the eastern region [79]. However, most of the increased water was converted from woodland, grassland or cultivated land [80], thus leading to a decline in carbon sequestration and, in turn, a reduction in the carbon budget. Due to the relatively backward economy in the western part of the study area, the local government still regards economic development as the primary task. This has resulted in many water areas being landfilled and developed for construction, reducing the carbon budget. Forests are important carbon sinks in terrestrial ecosystems [34,81]. In this study, forest and carbon budgets were positively correlated in most areas with significant spatial heterogeneity. The closer to the core area, the greater the impact of forest change on the carbon budget. This shows that compared with peripheral cities, forests in core areas play a more critical role in regulating the carbon budget. Grassland accounts for a small proportion of land use types in the YRD region and has little impact on the regional carbon budget as a whole. Grassland is mainly distributed in the southern area, and a large area of grassland was converted into forest during the study period [49,82]. The cultivated land is mainly distributed in the northern area. Due to the Grain for Green program, some cultivated land was converted into grassland. Therefore, the relationship between grassland and carbon budget changes was negatively correlated in the southern part of the YRD region and positively correlated in the northern part.

In terms of climate change factors, the overall impact of temperature on the carbon budget is greater than that of solar radiation and precipitation (Figure 6), which is consistent with the study by Zhang et al. [83]. We found that the effect of temperature on the carbon budget is the opposite to that of precipitation in many regions. For example, temperature and carbon budget are negatively correlated in southern Jiangsu, while precipitation and carbon budget are positively correlated; temperature and carbon budget are positively correlated in the northern area, Shanghai and most of Zhejiang Province, while precipitation and carbon budget are negatively correlated. Durand et al. demonstrated that increasing solar radiation can promote photosynthesis in vegetation, thereby increasing carbon sequestration [84]. It may also exacerbate the urban heat island effect and lead to an increase in carbon emissions [85]. In our study, the effect of radiation on the carbon budget showed an opposite situation in the east and the west. This may be due to the high degree of urbanization in the eastern region. High-density buildings, asphalt pavements, and cement pavements have larger heat absorption rates and smaller specific heat capacities, which will further enhance the urban heat island effect under the influence of increased radiation [86]. To mitigate the heat island effect, the eastern region consumes more energy to cool down. The increase in carbon sequestration caused by increased radiation is not enough to offset this part of carbon emissions. Due to the low degree of urbanization in the western area, the impact of radiation on carbon sequestration is greater than that of carbon emissions. Therefore, radiation in these regions is positively correlated with the carbon budget.

4.3. Implications, Limitations and Future Perspectives

Our study found that the spatial pattern of the carbon budget in the YRD region has a "core-edge" structural feature. As the core of the YRD region, Shanghai and its surrounding areas have high population density and rapid economic development and are under the pressure of an unbalanced carbon budget. The peripheral area farther from the core area assumed the main carbon sink function. Furthermore, we revealed that population density, GDP per capita, and unused land change have the largest impacts on the carbon budget. Low-carbon development not only requires effectively curbing carbon emissions from the perspective of carbon sources but also requires increasing carbon sinks as much as possible, thereby improving the level of carbon sequestration in the region. Based on these study results, we make recommendations from three aspects: population, industrial development, and land use. (1) Limit the continuous migration of the population from the peripheral area to the core area. The increase in the urban population will inevitably bring about a larger gathering of economic activities and an increase in built-up areas, which in turn will result in more energy consumption and carbon emissions. Therefore, this is a key factor in regulating the regional carbon budget. (2) Upgrade high-energy-consuming industries in the core area and further promote regional economic growth and low-carbon development through innovation; continue to carry out afforestation projects in peripheral areas and develop ecological industries. Based on the accounting results of the carbon budget, the government can formulate ecological compensation measures and improve the carbon-trading market to compensate for the carbon sink functional area. In this way, on the one hand, the counties in the peripheral area can use this part of the funds to improve the salaries, employment opportunities, and medical and education levels of residents to retain the local labor force and reduce the population pressure in the core area. On the other hand, the counties of peripheral area can also invest more funds in ecological protection projects. To support policy formulation, our research provided a list of counties with severe reductions and counties with increases in carbon budgets in the YRD region from 2000 to 2015 (Supplementary Tables S1 and S2). (3) Implement a compact and three-dimensional urban space development model in the core area to promote the intensive use of land, control the excessive growth of artificial surfaces in the peripheral area and reduce the occupation of natural open space.

Our results provide a scientific basis for local green and low-carbon development policies. However, this study still has some limitations that need to be improved. For example, the carbon budget of terrestrial ecosystems is scale-dependent. This study assumed that the YRD region is a closed area and did not consider its carbon cycle relationship with the larger external terrestrial ecosystem. Due to data availability limitations, soil carbon storage was not considered in this paper, which may lead to an underestimation of total carbon sequestration. In addition, policy formulation also has an impact on the carbon budget changes, but this study did not involve policy scenario simulation.

Despite the above limitations, the method used in this paper is highly maneuverable and can quickly and effectively reveal the spatiotemporal patterns of the regional carbon budget and its drivers. In the future, we plan to work on the following aspects to further improve the current research. First, we will conduct sampling of vegetation and soil quadrats to verify the accuracy of the CASA model and supplement the soil carbon storage component. Second, we will combine global carbon dioxide satellite-monitoring data with the IPCC inventory approach to improve the simulation accuracy of carbon emissions. Third, we plan to search for available regional policy data and analyze the impacts of policy factors on carbon budget changes.

5. Conclusions

This study explored the spatiotemporal pattern characteristics of the carbon budget and analyzed its drivers in the YRD region. The results showed that both carbon sequestration and carbon emissions in the study area increased from 2000 to 2015, but the increase in carbon emissions far exceeded the carbon sequestration. The study area changed from a carbon balance surplus in 2000 to a deficit in 2015. From the perspective of spatial patterns, the carbon budget in the YRD region has a "core-edge" structural feature. The closer it is to Shanghai, the core area, the more severe the carbon budget deficit; the farther from it, the greater the carbon budget surplus. We found that population density, GDP per capita and unused land change accounted for the top three impacts on the carbon budget in the YRD region. Overall, human activity has a larger impact on the carbon budget than climate change, and these impacts are mostly negative. Locally, the impact of each driver on the carbon budget showed obvious spatial heterogeneity. The different natural environments and socioeconomic development in different regions are the reasons for this spatial heterogeneity. Based on the results, we proposed limiting the continuous migration of the population from the peripheral area to the core area, carrying out industrial restructuring in the core area, and focusing on implementing ecological protection projects in the peripheral area, implementing a compact and three-dimensional urban space development model in the core area, and controlling the occupation of open natural spaces by built-up land in peripheral areas. Our study can provide a scientific basis for low-carbon development in the YRD region. Our methodology and findings can provide references for similar studies in other urbanized regions around the world.

Supplementary Materials: The following supporting information can be downloaded at: https://www.mdpi.com/article/10.3390/land11081230/s1, Figure S1: training and testing results of the relationship between carbon emissions and the sum of the Digital Number (DN) values of the counties in the YRD (a) 2000 and (b) 2015; Figure S2: the standard error coefficient of each driver; Table S1: counties with reductions in carbon budgets; Table S2: counties with increases in carbon budgets; Codes: codes used in Matlab for realizing the PSO-BP neural network model.

Author Contributions: Conceptualization, Q.F., X.B. and J.C.; methodology, Q.F. and M.G.; software, M.G.; formal analysis, Q.F.; resources, Y.W. and T.W.; data curation, M.G.; writing—original draft preparation, Q.F. and M.G.; writing—review and editing, Q.F. and X.B.; visualization, Q.F.; supervision, J.C.; project administration, Q.F.; funding acquisition, Q.F. All authors have read and agreed to the published version of the manuscript.

Funding: This research was funded by the National Natural Science Foundation of China (42101253), the Jiangsu Social Science Foundation (19GLC016), the Major Projects of Philosophical and Social Sciences Research in Colleges and Universities in Jiangsu Province (2019SJZDA043), and the Open fund of State Key Laboratory of urban and regional ecology (SKLURE2021-2-2).

Data Availability Statement: All data generated or analyzed during this study are included in this published article.

Acknowledgments: We would like to thank the anonymous reviewers for their valuable comments and suggestions.

Conflicts of Interest: The authors declare no conflict of interest.

References

1. Ali, G. Science of the Total Environment Climate Change and Associated Spatial Heterogeneity of Pakistan: Empirical Evidence Using Multidisciplinary Approach. *Sci. Total Environ.* **2018**, *634*, 95–108. [CrossRef] [PubMed]
2. Pata, U.K. Renewable Energy Consumption, Urbanization, Fi Nancial Development, Income and CO_2 Emissions in Turkey: Testing EKC Hypothesis with Structural Breaks. *J. Clean. Prod.* **2018**, *187*, 770–779. [CrossRef]
3. Adedoyin, F.; Ozturk, I.; Abubakar, I.; Kumeka, T.; Folarin, O. Structural Breaks in CO_2 Emissions: Are They Caused by Climate Change Protests or Other Factors? *J. Environ. Manag.* **2020**, *266*, 110628. [CrossRef]
4. Ebi, K.L.; Loladze, I. Elevated Atmospheric CO_2 Concentrations and Climate Change Will Affect Our Food's Quality and Quantity. *Lancet Planet. Health* **2019**, *3*, e283–e284. [CrossRef]
5. Zhang, Y.; Song, C.; Zhang, K.; Cheng, X.; Band, L.E.; Zhang, Q. Effects of Land Use/Land Cover and Climate Changes on Terrestrial Net Primary Productivity in the Yangtze River Basin, China, from 2001 to 2010. *J. Geophys. Res. Biogeosci.* **2014**, *119*, 1092–1109. [CrossRef]
6. Metcalf, C.J.E.; Graham, A.L.; Huijben, S.; Barclay, V.C.; Long, G.H.; Grenfell, B.T.; Read, A.F.; Bjørnstad, O.N. Partitioning Regulatory Mechanisms of Within-Host Malaria Dynamics Using the Effective Propagation Number. *Science* **2011**, *333*, 984–988. [CrossRef] [PubMed]

7. Le Quéré, C.; Raupach, M.R.; Canadell, J.G.; Marland, G.; Bopp, L.; Ciais, P.; Conway, T.J.; Doney, S.C.; Feely, R.A.; Foster, P.; et al. Trends in the Sources and Sinks of Carbon Dioxide. *Nat. Geosci.* **2009**, *2*, 831–836. [CrossRef]
8. Chmiola, J.; Yushin, G.; Gogotsi, Y.; Portet, C.; Simon, P.; Taberna, P.L. Anomalous Increase in Carbon at Pore Sizes Less than 1 Nanometer. *Science* **2006**, *313*, 1760–1763. [CrossRef]
9. Li, T.; Li, J.; Zhou, Z.; Wang, Y.; Yang, X.; Qin, K.; Liu, J. Taking Climate, Land Use, and Social Economy into Estimation of Carbon Budget in the Guanzhong-Tianshui Economic Region of China. *Environ. Sci. Pollut. Res.* **2017**, *24*, 10466–10480. [CrossRef] [PubMed]
10. Rogelj, J.; Forster, P.M.; Kriegler, E.; Smith, C.J.; Séférian, R. Estimating and Tracking the Remaining Carbon Budget for Stringent Climate Targets. *Nature* **2019**, *571*, 335–342. [CrossRef]
11. Matthews, H.D.; Tokarska, K.B.; Nicholls, Z.R.J.; Rogelj, J.; Canadell, J.G.; Friedlingstein, P.; Frölicher, T.L.; Forster, P.M.; Gillett, N.P.; Ilyina, T.; et al. Opportunities and Challenges in Using Remaining Carbon Budgets to Guide Climate Policy. *Nat. Geosci.* **2020**, *13*, 769–779. [CrossRef]
12. Houghton, R.A. Terrestrial Fluxes of Carbon in GCP Carbon Budgets. *Glob. Change Biol.* **2020**, *26*, 3006–3014. [CrossRef] [PubMed]
13. Yan, Y.; Zhao, B.; Chen, J.; Guo, H.; Gu, Y.; Wu, Q.; Li, B. Closing the Carbon Budget of Estuarine Wetlands with Tower-Based Measurements and MODIS Time Series. *Glob. Change Biol.* **2008**, *14*, 1690–1702. [CrossRef]
14. Haripriya, G.S. Carbon Budget of the Indian Forest Ecosystem. *Clim. Change* **2003**, *56*, 291–319. [CrossRef]
15. Li, J.; Yu, Q.; Sun, X.; Tong, X.; Ren, C.; Wang, J.; Liu, E.; Zhu, Z.; Yu, G. Carbon Dioxide Exchange and the Mechanism of Environmental Control in a Farmland Ecosystem in North China Plain. *Sci. China Ser. D Earth Sci.* **2006**, *49*, 226–240. [CrossRef]
16. Nagy, Z.; Pintér, K.; Czóbel, S.; Balogh, J.; Horváth, L.; Fóti, S.; Barcza, Z.; Weidinger, T.; Csintalan, Z.; Dinh, N.Q.; et al. The Carbon Budget of Semi-Arid Grassland in a Wet and a Dry Year in Hungary. *Agric. Ecosyst. Environ.* **2007**, *121*, 21–29. [CrossRef]
17. Kubo, A.; Kanda, J. Coastal Urbanization Alters Carbon Cycling in Tokyo Bay. *Sci. Rep.* **2020**, *10*, 20413. [CrossRef] [PubMed]
18. Piao, S.; He, Y.; Wang, X.; Chen, F. Estimation of China's Terrestrial Ecosystem Carbon Sink: Methods, Progress and Prospects. *Sci. China Earth Sci.* **2022**, *65*, 641–651. [CrossRef]
19. Tang, X.; Zhao, X.; Bai, Y.; Tang, Z.; Wang, W.; Zhao, Y.; Wan, H.; Xie, Z.; Shi, X.; Wu, B.; et al. Carbon Pools in China's Terrestrial Ecosystems: New Estimates Based on an Intensive Field Survey. *Proc. Natl. Acad. Sci. USA* **2018**, *115*, 4021–4026. [CrossRef] [PubMed]
20. Houghton, R.A. The Annual Net Flux of Carbon to the Atmosphere from Changes in Land Use 1850–1990. *Tellus B* **1999**, *51*, 298–313. [CrossRef]
21. Zhao, J.; Xie, H.; Ma, J.; Wang, K. Integrated Remote Sensing and Model Approach for Impact Assessment of Future Climate Change on the Carbon Budget of Global Forest Ecosystems. *Glob. Planet. Change* **2021**, *203*, 103542. [CrossRef]
22. Han, J.; Meng, X.; Liang, H.; Cao, Z.; Dong, L.; Huang, C. An Improved Nightlight-Based Method for Modeling Urban CO_2 Emissions. *Environ. Model. Softw.* **2018**, *107*, 307–320. [CrossRef]
23. Xu, Q.; Yang, R.; Dong, Y.X.; Liu, Y.X.; Qiu, L.R. The Influence of Rapid Urbanization and Land Use Changes on Terrestrial Carbon Sources/Sinks in Guangzhou, China. *Ecol. Indic.* **2016**, *70*, 304–316. [CrossRef]
24. Xiao, J.; Chevallier, F.; Gomez, C.; Guanter, L.; Hicke, J.A.; Huete, A.R.; Ichii, K.; Ni, W.; Pang, Y.; Rahman, A.F.; et al. Remote Sensing of the Terrestrial Carbon Cycle: A Review of Advances over 50 Years. *Remote Sens. Environ.* **2019**, *233*, 111383. [CrossRef]
25. Liu, X.; Zhang, Y.; Dong, G.; Jiang, M. Difference in Carbon Budget from Marshlands to Transformed Paddy Fields in the Sanjiang Plain, Northeast China. *Ecol. Eng.* **2019**, *137*, 60–64. [CrossRef]
26. Zhang, W.; Qiao, Y.; Lakshmanan, P.; Yuan, L.; Liu, J.; Zhong, C.; Chen, X. Combing Public-Private Partnership and Large-Scale Farming Increased Net Ecosystem Carbon Budget and Reduced Carbon Footprint of Maize Production. *Resour. Conserv. Recycl.* **2022**, *184*, 106411. [CrossRef]
27. Fu, Q.; Li, B.; Hou, Y.; Bi, X.; Zhang, X. Effects of Land Use and Climate Change on Ecosystem Services in Central Asia's Arid Regions: A Case Study in Altay Prefecture, China. *Sci. Total Environ.* **2017**, *607*, 633–646. [CrossRef] [PubMed]
28. Piao, S.; Fang, J.; Ciais, P.; Peylin, P.; Huang, Y.; Sitch, S.; Wang, T. The Carbon Balance of Terrestrial Ecosystems in China. *Nature* **2009**, *458*, 1009–1013. [CrossRef]
29. Kim, D.; Lim, C.H.; Song, C.; Lee, W.K.; Piao, D.; Heo, S.; Jeon, S. Estimation of Future Carbon Budget with Climate Change and Reforestation Scenario in North Korea. *Adv. Space Res.* **2016**, *58*, 1002–1016. [CrossRef]
30. Cui, X.; Wei, X.; Liu, W.; Zhang, F.; Li, Z. Spatial and Temporal Analysis of Carbon Sources and Sinks through Land Use/Cover Changes in the Beijing-Tianjin-Hebei Urban Agglomeration Region. *Phys. Chem. Earth* **2019**, *110*, 61–70. [CrossRef]
31. Liu, J.; Sleeter, B.M.; Zhu, Z.; Loveland, T.R.; Sohl, T.; Howard, S.M.; Key, C.H.; Hawbaker, T.; Liu, S.; Reed, B.; et al. Critical Land Change Information Enhances the Understanding of Carbon Balance in the United States. *Glob. Change Biol.* **2020**, *26*, 3920–3929. [CrossRef] [PubMed]
32. Zhao, S.; Liu, S.; Yin, R.; Li, Z.; Deng, Y.; Tan, K.; Deng, X.; Rothstein, D.; Qi, J. Quantifying Terrestrial Ecosystem Carbon Dynamics in the Jinsha Watershed, Upper Yangtze, China from 1975 to 2000. *Environ. Manag.* **2010**, *45*, 466–475. [CrossRef] [PubMed]
33. Zhou, X.; Luo, Y.; Gao, C.; Verburg, P.S.J.; Arnone, J.A.; Darrouzet-Nardi, A.; Schimel, D.S. Concurrent and Lagged Impacts of an Anomalously Warm Year on Autotrophic and Heterotrophic Components of Soil Respiration: A Deconvolution Analysis. *New Phytol.* **2010**, *187*, 184–198. [CrossRef] [PubMed]

34. Fu, Q.; Xu, L.; Zheng, H.; Chen, J. Spatiotemporal Dynamics of Carbon Storage in Response to Urbanization: A Case Study in the Su-Xi-Chang Region, China. *Processes* **2019**, *7*, 836. [CrossRef]
35. Friedlingstein, P.; Jones, M.W.; O'Sullivan, M.; Andrew, R.M.; Bakker, D.C.E.; Hauck, J.; le Quéré, C.; Peters, G.P.; Peters, W.; Pongratz, J.; et al. Global Carbon Budget 2021. *Earth Syst. Sci. Data* **2022**, *14*, 1917–2005. [CrossRef]
36. Grau, H.R.; Aide, T.M.; Zimmerman, J.K.; Thomlinson, J.R. Trends and Scenarios of the Carbon Budget in Postagricultural Puerto Rico (1936–2060). *Glob. Change Biol.* **2004**, *10*, 1163–1179. [CrossRef]
37. Tan, S.; Zhang, M.; Wang, A.; Zhang, X.; Chen, T. How Do Varying Socio-Economic Driving Forces Affect China's Carbon Emissions? New Evidence from a Multiscale Geographically Weighted Regression Model. *Environ. Sci. Pollut. Res.* **2021**, *28*, 41242–41254. [CrossRef]
38. Han, P.; Lin, X.; Zhang, W.; Wang, G.; Wang, Y. Projected Changes of Alpine Grassland Carbon Dynamics in Response to Climate Change and Elevated CO_2 Concentrations under Representative Concentration Pathways (RCP) Scenarios. *PLoS ONE* **2019**, *14*, e0215261. [CrossRef] [PubMed]
39. Chen, Z.; Yu, G.; Ge, J.; Sun, X.; Hirano, T.; Saigusa, N.; Wang, Q.; Zhu, X.; Zhang, Y.; Zhang, J.; et al. Temperature and Precipitation Control of the Spatial Variation of Terrestrial Ecosystem Carbon Exchange in the Asian Region. *Agric. For. Meteorol.* **2013**, *182*, 266–276. [CrossRef]
40. Svirejeva-Hopkins, A.; Schellnhuber, H.J. Modelling Carbon Dynamics from Urban Land Conversion: Fundamental Model of City in Relation to a Local Carbon Cycle. *Carbon Balance Manag.* **2006**, *1*, 8. [CrossRef]
41. Zhang, C.; Tian, H.; Chen, G.; Chappelka, A.; Xu, X.; Ren, W.; Hui, D.; Liu, M.; Lu, C.; Pan, S.; et al. Impacts of Urbanization on Carbon Balance in Terrestrial Ecosystems of the Southern United States. *Environ. Pollut.* **2012**, *164*, 89–101. [CrossRef] [PubMed]
42. Li, H.; Wei, Y.M. Is It Possible for China to Reduce Its Total CO_2 Emissions? *Energy* **2015**, *83*, 438–446. [CrossRef]
43. Chen, C.; Park, T.; Wang, X.; Piao, S.; Xu, B.; Chaturvedi, R.K.; Fuchs, R.; Brovkin, V.; Ciais, P.; Fensholt, R.; et al. China and India Lead in Greening of the World through Land-Use Management. *Nat. Sustain.* **2019**, *2*, 122–129. [CrossRef] [PubMed]
44. Li, H.; Qin, Q. Challenges for China's Carbon Emissions Peaking in 2030: A Decomposition and Decoupling Analysis. *J. Clean. Prod.* **2019**, *207*, 857–865. [CrossRef]
45. Yu, X.; Wu, Z.; Zheng, H.; Li, M.; Tan, T. How Urban Agglomeration Improve the Emission Efficiency? A Spatial Econometric Analysis of the Yangtze River Delta Urban Agglomeration in China. *J. Environ. Manag.* **2020**, *260*, 110061. [CrossRef]
46. Shao, S.; Chen, Y.; Li, K.; Yang, L. Market Segmentation and Urban CO_2 Emissions in China: Evidence from the Yangtze River Delta Region. *J. Environ. Manag.* **2019**, *248*, 109324. [CrossRef] [PubMed]
47. Zhang, S.; Bai, X.; Zhao, C.; Tan, Q.; Luo, G.; Wu, L.; Xi, H.; Li, C.; Chen, F.; Ran, C.; et al. China's Carbon Budget Inventory from 1997 to 2017 and Its Challenges to Achieving Carbon Neutral Strategies. *J. Clean. Prod.* **2022**, *347*, 130966. [CrossRef]
48. Gao, J.; Wang, L. Embedding Spatiotemporal Changes in Carbon Storage into Urban Agglomeration Ecosystem Management—A Case Study of the Yangtze River Delta, China. *J. Clean. Prod.* **2019**, *237*, 117764. [CrossRef]
49. Pei, F.; Zhong, R.; Liu, L.-A.; Qiao, Y. Decoupling the Relationships between Carbon Footprint and Economic Growth within an Urban Agglomeration—A Case Study of the Yangtze River Delta in China. *Land* **2021**, *10*, 923. [CrossRef]
50. Han, J.; Meng, X.; Zhou, X.; Yi, B.; Liu, M.; Xiang, W.N. A Long-Term Analysis of Urbanization Process, Landscape Change, and Carbon Sources and Sinks: A Case Study in China's Yangtze River Delta Region. *J. Clean. Prod.* **2017**, *141*, 1040–1050. [CrossRef]
51. Cai, W.; Peng, W. Exploring Spatiotemporal Variation of Carbon Storage Driven by Land Use Policy in the Yangtze River Delta Region. *Land* **2021**, *10*, 1120. [CrossRef]
52. Liu, C.; Sun, W.; Li, P. Characteristics of Spatiotemporal Variations in Coupling Coordination between Integrated Carbon Emission and Sequestration Index: A Case Study of the Yangtze River Delta, China. *Ecol. Indic.* **2022**, *135*, 108520. [CrossRef]
53. Tao, Y.; Tao, Q.; Sun, X.; Qiu, J.; Pueppke, S.G.; Ou, W.; Guo, J.; Qi, J. Mapping Ecosystem Service Supply and Demand Dynamics under Rapid Urban Expansion: A Case Study in the Yangtze River Delta of China. *Ecosyst. Serv.* **2022**, *56*, 101448. [CrossRef]
54. Zhu, W.; Pan, Y.; He, H.; Yu, D.; Hu, H. Simulation of Maximum Light Use Efficiency for Some Typical Vegetation Types in China. *Chin. Sci. Bull.* **2006**, *51*, 457–463. [CrossRef]
55. Yang, H.; Zhong, X.; Deng, S.; Xu, H. Assessment of the Impact of LUCC on NPP and Its Influencing Factors in the Yangtze River Basin, China. *Catena* **2021**, *206*, 105542. [CrossRef]
56. Chen, J.; Fan, W.; Li, D.; Liu, X.; Song, M. Driving Factors of Global Carbon Footprint Pressure: Based on Vegetation Carbon Sequestration. *Appl. Energy* **2020**, *267*, 114914. [CrossRef]
57. Clarke-Sather, A.; Qu, J.; Wang, Q.; Zeng, J.; Li, Y. Carbon Inequality at the Sub-National Scale: A Case Study of Provincial-Level Inequality in CO_2 Emissions in China 1997–2007. *Energy Policy* **2011**, *39*, 5420–5428. [CrossRef]
58. Zhang, M.; Mu, H.; Ning, Y. Accounting for Energy-Related CO_2 Emission in China, 1991–2006. *Energy Policy* **2009**, *37*, 767–773. [CrossRef]
59. Xi, J.; Gong, H.; Zhang, Y.; Dai, X.; Chen, L. The Evaluation of GHG Emissions from Shanghai Municipal Wastewater Treatment Plants Based on IPCC and Operational Data Integrated Methods (ODIM). *Sci. Total Environ.* **2021**, *797*, 148967. [CrossRef] [PubMed]
60. Chen, J.; Gao, M.; Cheng, S.; Hou, W.; Song, M.; Liu, X.; Liu, Y.; Shan, Y. County-Level CO_2 Emissions and Sequestration in China during 1997–2017. *Sci. Data* **2020**, *7*, 391. [CrossRef] [PubMed]
61. Zhao, R.; Liu, Y. *Theoretical and Empirical Study on Regional Carbon Budget Accounting*; Science Press: Beijing, China, 2015.

62. Li, K.; Hou, Y.; Andersen, P.S.; Xin, R.; Rong, Y.; Skov-Petersen, H. An Ecological Perspective for Understanding Regional Integration Based on Ecosystem Service Budgets, Bundles, and Flows: A Case Study of the Jinan Metropolitan Area in China. *J. Environ. Manag.* **2022**, *305*, 114371. [CrossRef] [PubMed]
63. Sun, X.; Tang, H.; Yang, P.; Hu, G.; Liu, Z.; Wu, J. Spatiotemporal Patterns and Drivers of Ecosystem Service Supply and Demand across the Conterminous United States: A Multiscale Analysis. *Sci. Total Environ.* **2020**, *703*, 135005. [CrossRef]
64. Wang, J.F.; Zhang, T.L.; Fu, B.J. A Measure of Spatial Stratified Heterogeneity. *Ecol. Indic.* **2016**, *67*, 250–256. [CrossRef]
65. Kashki, A.; Karami, M.; Zandi, R.; Roki, Z. Evaluation of the Effect of Geographical Parameters on the Formation of the Land Surface Temperature by Applying OLS and GWR, A Case Study Shiraz City, Iran. *Urban Clim.* **2021**, *37*, 100832. [CrossRef]
66. Zhang, Y.; Cao, W.; Zhang, K. Mapping Urban Networks through Inter-Firm Linkages: The Case of Listed Companies in Yangtze River Delta, China. *J. Geosci. Environ. Prot.* **2020**, *8*, 23–36. [CrossRef]
67. Yang, J.; Zhang, J.; Lu, J.; Sun, D. The Core-Periphery Structure in the Yangtze River Delta: An Enterprise Linkage Perspective, 1978–2019. *Complexity* **2021**, *2021*, 9351741. [CrossRef]
68. Wang, J.; Deng, Y.; Qalati, S.A.; Qureshi, N.A. Urban Resilience and Transportation Infrastructure Level in the Yangtze River Delta. *Front. Environ. Sci.* **2022**, *10*, 893964. [CrossRef]
69. Sun, X.; Ma, Q.; Fang, G. Spatial Scaling of Land Use/Land Cover and Ecosystem Services across Urban Hierarchical Levels: Patterns and Relationships. *Landsc. Ecol.* **2022**. Available online: https://link.springer.com/article/10.1007/s10980-021-01387-4#citeas (accessed on 5 June 2021). [CrossRef]
70. Liu, C.; Tang, R.; Guo, Y.; Sun, Y.; Liu, X. Research on the Structure of Carbon Emission Efficiency and Influencing Factors in the Yangtze River Delta Urban Agglomeration. *Sustainability* **2022**, *14*, 6114. [CrossRef]
71. Xu, Q.; Dong, Y.; Yang, R.; Zhang, H.; Wang, C.; Du, Z. Temporal and Spatial Differences in Carbon Emissions in the Pearl River Delta Based on Multi-Resolution Emission Inventory Modeling. *J. Clean. Prod.* **2019**, *214*, 615–622. [CrossRef]
72. Zhou, Y.; Shan, Y.; Liu, G.; Guan, D. Emissions and Low-Carbon Development in Guangdong-Hong Kong-Macao Greater Bay Area Cities and Their Surroundings. *Appl. Energy* **2018**, *228*, 1683–1692. [CrossRef]
73. Oda, T.; Maksyutov, S. A Very High-Resolution (1 km $\times$ 1 km) Global Fossil Fuel CO2 Emission Inventory Derived Using a Point Source Database and Satellite Observations of Nighttime Lights. *Atmos. Chem. Phys.* **2011**, *11*, 543–556. [CrossRef]
74. Rogger, J.; Hörtnagl, L.; Buchmann, N.; Eugster, W. Carbon Dioxide Fluxes of a Mountain Grassland: Drivers, Anomalies and Annual Budgets. *Agric. For. Meteorol.* **2022**, *314*, 108801. [CrossRef]
75. Dong, K.; Hochman, G.; Zhang, Y.; Sun, R.; Li, H.; Liao, H. CO_2 Emissions, Economic and Population Growth, and Renewable Energy: Empirical Evidence across Regions. *Energy Econ.* **2018**, *75*, 180–192. [CrossRef]
76. Hao, Y.; Zhang, Z.Y.; Yang, C.; Wu, H. Does Structural Labor Change Affect CO_2 Emissions? Theoretical and Empirical Evidence from China. *Technol. Forecast. Soc. Change* **2021**, *171*, 120936. [CrossRef]
77. Zhang, F.; Xu, N.; Wang, C.; Wu, F.; Chu, X. Effects of Land Use and Land Cover Change on Carbon Sequestration and Adaptive Management in Shanghai, China. *Phys. Chem. Earth* **2020**, *120*, 102948. [CrossRef]
78. Anderson, V.; Gough, W.A. Evaluating the Potential of Nature-Based Solutions to Reduce Ozone, Nitrogen Dioxide, and Carbon Dioxide through a Multi-Type Green Infrastructure Study in Ontario, Canada. *City Environ. Interact.* **2020**, *6*, 100043. [CrossRef]
79. Wu, C.; Li, J.; Wang, C.; Song, C.; Chen, Y.; Finka, M.; la Rosa, D. Understanding the Relationship between Urban Blue Infrastructure and Land Surface Temperature. *Sci. Total Environ.* **2019**, *694*, 133742. [CrossRef] [PubMed]
80. Han, L.; Xu, Y.; Yang, L.; Deng, X.; Hu, C.; Xu, G. Temporal and Spatial Change of Stream Structure in Yangtze River Delta and Its Driving Forces during 1960s–2010s. *Acta Geogr. Sin.* **2015**, *70*, 819–827. [CrossRef]
81. Liu, W.; Guo, Z.; Lu, F.; Wang, X.; Zhang, M.; Liu, B.; Wei, Y.; Cui, L.; Luo, Y.; Zhang, L.; et al. The Influence of Disturbance and Conservation Management on the Greenhouse Gas Budgets of China's Forests. *J. Clean. Prod.* **2020**, *261*, 121000. [CrossRef]
82. Li, X.; Fang, B.; Yin, M.; Jin, T.; Xu, X. Multi-Dimensional Urbanization Coordinated Evolution Process and Ecological Risk Response in the Yangtze River Delta. *Land* **2022**, *11*, 723. [CrossRef]
83. Zhang, L.; Zhou, G.; Ji, Y.; Bai, Y. Grassland Carbon Budget and Its Driving Factors of the Subtropical and Tropical. *Sci. Rep.* **2017**, *7*, 14717. [CrossRef] [PubMed]
84. Durand, M.; Murchie, E.H.; Lindfors, A.V.; Urban, O.; Aphalo, P.J.; Robson, T.M. Diffuse Solar Radiation and Canopy Photosynthesis in a Changing Environment. *Agric. For. Meteorol.* **2021**, *311*, 108684. [CrossRef]
85. Roxon, J.; Ulm, F.J.; Pellenq, R.J.M. Urban Heat Island Impact on State Residential Energy Cost and CO_2 Emissions in the United States. *Urban Clim.* **2020**, *31*, 100546. [CrossRef]
86. Li, Y.; Schubert, S.; Kropp, J.P.; Rybski, D. On the Influence of Density and Morphology on the Urban Heat Island Intensity. *Nat. Commun.* **2020**, *11*, 2647. [CrossRef] [PubMed]

land

Article

Coupling Coordination Analysis of Ecosystem Services and Urbanization in Inner Mongolia, China

Li Na, Yangling Zhao and Luo Guo *

College of Life and Environmental Sciences, Minzu University of China, Beijing 100081, China
* Correspondence: guoluo@muc.edu.cn

Abstract: Given that ecological and environmental functions are greatly influenced by rapid urbanization, a clear understanding of the relationship between ecosystem services (ESs) and urbanization is urgently needed to improve sustainable development in Inner Mongolia. In this study, we first carried out ecosystem service valuation (ESV) using the value coefficient method. We then examined the urbanization level using a comprehensive indicator system. Finally, we applied the coupling coordination degree model to analyze the coordination relationship between ecosystem services and urbanization from 1995 to 2020 in Inner Mongolia. The results showed that there was an increase in both the urbanization level and all ecosystem services excluding climate regulation, environmental purification, and biodiversity services. The coupling coordination degree (CCD) of Inner Mongolia is not ideal, and most counties remain at a low level of coordination degree. Furthermore, spatiotemporal heterogeneity was evident in the CCD of ecosystem services and urbanization as it was higher in the center and east of the country, but lower in the north and west regions. Relevant policies should be implemented to strengthen the advantages of local ecology, encourage environmentally friendly industrialization, and promote ecologically and economically sustainable development.

Keywords: ecosystem services; urbanization; coupling coordination relationship; Inner Mongolia

Citation: Na, L.; Zhao, Y.; Guo, L. Coupling Coordination Analysis of Ecosystem Services and Urbanization in Inner Mongolia, China. *Land* **2022**, *11*, 1870. https://doi.org/10.3390/land11101870

Academic Editor: Luca Congedo

Received: 23 September 2022
Accepted: 17 October 2022
Published: 21 October 2022

1. Introduction

Since the beginning of the twenty-first century, China has undergone a significant increase in urban development. The urbanization rate of China surpassed 60% at the end of 2020 [1] and, based on Northam's classification of urbanization stages, is considered to be in the accelerated development stage [2,3]. China's rapid urbanization is not only illustrated by fast population urbanization, defined as the population transfer from rural to urban districts, but is also characterized by land and economic urbanization [4]. Urbanization can drive natural ecosystems to become human–nature combined or human-dominated ecosystems [5]. Recent rapid urbanization has brought about an increasing number of environmental problems including degradation of environmental quality, scarcity of arable land, and most importantly, the decline of ecosystem provision which has affected both the structure and function of ecosystems in regional areas, thereby destroying ESs [6]. ESV can be utilized to assess the human services potential provided by regional ecosystems and to analyze the environmental transformation induced by human disturbances. In many studies, experts have considered ESV to be a key indicator used for measuring changes in the ecological environment [7].

ESs are defined as the benefits and welfare that humans derive from ecosystems, either directly or indirectly [8]. More specifically, ecosystem services encompass provision services, regulation services, support services, and recreation and culture services [9]. The integration of provision, regulation, support, and cultural services contributes significantly to the synergistic achievements of the United Nations Sustainable Development Goals [10]. Globally, however, ESs have declined substantially over the past few decades, and it is predicted that these declines will continue in the decades to come [11]. Additionally, the structure (for example vegetation coverage and soil quality) as well as the processes (for example animal mitigation) within ecosystems are profoundly altered [12]. Intensifying urbanization is putting increasing pressure on ecosystems. With their vital role in connecting human society to nature, ecosystem services are usually impacted by urbanization either directly or indirectly [13]. Quantifying the relationship between ecosystem services (ESs) and urbanization is of utmost importance to identify the underlying mechanisms between two variables and achieve sustainable development between man and nature [14]. In this context, scholars, policymakers, and stakeholders worldwide are closely focused on investigating the interactions and coordination mechanisms between urbanization and ESs. A solution to this common and challenging issue is urgently needed to enhance regional sustainability [15,16]. It is imperative to better understand the coupling relationship and assess the coordination and conflict between urbanization and ESs to strike a long-term balance between urbanization and ecosystem protection.

Scholars have conducted numerous studies on ESs and urbanization. Studies of ESs rely primarily on the invest model or the value scale developed by Costanza et al. [17] for simulating and exploring ecosystem services, and ESV is calculated by continually optimizing the value equivalent factor [18]. In addition, investigations on urbanization have been conducted primarily based on LUCC change [19], nighttime lighting data, urbanization rate, and single-dimensional estimates [20]. A simple linear relationship usually does not exist between ecosystem services and urbanization. The relationship may involve a correlation mechanism that is dynamic and nonlinear [21]. There are different approaches currently being applied to examine such nonlinear correlations. Wang et al. [22] analyzed the interaction mechanism between ESs and urbanization utilizing curve estimation. Other researchers have used piecewise linear regression to explain the relationship between two variables in Beijing suburbs [23]. The ecological environment and urbanization are generally interconnected. Studies have been conducted on this topic specifically for metropolitan regions of high urbanization or other administrative regions [24–26]. They are however not able to provide a coherent assessment of the coupling coordination relationship between ecosystem services and urbanization for the ecological environments of relatively fragile arid and semi-arid regions and a more precise resolution for the coordinated development between two variables in the typical arid and semi-arid regions. Nevertheless, insufficient research is being done on coupling coordination between urbanization and ESs, particularly in ecologically fragile zones, e.g., arid regions and plateaus, which play an essential role in long-term sustainable development and ecological management in these regions [27]. This study examines the spatio-temporal patterns of the coordination relationships between ESs and urbanization in the typical arid and semi-arid region using a coupling coordination degree (CCD) model. Further, scholars from different backgrounds and subjects have employed different methods and models including ordinary least square model (OLS) [28], geographical weighted regression model (GWR) [29], and panel data regression [30] model to investigate the coupling coordination relationship between socioeconomic and ecological systems. Nonetheless, these models can only explain urbanization's unidirectional effect on the natural environment and fail to take into account the spatial-temporal effect. The coupling coordination degree (CCD) model fills the gaps mentioned above [31]. Additionally, the existing literature is primarily concentrated on the regional, provincial, and municipal scales with few studies conducted at the county level in the entire autonomous region. To fill this research gap, the current study was implemented at the county level. Referring to administrative units, the evaluation results are more targeted and provide more effective guidance for actions taken by the local government [32]. Therefore, conducting evaluations at the county level is a useful way of making sustainable development a reality and improving the coupling coordination development of urbanization and ecosystem services.

As a typical resource-based area, the Inner Mongolian economy is heavily dependent on coal mining and extraction of other natural resources to promote its economic prosperity, which has led to extensive resource consumption and fragile environmental conditions in the entire region. The rapidly degrading environment in turn hinders local urbanization development, and Inner Mongolia faces severe challenges in its attempt to achieve sustainable development. With the above considerations, we developed a comprehensive assessment indicator system that includes the multiple aspects of population, economy, and land to identify urbanization. In addition, the spatial-temporal evolution characteristics of ESs and urbanization were identified in Inner Mongolia over the period 1995 to 2020. Specifically, this article aimed to (1) investigate the spatio-temporal distribution pattern of individual ESs in Inner Mongolia from 1995 to 2020; (2) evaluate the urbanization level of Inner Mongolia from 1995 to 2020 based on three dimensions relating to population, economy, and land; and (3) utilize a CCDM model to get a clear understanding of the coordination interaction between ESs and urbanization. The results serve as a policy-making basis for achieving environmental protection as well as sustainable development in Inner Mongolia and a reference for improving the coordination relationship between ESs and urbanization in similar areas worldwide.

2. Study Area, Data Sources, and Research Methods

2.1. Study Area

Inner Mongolia stretches across Northern China covering 1.183 million km^2 and containing 25 million people [33]. It consists of 101 counties and occupies 12.3% of China's total land area (Figure 1). The climate in this area is largely dominated by the continental monsoon due to its high latitude and altitude and distance from the ocean [34]. The average annual temperature is 0–8 °C declining gradually from south to north. Annual precipitation decreases progressively from northeast to southwest ranging from 50 to 550 mm, with 75% occurring between July and September [35]. The area consists of six major vegetation ecological zones: shrub desert, desert steppe, forest steppe, typical steppe, meadow steppe, and deciduous forest. Xilamuren, Xilingol, Horqin, and Hulunbuir are four major prairies in the region, and Badain Jaran, Ulanbuh, Kubuqi, and Tengger are four major deserts. According to LUCC in 2020, grassland made up 46.13%of Inner Mongolia's total area, followed by desert or barren land (25.37%), forested land (14.46%), cultivated land (9.99%), wetland (3.14%), and construction land (1.02%). In recent years, Inner Mongolia has seen rapid urbanization due to its abundant energy resources, which also contributed to the region's social and economic progress. During the past few years, the region's social economic development and the construction of infrastructure have made great progress due to its energy resources advantages. The urbanization level has greatly improved, and now it has entered a phase of rapid urbanization development. Rapid urbanization not only occupies the ecological environment directly, but also exerts a coercive influence on the environment, energy, and resources. Therefore, conflicts between the ecological environment and urban development are becoming more dramatic.

2.2. Data Sources

In this article, LUCC datasets were obtained from the Chinese Academy of Sciences Resource Environmental Data Center (RESDC). These LUCC data with a spatial resolution of 1 km and accuracy of 90% were considered appropriate for this research [2]. Then, by combining supervision classifications and manual visual interpretations as well as field surveys, land use types were classified into grassland, forest land, cultivated land, water body, construction land, and other land (Figure 2). Additionally, population density and gross domestic product with a resolution of 1 × 1 km from 1995 to 2020 were also available on the website of RESDC. For the purpose of facilitating the exploration of the spatial distribution characteristics of Urbanization data and ESV, a grid of 2 × 2 km was applied to spatial sampling, and using ArcGIS software, the mean values in a total of 286,793 grid squares of the 6-phase data were calculated.

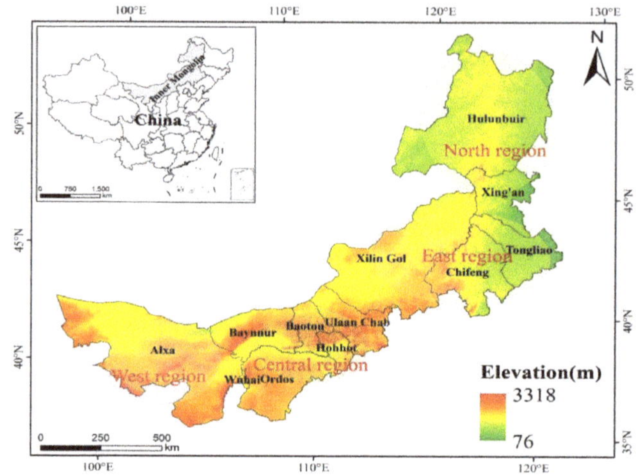

Figure 1. Location of the study area.

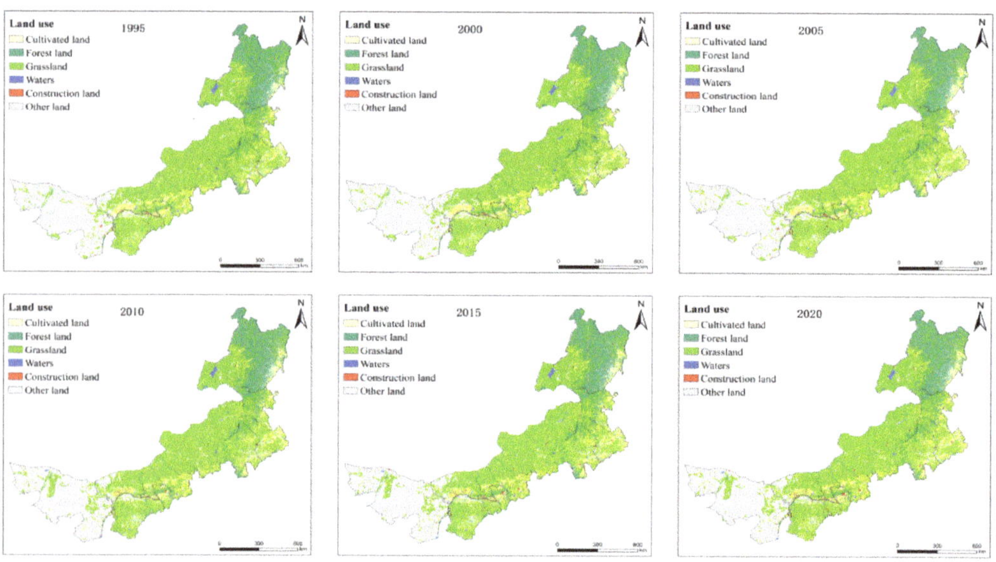

Figure 2. Land use types in Inner Mongolia from 1995 to 2020.

2.3. *Study Methods*

A total of 286,793 grids were generated over the six-year period (i.e., 1995, 2000, 2005, 2010, 2015, and 2020) using the "Create fishnet" function tool [36]. The spatial resolution of each fishnet cell was set to 2 × 2 km applying the ArcGIS 10.4 spatial analysis tool.

2.3.1. Assessing Ecosystem Service Values

ESV are widely evaluated by applying the equivalent factor methodology on the basis of the value per unit area [37,38]. The ESV is evaluated here according to the improved method of the equivalent factor [39], and the value equivalent factor is also adjusted according to specific circumstances. The equivalent factor of ESV according to Xie et al. [40] in a unit of measurement that is standard equals the economic value of natural grain produced annually on one hectare of farmland, equivalent to approximately one-seventh of the national grain average value of that year the same. The inter-annual variations in grain production's economic value were considered to calculate grain value on average in the area and the adjusted ESV of each standardized equivalent factor in the research field referred to 2172.62 yuan/hm^2. The ESV value coefficients for each unit area are presented in Table 1. The ESV is estimated using the formula below:

$$\mathrm{ESV} = \sum_i \sum_j A_i \times VC_{ij} \quad (1)$$

where ESV represents the total ecosystem services value, A_i represents land use type i's area, and VC_{ij} is the land use type i value coefficient and ecosystem service function type f.

Table 1. Value coefficients of ESV each unit area (yuan/hm^2·year).

Ecosystem Services Functions		Cultivated Land	Forestland	Grassland	Waters	Other Land
Supply services	Food production	263.5	83.7	71.3	248.0	3.1
	Raw material production	124.0	195.3	105.4	71.3	9.3
	Water supply	6.2	102.3	58.9	2569.9	6.2
Regulation services	Gas regulation	207.7	641.7	375.1	238.7	34.1
	Climate regulation	111.6	1922.0	988.9	709.9	31.0
	Environmental purification	31.0	558.0	325.5	1720.5	96.1
	Hydrological regulation	83.7	1196.6	725.4	31,694.4	65.1
Support services	Soil conservation	319.3	781.2	455.7	288.3	40.3
	Nutrient cycling	37.2	58.9	34.1	21.7	3.1
	Biodiversity	40.3	713.0	415.4	790.5	37.2
Cultural services	Aesthetic landscape	18.6	310.0	182.9	585.9	15.5

2.3.2. Comprehensive Urbanization Indicators Assessment

Urbanization is categorized into three-dimensions, which can be further broken down into three categories: population, economy, and land urbanization [41]. Specifically, the three key metrics were the population density (POP), the GDP density (GDP), and the urban land percentage (UBP). Afterwards, the urbanization factor layers were obtained by aggregating the 1 × 1 km resolution data of *PD*, *GDPD*, and *BLP* into each 2 × 2 km evaluation grid. Finally, the comprehensive urbanization level (CUL) was obtained by superimposing *PD*, *GDPD*, and *BLP*. The formula is as follows:

$$CUL_j = \frac{1}{3}\left(PD'_j + GDPD'_j + BLP'_j\right) \quad (2)$$

where CUL_j refers to the comprehensive urbanization level in unit grid j, and PD'_j, $GDPD'_j$, and BLP'_j correspond to the standardized values of population density (PD), gross domestic product density (GDPD), and build-up land percentage (BLP).

2.3.3. Coupling Coordination Relationship between ESs and Urbanization

The coupling associations between the urbanization and ecosystem services originated from a measure that exists in physics [42–44]. It is illustrated that a positive dynamic relationship forms from interactions between two or more systems in which both systems benefit mutually [45]. Using the theory of coupling in physics, we formulate a coupling model calculation of ecosystem services and urbanization as shown below:

$$C = 2\left\{\frac{W_{UD} \times W_{ES}}{(W_{UD} + W_{ES})^2}\right\}^{\frac{1}{2}}, C \in [0,1] \quad (3)$$

where C represents the coupling level of the ecosystem services and urbanization, which has the value of 0–1; W_{UD} represents the urbanization assessment value; and W_{ES} refers to the comprehensive ecosystem services assessment value. Higher C values indicate stronger interactions between urbanization and ecosystems. The coupling degree only reflects their correlation level in contrast to the actual development status between the two systems. Possibly, the two systems have low values, but high coupling degrees. Thus, it is important to note that the high coupling value is different from the high horizontal coupling. Thus, the CCDM model was applied to measure the level of coordination between ecosystem services and urbanization using the following formulas:

$$\begin{cases} D = \sqrt{C \times T} \\ T = \alpha W_{UD} + \beta W_{ES} \end{cases} \tag{4}$$

where D refers to the CCD of the urbanization and ecosystem; T indicate the overall synergy effect represented in the two systems' comprehensive evaluation index; and α and β refer to the two systems' weight coefficients, most commonly in 0–1. The current study presumes that urbanization and the ecosystem are equally important; therefore, both α and β were given the same value of 0.5 [46].

We used the CCD division standard below to analyze the coordination development of ecosystem services and urbanization, and the divisional results are presented in Table 2 [47,48].

Table 2. CCD standard between ESV and urbanization.

Range of D Value	Level of CCD	Type of CCD
$0.8 < D \leq 1$	Quality coordination (L1)	$W_{UD} > W_{ES}$ Quality coordination with lagging ecosystem $W_{UD} \approx W_{ES}$ Quality coordination of urbanization and ecosystem $W_{UD} < W_{ES}$ Quality coordination with lagging urbanization
$0.6 < D \leq 0.8$	Intermediate coordination (L2)	$W_{UD} > W_{ES}$ Intermediate coordination with lagging ecosystem $W_{UD} \approx W_{ES}$ Intermediate coordination of urbanization and ecosystem $W_{UD} < W_{ES}$ Intermediate coordination with lagging urbanization
$0.5 < D \leq 0.6$	Primary coordination (L3)	$W_{UD} > W_{ES}$ Primary coordination with lagging ecosystem $W_{UD} \approx W_{ES}$ Primary coordination of urbanization and ecosystem $W_{UD} < W_{ES}$ Primary coordination with lagging urbanization
$0.4 < D \leq 0.5$	Basic incoordination (L4)	$W_{UD} > W_{ES}$ Basic incoordination with hindered ecosystem $W_{UD} \approx W_{ES}$ Basic incoordination of urbanization and ecosystem $W_{UD} < W_{ES}$ Basic incoordination with hindered urbanization
$0.2 < D \leq 0.4$	Intermediate incoordination (L5)	$W_{UD} > W_{ES}$ Intermediate incoordination with hindered ecosystem $W_{UD} \approx W_{ES}$ Intermediate incoordination of urbanization and ecosystem $W_{UD} < W_{ES}$ Intermediate incoordination with hindered urbanization
$0 < D \leq 0.2$	Extreme incoordination (L6)	$W_{UD} > W_{ES}$ Extreme incoordination with hindered ecosystem $W_{UD} \approx W_{ES}$ Extreme incoordination of urbanization and ecosystem $W_{UD} < W_{ES}$ Extreme incoordination with hindered urbanization

3. Results

3.1. Spatiotemporal Change of the ESV

The distribution pattern of individual ESs in Inner Mongolia showed significant spatial heterogeneity from 1995 to 2020 (Figure 3). Natural conditions across the study area varied greatly, and districts varied significantly in their ability to provide ecosystem services. The spatial pattern of water supply services in most districts has remained mostly unchanged over the past 25 years. WY displayed the highest high–low spatial distribution pattern attributed to regional differences in precipitation from the northeast to the southwest. The most significant changes in WY were observed in the central region between 2000 and 2015, with initial decreases followed by increases, mainly due to precipitation and vegetation evapotranspiration. SR values were lower in the western region, where bare and unused land make up most of the land use type. Higher SR districts were concentrated in the area with abundant precipitation and dense vegetation coverage, making it evident that forests contributed significantly to soil retention, alleviating urbanization's negative effect on soil retention. The spatial pattern of SR in the west has changed dramatically since 2000. Climate regulation services are located high in the northeastern region. It becomes clear that high-value areas are distributed in the same way as forest types when superimposed on the land use map, indicating that this area is highly climate-regulating. Overall, climate regulation value decreased during the study period. FP is influenced by a variety of factors, including the amount of farmland. The improvement of agrotechnology contributes to an average FP increase of 1.76% despite the GFGP implementation. Recreation potential did not show significant changes over the studied time interval. Nevertheless, this service displayed clear spatial heterogeneity as high recreation values were mainly found near water sources and mountains.

The spatial pattern of ESV in Inner Mongolia was relatively stable, but there was an evident spatial distribution pattern owing to differences in land use structure and geographic location (Figure 4). Forest coverage in hilly and mountainous areas in northeastern Inner Mongolia contributed to increased ESV. ESV in the west region was low on account of construction land expansion under rapid urbanization and barren land distribution in this area. Farmland and grassland mainly dominate the medium-ESV regions. In general, ES was spatially distributed with the highest values in forests and wetlands and the lowest values in urban areas. Total ESV had a small, fluctuating upward trend from 1995 to 2020. It increased from nearly 384.27 billion in 1995 to 386.3.7 billion in 2020, indicating a slight improvement of approximately 0.55% in ecosystem service functions (Table 3). Across eleven ecosystem service categories, regulation services accounted for the greatest percentage of contributions with 67.27% of ESV in 1995 and 67.35% in 2020. Individual ecosystem services such as environmental purification and biodiversity decreased. The climate regulation service was the most severely impacted regulatory service, with its value decreasing by 0.17% over the 25 years from 1995 to 2020. The total ESV increased by approximately 2.95 billion yuan due to increased hydrological regulation.

Table 3. ESV of different ecosystem services in Inner Mongolia (10^8 yuan).

Ecosystem Services Functions	ESV					
	1995	2000	2005	2010	2015	2020
Food production	84.41	85.64	85.45	85.92	85.93	85.89
Raw material production	105.37	105.21	105.00	105.87	105.83	105.77
Water supply	86.29	87.26	84.81	86.49	87.47	87.78
Gas regulation	340.87	339.01	338.03	341.41	341.30	341.13
Climate regulation	872.14	864.00	860.91	870.96	870.99	870.63
Environmental purification	322.08	320.39	318.23	321.42	321.78	321.83
Hydrological regulation	1049.90	1062.09	1031.89	1052.72	1064.77	1068.61
Soil conservation	421.26	419.41	418.28	422.29	422.16	421.94
Nutrient cycling	33.07	33.01	32.94	33.22	33.21	33.189
Biodiversity	364.43	361.50	359.76	363.99	364.12	364.03
Aesthetic landscape	162.90	161.77	160.79	162.76	162.91	162.90
Total	3842.73	3839.30	3796.11	3847.07	3860.47	3863.71

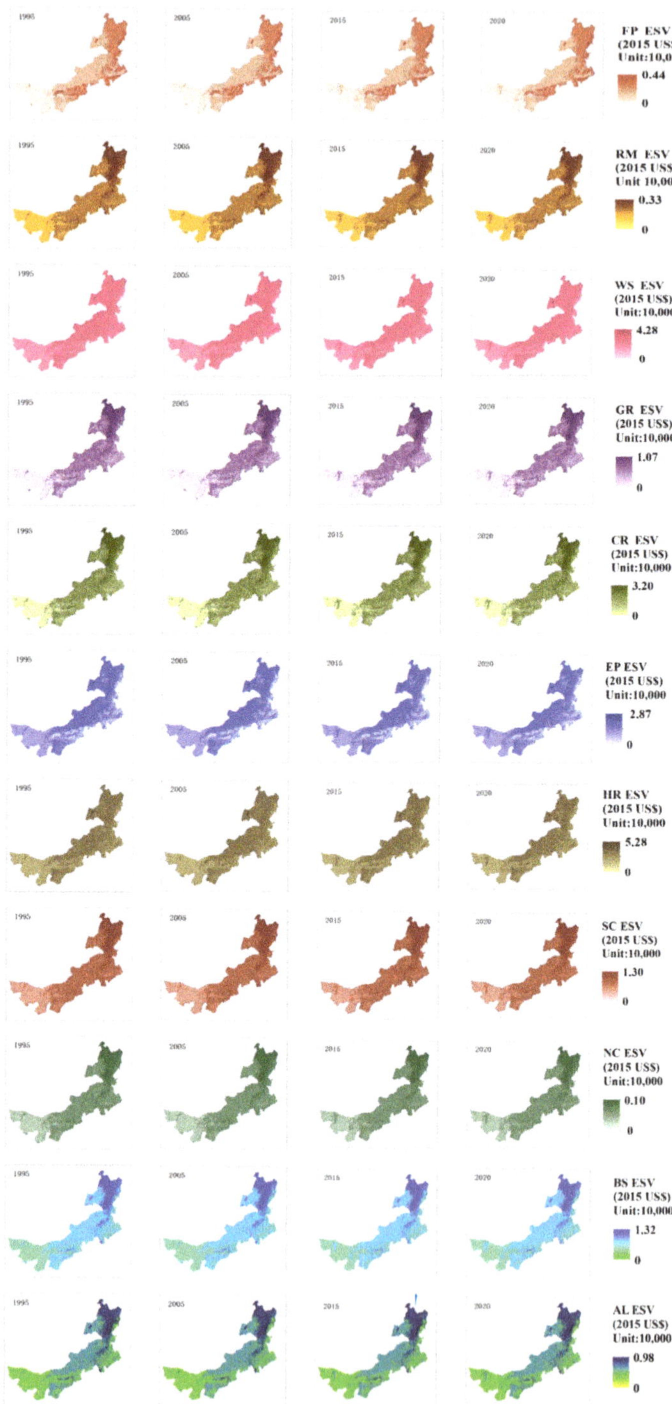

Figure 3. Spatial pattern of individual ESV in Inner Mongolia from 1995 to 2020.

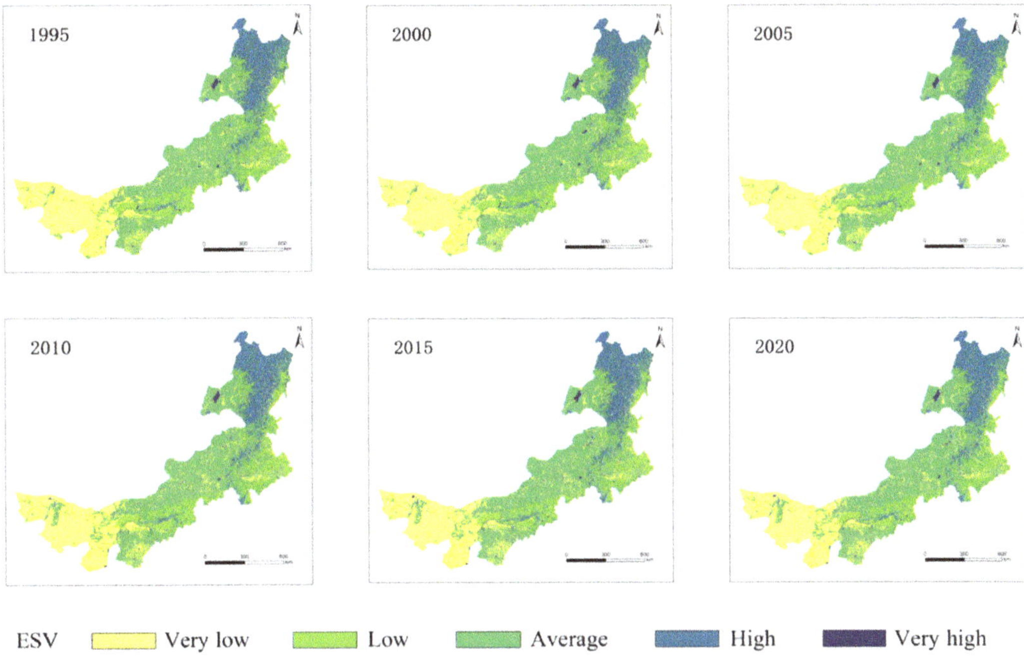

Figure 4. Spatial pattern of total ESV in Inner Mongolia from 1995 to 2020.

3.2. The Dynamics of Urbanization in Inner Mongolia

The comprehensive evaluation of Inner Mongolia's urbanization level in this study was conducted using economic, population, and land urbanization factors. The spatial pattern of urbanization in Inner Mongolia from 1995 to 2020 has experience rapid growth, especially since 2000, as shown in Figure 5. The urbanization level of all districts in Inner Mongolia increased from 1995 to 2020 despite regional variations. The Hubao-E urban agglomeration saw the greatest increase among all districts. The Hubao-E urban agglomeration had the highest level of urbanization with a maximum value of 0.65 in 1995. There were many areas of the region with low levels of urbanization, such as Hulun Buir, Xilin Gol, and Alxa (CUB: 0–0.12). Since the grasslands and mountains of Northern Inner Mongolia create an ecological barrier, many ecological restoration projects have been implemented there, resulting in a relatively low level of urbanization development in this region. Hubao-E had the highest urbanization score, still higher than peripheral areas in 2015. The urbanization level in Inner Mongolia progressively decreased from the urban center to peripheral areas. In other words, the highest level of urbanization was found in the extended urban function zone, following the new urbanization zone, with the lowest level occurring in the ecological conservation zone.

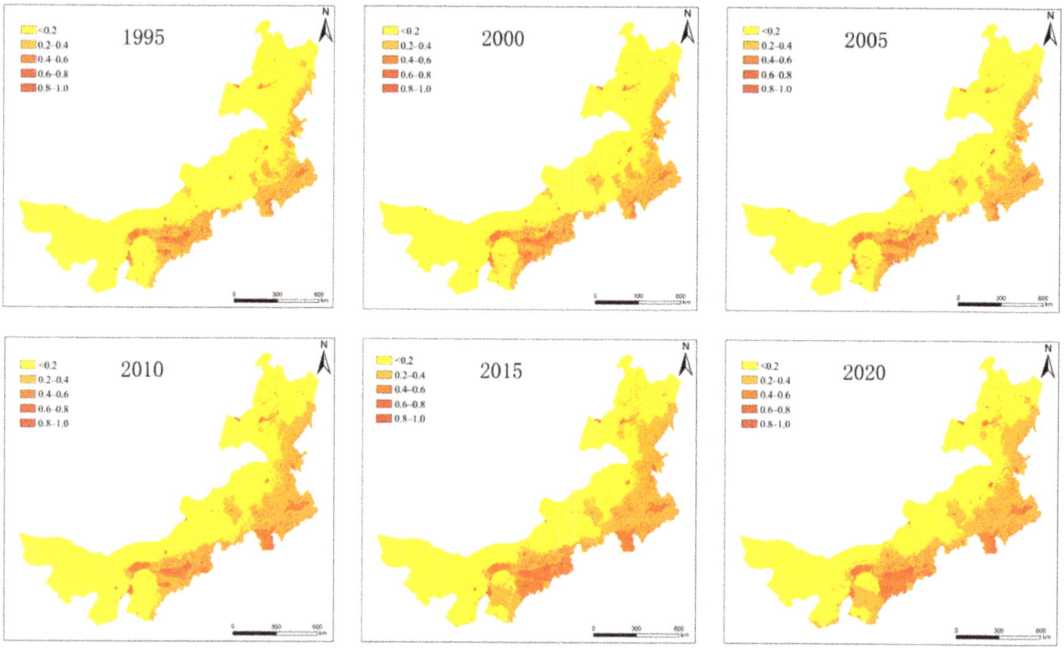

Figure 5. Spatial distribution of urbanization from 1995 to 2020 in Inner Mongolia.

3.3. *CCD between Urbanization and Ecosystem Services*

3.3.1. Coupling Degree of ESs and Urbanization

Figure 6 shows the CD of ESs and urbanization in different counties of Inner Mongolia from 1995 to 2020. The CD consists mainly of four categories: low-level coupling, antagonism, running-in, and high-level coupling. The coupling degree across Inner Mongolia shows a clear spatial distribution and wide differences among counties. In 1995, approximately 38% of the counties exhibited a low-level coupling between ESs and urbanization. High-level coupling and running-in coupling between ESs and urbanization occurred in 48% and 14% of the counties, respectively. By 2020, 4% of the counties had transitioned to high-level coupling from the running-in coupling. The coupling degree in the Horqin Right Middle Banner decreased from the running-in to low-level from 1995 to 2010 and has been maintained at running-in coupling levels since 2010. Ecosystem service provision exceeded the urbanization level of Ejin Horo Banner after 2005, and the gap between the two progressively increased, leading to an increase in its coupling degree. The Horqin District coupling degree continually declined. The coupling degrees of Bairin Youqi, Hainan District, Tumote Right Banner, and Dongsheng District steadily increased, while in Alxa Zuoqi, it decreased from 1995 to 2020. The coupling degrees of other counties in Inner Mongolia slightly changed but remained at the same coupling stage between 1995 and 2020.

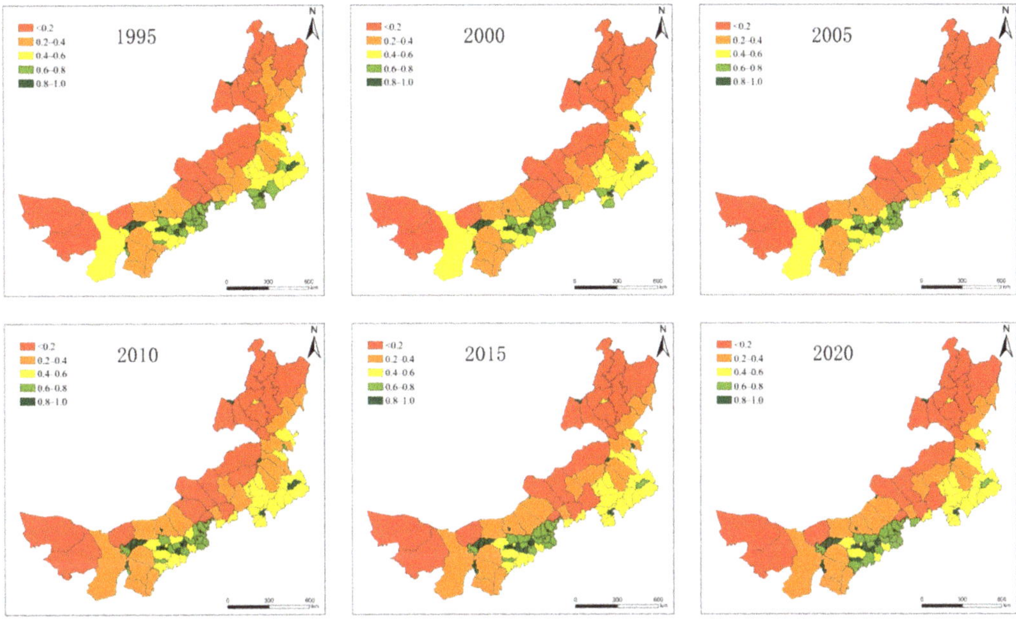

Figure 6. Coupling degree of ESs and urbanization in Inner Mongolia from 1995 to 2020.

3.3.2. Coupling Coordination Analysis of Urbanization and Ecosystem Services

The spatial distribution of CCD between ecosystem services and urbanization on a county level in Inner Mongolia was analyzed using ArcGIS software (Figure 7). Regarding the spatial pattern, the coordinative relationship performed better in the central urban agglomeration, namely Baotou, Hohhot, and Ordos counties, which is consistently in the primary or quality coordination development stage. In recent years, Ordos and Baotou faced industrial transformation and labor shortages, actively developed tourism as a means of improving their industrial structure, and invested in technological innovation to increase enterprise productivity and decrease environmental pollution produced by traditional industries [49]. In contrast, the cities located in the east region including Tongliao, Chifeng, and Xilin Gol have not yet achieved a coordination relationship between ecosystem services and urbanization, remaining in the basic or extreme incoordination phase. This is attributed to the large amount of grassland in these cities, as well as their rapid urban development in recent years. Intensive human activities have negatively affected ecosystem structure, process, and function and reduced the supply of ESs, resulting in incoordination between the two variables. The overall CCD in the western zones (Alxa) was unchanged, remaining at the extreme incoordination level. In the north region, containing the cities of Hulunbuir and Xing'an, the coordination relationship between ecosystem services and urbanization failed to perform well, remaining in the intermediate incoordination stage. In conclusion, urbanization in Inner Mongolia significantly impacted the ecosystem during the research period. Aside from a few counties, the coordination relationship between ecosystem services and urbanization was unfavorable, and there was a significant regional difference with a higher value in the northeastern than the southwestern parts.

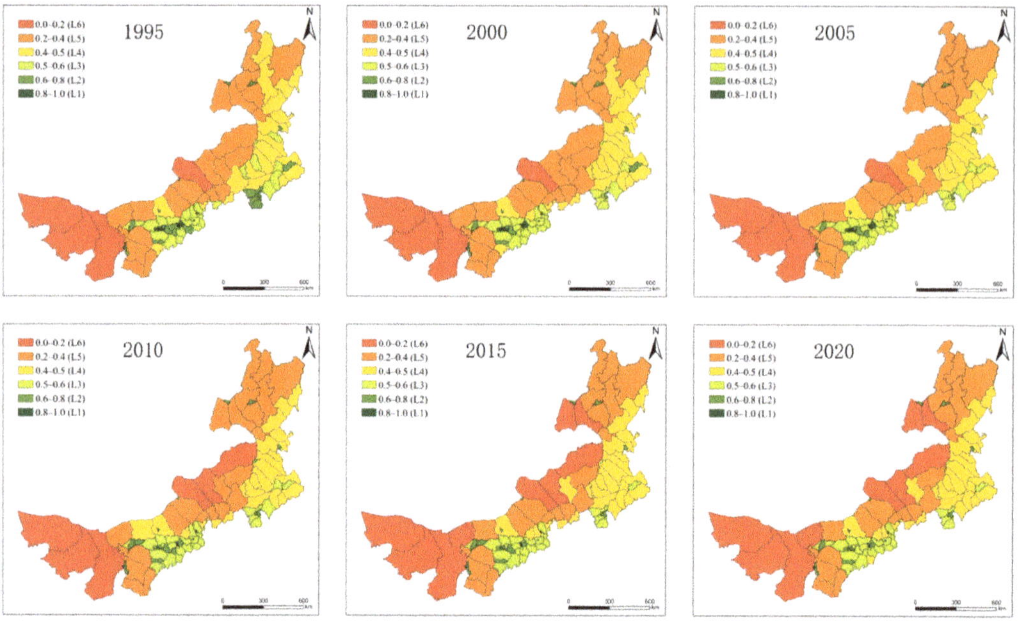

Figure 7. Spatial pattern of CCD in Inner Mongolia.

Figure 8 illustrates the average coupling coordination value of temporal scale variations in 101 counties from 1995 to 2020, which shows the coordination level of each county except for Alxa county. Since Alxa's temporal changes from 1995 to 2020 are not significant, we excluded them from Figure 8. The CCD values in approximately 66% of the counties exhibit a fluctuating downward trend. Located primarily in city centers and surrounding areas, these counties are mostly concentrated in the regions with the highest level of economic development and urbanization. Because of rapid industrialization and urbanization, land use patterns are largely disordered (construction land increases, and forestland and grassland decrease) and natural ecosystem foundations are changed, which adversely impacts the supply of core ESs. In contrast, the CCD changes in 23% of the counties showed an upward trend, including Chahar Right Front Banner, Dengkou, Dongsheng, Dong Ujimqin, Erenhot, Haibowan District, Hangjinhouqi, Hangjinqi, Hongshan District, Hollingol, Jining District, Linhe District, Sunite Right Banner, Wuda District, Urad Middle Qi, Wushenqi, Wuyuan Banner, West Ujimqin, Xilinhot, Ejin Horo, Yuquan District, and Jungar. The Grain for Green Program and the Natural Forest Conservation Program have contributed greatly to this growth. The implementation of these projects has considerably increased vegetation coverage, improving the CCD. The remaining 11% of counties, including Darhan Muminggan, Xin Barag Youqi in Hulunbuier, and Jarud in Xingan League, had no clear changes. Urbanization and ecosystem services coexist well in these counties on account of the low land use intensity, the small influence of human disturbances, and the relatively dense vegetation coverage.

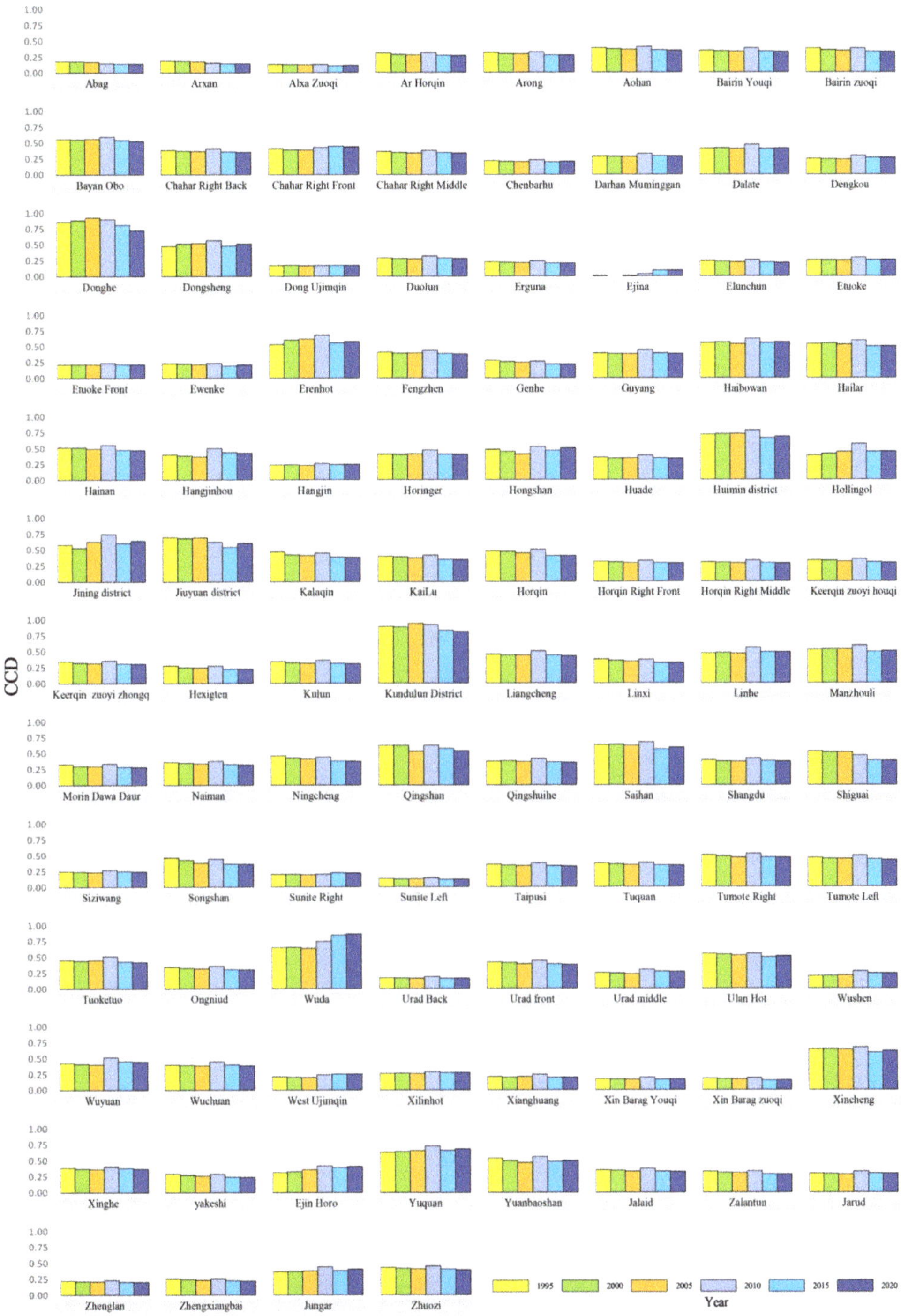

Figure 8. The temporal changes of the CCD between ESs and urbanization by county.

4. Discussion

4.1. Coupling Coordination Analysis between ESs and Urbanization

Inner Mongolia is a large arid and semi-arid region and a significant part of China's energy and food production base. The region has undergone dramatic changes in both socioeconomic and ecological environments during the research period. Urbanization in Inner Mongolia has grown rapidly: The Inner Mongolian urban population increased from approximately 8 million in the late 1980s to 15.14 million by 2015, and built-up space as a percentage of the provincial land area grew from 0.1% to 0.45% [33]. Growing prosperity came at an enormous cost to the environment. Generally, urbanization can negatively impact the ecological environment through economic development, population growth, traffic expansions, and energy consumption [50]. In contrast, the ecological environment can hinder urbanization development in many ways, including capital competition, population expulsion, policy interventions, and capital exclusion [51]. The rapid urbanization intensified the loss of ecosystem services, which adversely affected the ecological environment [52]. Thus, it is imperative to coordinate the relationship between the two factors reasonably and ensure the long-term development and sustainability of arid and semi-arid regions. If the contradiction between two variables decreases and both move in a positive direction, the CCD value will increase.

High-value ecosystem services are predominantly distributed in environmentally protected areas and low-urbanized areas, whereas low-value ecosystem services are mainly located in highly urbanized areas. Urbanization interacts with the benefits derived from land use types, which supports the findings of Zhou et al. [53]. Hulunbuir has the highest value of ecosystem services owing to the rich forest coverage, but coupled situation between ESV and urbanization has not yet reached a balanced and coordinated stage because of the low urbanization development. Chifeng City, as a highly urbanized region, expands its built-up land by occupying other types of arable land, which has resulted in a corresponding decline in ecosystem service value. Therefore, an increase in built-up land area per unit can negatively affects the total value of ecosystem services. In accordance with Tian's findings, a higher level of urbanization leads to greater pressure on ESV [54]. There are some studies suggesting that population density may also relate to this phenomenon. Our results further show that most counties' ecosystem services-urbanization CCDs are at the preliminary incoordination stage. While some counties are at the intermediate coordination stage, only a few were either not coordinated or at the quality coordination stage. The majority of Inner Mongolia counties were in the coordinate development ecological–economic coupling transition stage and the coordination relationship between ESs and urbanization in Inner Mongolia overall failed to reach a satisfactory level. Moreover, there were noticeable distinctions between different counties, with an evident spatial distribution pattern of lower CCD values in the northeast and west, with higher values in the center. Most counties with urban agglomeration were primarily in the quality, intermediate, and primary coordination stages, particularly the counties of Ordos, Baotou, and Hohhot. Despite having higher levels of urbanization, these counties still realized coordination development between ESs and urbanization. Hohhot, Baotou, and Ordos, as part of the Golden Triangle, have become the economic and social hubs of the autonomous region owing to their rich mineral resources, excellent geographical conditions, unique location advantages, and good industry matching. Recently, responding to central and local policies, these regions have an increased focus on green development, which is defined as a gradual shift from traditional regional development modes to a sustainable development mode. This development is in accordance with the global acceptance of reasonable protection and orderly utilization of limited resources. However, the CCD level was not high in most prefecture-level cities in Inner Mongolia, including Xilin Gol League, Hulunbuir, Ulaan Chab, Chifeng, and Tongliao, which have a high population but low per capita economic indicators. Secondary and tertiary industries are not prevalent, and the overall urbanization rate is slower than in the urban agglomeration area. Both Baynnur City and Alxa League have a low level of urbanization and development owing to their special geographical location and harsh environmental conditions. The Xing'an League is a region rich in agricultural and animal husbandry production. Agriculture accounts for 30% of the region's GDP, and the agricultural population is high. However, many other economic indicators are at low levels, resulting in a backward urbanization process. The above analysis illustrates that Inner Mongolia is still facing the double pressure of ecological environmental protection and socioeconomic growth.

Therefore, it is necessary to adopt different urbanization and ecological environmental policies in different economic regions to promote the coordinative development between ESs and urbanization in Inner Mongolia. Furthermore, the results of this study can serve as a useful reference for coordinative development between ESs and urbanization in other arid and semi-arid areas.

4.2. Policy Implications

We explored the coordination relationship between ESs and urbanization and found regional differences between the center and the rest of Inner Mongolia. Generally speaking, it is imperative to realize that the relationship between ecosystem service and urbanization is a coordinated process, not an unbalanced one [55]. Efforts should be made to protect land conservation and ecosystem health to ensure the synergistic development of ecosystem services and urbanization. Specifically, it is therefore more important to put city-specific policies into practice instead of implementing one-size-fits-all ecological restoration and protection measures. Cities situated in the northern, eastern, and western parts of the region must further enhance their development infrastructure and technical personnel to turn ecological environmental benefits into benefits of regional development and achieve green and sustainable development. The most efficient approach to increasing ESV is to protect ecosystems with high ESV equivalence factors (forest lands, wetlands, water bodies) [56]. Nevertheless, the government should not expand these ecosystems blindly when setting land use policies but rather should protect the original ecological state and ensure the integrity of the existing ecosystem. Several large-scale engineering projects implemented in Inner Mongolia have contributed to improving the overall level of the ecological environment, which fosters a better coordination relationship between socioeconomic development and the ecological environment. As for the cities in the central part of Inner Mongolia, the intensity and quantity of new built-up land in urban areas with extensive human disturbance need to be under strict control, and the conversion of existing built-up land into ecological land should be promoted. Additionally, it is essential to rationally organize and optimize green spaces in cities, including water bodies and parks, to balance the ecological environment and human needs. Unreasonable resource extraction and urban construction destroy ecosystem services by disrupting the supply-and-demand balance. It is thus important for these areas to close down industries that consume resources, develop green service industries, set up nature reserves, and form a system of ecosystem service that is low-consumption, green, and low-carbon, rather than one which consumes resources.

4.3. Limitations and Suggestions

This article identified the coordination interaction between ESs and urbanization by using an arid and semi-arid region as a case study. However, the current study has some deficiencies that need to be taken into account in future studies. First, no validation analysis was conducted; a greater quantity of field observation data is needed to confirm the reliability of the ES evaluation. Additionally, input data resolution greatly affects the accuracy of the estimated result. The data pertinent to the research area are largely on the national scale with a resolution of 1 km, leading to limitations and uncertainties in measuring ESs. Furthermore, the developed land equivalent factor was set to 0. While ESs provided by developed land have been viewed by many studies as having no value [57], some studies have argued that entertainment, tourism, and culture are valuable ESs provided by developed land. That semi-artificial or artificial ecosystems replacing natural ecosystems in the process of urbanization will undoubtedly result in ecosystem destruction, and this eventuality coupled with the hysteresis effect of ESs leads us to suggest setting the developed land value to 0. Lastly, the urbanization evaluation index systems were based only on three indices representing the aspects of the economy, population, and land. To obtain more reliable results, future studies should consider a more accurate and comprehensive urbanization assessment indicator system.

5. Conclusions

This study analyzed the spatio-temporal evolution of ESs and urbanization level in Inner Mongolia from 1995 to 2020, then utilized a CCDM model to explore the coordination relationship between the two variables. This study serves as a scientific foundation for ecological environment management and sustainable development strategies. The results indicate that (1) climate regulation, environmental purification, and biodiversity services decreased while other types of services increased. In terms of ecosystem service value categories, the regulation function has the greatest value followed by supply services and support services, and the cultural services are considered to be the lowest value function. There is evident spatial heterogeneity in the capacity of ES provision in different regions. High ESVs generally were found in the forest-rich cities located in northeastern Inner Mongolia, while medium-ESVs were found in the central and eastern plains dominated by grassland and cultivated land. ESVs in the West Region were low due to barren land distribution in this area. The above changes are mainly caused by the GFGP policy and rapid urbanization, both of which have led to vegetation coverage growth and built-up land expansion. (2) The urbanization level of Inner Mongolia had a noticeable increasing trend. Furthermore, the urbanization degree of most districts presented an upward trend, and urban expansion was concentrated in the developed city centers. An obvious hierarchy of urbanization levels exists in the study region, with the higher administrative level corresponding to the higher level of urbanization. Taking Hohhot as the core city and Baotou as the sub-core city, the spatial distribution characteristics of urbanization decrease from the center to the outside as well as from the south to the north. (3) Overall, the CCDs between ecosystem services and urbanization in most counties were too low. The CCD showed a spatial trend of higher values in the center and east, and lower values in the west and north. Relating to the spatial distribution of coupling and coordination between ecosystem services and urbanization, large areas of the Inner Mongolian Autonomous Region are still dominated by imbalances.

Author Contributions: Conceptualization, L.G.; methodology, L.N. and Y.Z.; formal analysis, L.N.; investigation, L.N.; data curation, L.N. and Y.Z.; writing—original draft preparation, L.N.; writing—review and editing, L.G. and L.N.; visualization, L.N.; supervision, L.G. All authors have read and agreed to the published version of the manuscript.

Funding: This research was funded by the Ministry of Science and Technology of the People's Republic of China (2022YFF1303001).

Data Availability Statement: The sources and preprocessing of data are in Sections 2.2 and 2.3. Other relevant data to support this study are available from the authors upon request.

Conflicts of Interest: The authors declare no conflict of interest.

References

1. NBSC. *Statistical Communique of the People's Republic Of China on the 2020 National Economic and Social Development*; China Statistics Press: Beijing, China, 2020.
2. Liu, N.; Liu, C.; Xia, Y.; Da, B. Examining the coordination between urbanization and eco-environment using coupling and spatial analyses: A case study in China. *Ecol. Indic.* **2018**, *93*, 1163–1175. [CrossRef]
3. Liu, W.; Jiao, F.; Ren, L.; Xu, X.; Wang, J.; Wang, X. Coupling coordination relationship between urbanization and atmospheric environment security in Jinan City. *J. Clean. Prod.* **2018**, *204*, 1–11. [CrossRef]

4. Shi, Y.; Feng, C.C.; Yu, Q.R.; Guo, L. Integrating supply and demand factors for estimating ecosystem services scarcity value and its response to urbanization in typical mountainous and hilly regions of south China. *Sci. Total Environ.* **2021**, *796*, 149032. [CrossRef]
5. Qiu, B.; Li, H.; Zhou, M.; Zhang, L. Vulnerability of ecosystem services provisioning to urbanization: A case of China. *Ecol. Indic.* **2015**, *57*, 505–513. [CrossRef]
6. Huang, Q.; Peng, B.; Elahi, E.; Wan, A. Evolution and driving mechanism of ecological security pattern: A case study of Yangtze River urban agglomeration. *Integr. Environ. Assess. Manag.* **2021**, *17*, 573–583. [CrossRef] [PubMed]
7. Wang, L.; Li, Q.; Bi, H.; Mao, X.Z. Human impacts and changes in the coastal waters of south China. *Sci. Total Environ.* **2016**, *562*, 108–114. [CrossRef]
8. Guerry, A.D.; Polasky, S.; Lubchenco, J.; Chaplin-Kramer, R.; Daily, G.C.; Griffin, R.; Ruckelshaus, M.; Bateman, I.J.; Duraiappah, A.; Elmqvist, T.; et al. Natural capital and ecosystem services informing decisions: From promise to practice. *Proc. Natl. Acad. Sci. USA* **2015**, *112*, 7348–7355. [CrossRef] [PubMed]
9. Arowolo, A.O.; Deng, X.; Olatunji, O.A.; Obayelu, A.E. Assessing changes in the value of ecosystem services in response to land-use/land-cover dynamics in Nigeria. *Sci. Total Environ.* **2018**, *636*, 597–609. [CrossRef]
10. Yuan, M.H.; Lo, S.L. Ecosystem services and sustainable development: Perspectives f1 rom the food-energy-water Nexus. *Ecosyst. Serv.* **2020**, *46*, 101217. [CrossRef]
11. Chen, W.; Chi, G.; Li, J. The spatial association of ecosystem services with land use and land cover change at the county level in China, 1995–2015. *Sci. Total Environ.* **2019**, *669*, 459–470. [CrossRef]
12. Long, H.; Liu, Y.; Hou, X.; Li, T.; Li, Y. Effects of land use transitions due to rapid urbanization on ecosystem services: Implications for urban planning in the new developing area of China. *Habitat Int.* **2014**, *44*, 536–544. [CrossRef]
13. Ai, J.; Feng, L.; Dong, X.; Zhu, X.; Li, Y. Exploring coupling coordination between urbanization and ecosystem quality (1985–2010): A case study from Lianyungang City, China. *Front. Earth Sci.* **2016**, *10*, 527–545. [CrossRef]
14. Al-Mulali, U.; Solarin, S.A.; Ozturk, I. Investigating the presence of the environmental Kuznets curve (EKC) hypothesis in Kenya: An autoregressive distributed lag (ARDL) approach. *Nat. Hazards* **2016**, *80*, 1729–1747. [CrossRef]
15. Li, C.; Li, J.; Wu, J. Quantifying the speed, growth modes, and landscape pattern changes of urbanization: A hierarchical patch dynamics approach. *Landsc. Ecol.* **2013**, *28*, 1875–1888. [CrossRef]
16. Li, B.; Chen, D.; Wu, S.; Zhou, S.; Wang, T.; Chen, H. Spatio-temporal assessment of urbanization impacts on ecosystem services: Case study of Nanjing City, China. *Ecol. Indic.* **2016**, *71*, 416–427. [CrossRef]
17. Costanza, R.; D'Arge, R.; De Groot, R.; Farber, S.; Grasso, M.; Hannon, B.; Limburg, K.; Naeem, S.; O'Neill, R.V.; Paruelo, J.; et al. The value of the world's ecosystem services and natural capital. *Nature* **1997**, *387*, 253–260. [CrossRef]
18. Zamboni, N.S.; Noleto Filho, E.M.; Carvalho, A.R. Unfolding differences in the distribution of coastal marine ecosystem services values among developed and developing countries. *Ecol. Econ.* **2021**, *189*, 107151. [CrossRef]
19. Rahman, M.M.; Szabó, G. Impact of land use and land cover changes on urban ecosystem service value in Dhaka, Bangladesh. *Land* **2021**, *10*, 793. [CrossRef]
20. Hou, X.; Liu, J.; Zhang, D.; Zhao, M.; Xia, C. Impact of urbanization on the eco-efficiency of cultivated land utilization: A case study on the Yangtze River Economic Belt, China. *J. Clean. Prod.* **2019**, *238*, 117916. [CrossRef]
21. Zhang, W.; Yang, D.; Huo, J. Studies of the relationship between city size and urban benefits in China based on a panel data model. *Sustainability* **2016**, *8*, 554. [CrossRef]
22. Wang, J.; Zhou, W.; Pickett, S.T.A.; Yu, W.; Li, W. A multiscale analysis of urbanization effects on ecosystem services supply in an urban megaregion. *Sci. Total Environ.* **2019**, *662*, 824–833. [CrossRef] [PubMed]
23. Peng, J.; Tian, L.; Liu, Y.; Zhao, M.; Hu, Y.N.; Wu, J. Ecosystem services response to urbanization in metropolitan areas: Thresholds identification. *Sci. Total Environ.* **2017**, *607–608*, 706–714. [CrossRef] [PubMed]
24. Cortinovis, C.; Geneletti, D. Ecosystem services in urban plans: What is there, and what is still needed for better decisions. *Land Use Policy* **2018**, *70*, 298–312. [CrossRef]
25. Ding, T.; Chen, J.; Fang, Z.; Chen, J. Assessment of coordinative relationship between comprehensive ecosystem service and urbanization: A case study of Yangtze River Delta urban Agglomerations, China. *Ecol. Indic.* **2021**, *133*, 108454. [CrossRef]
26. Xiao, R.; Lin, M.; Fei, X.; Li, Y.; Zhang, Z.; Meng, Q. Exploring the interactive coercing relationship between urbanization and ecosystem service value in the Shanghai–Hangzhou Bay Metropolitan Region. *J. Clean. Prod.* **2020**, *253*, 119803. [CrossRef]
27. Sun, Y.; Liu, S.; Dong, Y.; An, Y.; Shi, F.; Dong, S.; Liu, G. Spatio-temporal evolution scenarios and the coupling analysis of ecosystem services with land use change in China. *Sci. Total Environ.* **2019**, *681*, 211–225. [CrossRef]
28. Yu, Y.; Tong, Y.; Tang, W.; Yuan, Y.; Chen, Y. Identifying spatiotemporal interactions between urbanization and eco-environment in the Urban agglomeration in the middle reaches of the Yangtze River, China. *Sustainability* **2018**, *10*, 290. [CrossRef]
29. Sannigrahi, S.; Zhang, Q.; Pilla, F.; Joshi, P.K.; Basu, B.; Keesstra, S.; Roy, P.S.; Wang, Y.; Sutton, P.C.; Chakraborti, S.; et al. Responses of ecosystem services to natural and anthropogenic forcings: A spatial regression based assessment in the world's largest mangrove ecosystem. *Sci. Total Environ.* **2020**, *715*, 137004. [CrossRef]
30. Yao, X.; Kou, D.; Shao, S.; Li, X.; Wang, W.; Zhang, C. Can urbanization process and carbon emission abatement be harmonious? New evidence from China. *Environ. Impact Assess. Rev.* **2018**, *71*, 70–83. [CrossRef]

31. Zheng, J.; Hu, Y.; Boldanov, T.; Bazarzhapov, T.; Meng, D.; Li, Y.; Dong, S. Comprehensive assessment of the coupling coordination degree between urbanization and ecological environment in the Siberian and Far East Federal Districts, Russia from 2005 to 2017. *PeerJ* **2020**, *8*, e9125. [CrossRef]
32. Liu, S.; Bai, J.; Chen, J. Measuring SDG 15 at the county scale: Localization and practice of SDGs indicators based on geospatial information. *ISPRS Int. J. Geo-Inf.* **2019**, *8*, 515. [CrossRef]
33. Inner Mongolia Autonomous Region Bureau of Statistics. *Inner Mongolia Statistical Yearbooks 2020*; China Statistic Press: Beijing, China, 2020.
34. Xu, G.H.; Wu, J.G. Social-ecological transformations of Inner Mongolia: A sustainability perspective. *Ecol. Process.* **2016**, *5*, 23. [CrossRef]
35. Zhang, H.; Fan, J.; Cao, W.; Harris, W.; Li, Y.; Chi, W.; Wang, S. Response of wind erosion dynamics to climate change and human activity in Inner Mongolia, China during 1990 to 2015. *Sci. Total Environ.* **2018**, *639*, 1038–1050. [CrossRef] [PubMed]
36. Zhou, Y.; Ning, L.; Bai, X. Spatial and temporal changes of human disturbances and their effects on landscape patterns in the Jiangsu coastal zone, China. *Ecol. Indic.* **2018**, *93*, 111–122. [CrossRef]
37. de Groot, R.S.; Alkemade, R.; Braat, L.; Hein, L.; Willemen, L. Challenges in integrating the concept of ecosystem services and values in landscape planning, management and decision making. *Ecol. Complex.* **2010**, *7*, 260–272. [CrossRef]
38. Millennium Ecosystem Assessment. *Living Beyond Our Means: Natural Assets and Human Well-Being: Statement from the Board*; Millennium Ecosystem Assessment: Washington, DC, USA, 2005.
39. Xie, G.D.; Zhang, C.X.; Zhang, L.M.; Chen, W.H.; Li, S.M. Improvement of the evaluation method for ecosystem ser-vice value based on per unit area. *J. Nat. Resour.* **2015**, *30*, 1243–1254.
40. Xie, G.D.; Lu, C.X.; Leng, Y.F.; Zheng, D.; Li, S.C. Ecological assets valuation of the Tibetan Plateau. *J. Nat. Resour.* **2003**, *18*, 189–196.
41. Zhang, Y.; Liu, Y.; Zhang, Y.; Liu, Y.; Zhang, G.; Chen, Y. On the spatial relationship between ecosystem services and urbanization: A case study in Wuhan. *China. Sci. Total Environ.* **2018**, *637–638*, 780–790. [CrossRef]
42. Morzillo, A.T.; de Beurs, K.M.; Martin-Mikle, C.J. A conceptual framework to evaluate human-wildlife interactions within coupled human and natural systems. *Ecol. Soc.* **2014**, *19*, 44. [CrossRef]
43. Kurniawan, F.; Adrianto, L.; Bengen, D.G.; Prasetyo, L.B. The social-ecological status of small islands: An evaluation of island tourism destination management in Indonesia. *Tour. Manag. Perspect.* **2019**, *31*, 136–144. [CrossRef]
44. Zhu, S.; Huang, J.; Zhao, Y. Coupling coordination analysis of ecosystem services and urban development of resource-based cities: A case study of Tangshan city. *Ecol. Indic.* **2022**, *136*, 108706. [CrossRef]
45. Dong, F.; Li, W. Research on the coupling coordination degree of "upstream-midstream-downstream" of China's wind power industry chain. *J. Clean. Prod.* **2021**, *283*, 124633. [CrossRef]
46. Zhang, Z.; Li, Y. Coupling coordination and spatiotemporal dynamic evolution between urbanization and geological hazards–A case study from China. *Sci. Total Environ.* **2020**, *728*, 138825. [CrossRef] [PubMed]
47. Shi, T.; Yang, S.; Zhang, W.; Zhou, Q. Coupling coordination degree measurement and spatiotemporal heterogeneity between economic development and ecological environment—Empirical evidence from tropical and subtropical regions of China. *J. Clean. Prod.* **2020**, *244*, 118739. [CrossRef]
48. Ariken, M.; Zhang, F.; Chan, N.W.; Kung, H.T. Coupling coordination analysis and spatio-temporal heterogeneity between urbanization and eco-environment along the Silk Road Economic Belt in China. *Ecol. Indic.* **2021**, *121*, 107014. [CrossRef]
49. Zheng, H.; Miao, C.; Wu, J.; Lei, X.; Liao, W.; Li, H. Temporal and spatial variations in water discharge and sediment load on the Loess Plateau, China: A high-density study. *Sci. Total Environ.* **2019**, *666*, 875–886. [CrossRef]
50. Salvati, L.; Zambon, I.; Chelli, F.M.; Serra, P. Do spatial patterns of urbanization and land consumption reflect different socioeconomic contexts in Europe? *Sci. Total Environ.* **2018**, *625*, 722–730. [CrossRef]
51. Feng, X.; Fu, B.; Piao, S.; Wang, S.; Ciais, P.; Zeng, Z.; Lü, Y.; Zeng, Y.; Li, Y.; Jiang, X.; et al. Revegetation in China's Loess Plateau is approaching sustainable water resource limits. *Nat. Clim. Chang.* **2016**, *6*, 1019–1022. [CrossRef]
52. Grimm, N.B.; Faeth, S.H.; Golubiewski, N.E.; Redman, C.L.; Wu, J.; Bai, X.; Briggs, J.M. Global change and the ecology of cities. *Science* **2008**, *319*, 756–760. [CrossRef]
53. Zhou, D.; Tian, Y.; Jiang, G. Spatio-temporal investigation of the interactive relationship between urbanization and ecosystem services: Case study of the Jingjinji urban agglomeration, China. *Ecol. Indic.* **2018**, *95*, 152–164. [CrossRef]
54. Tian, Y.; Zhou, D.; Jiang, G. Conflict or coordination? Multiscale assessment of the spatio-temporal coupling relationship between urbanization and ecosystem services: The case of the Jingjinji Region, China. *Ecol. Indic.* **2020**, *117*, 106543. [CrossRef]
55. Cao, Y.; Kong, L.; Zhang, L.; Ouyang, Z. The balance between economic development and ecosystem service value in the process of land urbanization: A case study of China's land urbanization from 2000 to 2015. *Land Use Policy* **2021**, *108*, 105536. [CrossRef]
56. Chuai, X.; Huang, X.; Wang, W.; Wu, C.; Zhao, R. Spatial simulation of land use based on terrestrial ecosystem carbon storage in Coastal Jiangsu, China. *Sci. Rep.* **2014**, *4*, 5667. [CrossRef]
57. Xiong, Y.; Zhang, F.M.; Gong, C.A.; Luo, P. Spatial-temporal evolvement of ecosystem service value in Hunan Province based on LUCC. *Resour. Environ. Yangtze Basin* **2018**, *27*, 1397–1408.

land

MDPI

Article

Land Consumption Mapping with Convolutional Neural Network: Case Study in Italy

Giulia Cecili [1,*], Paolo De Fioravante [2], Luca Congedo [2], Marco Marchetti [1] and Michele Munafò [2]

[1] Department of Biosciences and Territory, University of Molise, C/da Fonte Lappone, 86090 Pesche, Italy
[2] Italian Institute for Environmental Protection and Research (ISPRA), Via Vitaliano Brancati 48, 00144 Rome, Italy
* Correspondence: g.cecili@studenti.unimol.it

Abstract: In recent years, deep learning (DL) algorithms have been widely integrated for remote sensing image classification, but fewer studies have applied it for land consumption (LC). LC is the main factor in land transformation dynamics and it is the first cause of natural habitat loss; therefore, monitoring this phenomenon is extremely important for establishing effective policies and sustainable planning. This paper aims to test a DL algorithm on high-resolution aerial images to verify its applicability to land consumption monitoring. For this purpose, we applied a convolutional neural networks (CNNs) architecture called ResNet50 on a reference dataset of six high-spatial-resolution aerial images for the automatic production of thematic maps with the aim of improving accuracy and reducing costs and time compared with traditional techniques. The comparison with the National Land Consumption Map (LCM) of ISPRA suggests that although deep learning techniques are not widely exploited to map consumed land and to monitor land consumption, it might be a valuable support for monitoring and reporting data on highly dynamic peri-urban areas, especially in view of the rapid evolution of these techniques.

Keywords: deep learning; convolutional neural networks; land consumption mapping; land cover; remote sensing; semantic segmentation

Citation: Cecili, G.; De Fioravante, P.; Congedo, L.; Marchetti, M.; Munafò, M. Land Consumption Mapping with Convolutional Neural Network: Case Study in Italy. *Land* **2022**, *11*, 1919. https://doi.org/10.3390/land11111919

Academic Editor: Rohan Bennett

Received: 29 September 2022
Accepted: 24 October 2022
Published: 28 October 2022

Publisher's Note: MDPI stays neutral with regard to jurisdictional claims in published maps and institutional affiliations.

1. Introduction

1.1. Background

Soil is one of the most complex and extremely fragile natural resources which provides fundamental ecosystem services for human wellbeing, society, and the environment, such as food production, biomass, raw materials, climate regulation, carbon storage, etc. [1–5]. Soil health is essential to meet the needs of current and future generations; therefore, it is a central topic in sustainable development strategies [2].

In recent years, rapid urbanization, industrialization, and infrastructure development are among the global trends which determined many land use and land cover changes [6]. Landscape changes are shaping society, the biosphere, and endangering soil health: about three-quarters of the Earth's surface has undergone anthropic pressures during the last centuries [7], becoming the most widespread cause of arable land loss, habitat destruction, and climate changes [2,8].

Land consumption (LC) is the land transformation process related to the construction of new buildings, the expansion of cities, the infrastructure of the territory, and other soil sealing interventions [2,9,10]. Land consumption is defined as the loss of agricultural, natural, and semi-natural areas due to the new artificial cover [2]. Due to LC, the soil loses its ability to provide ecosystem services and sustain biodiversity; it negatively affects our wellbeing and quality of our lives [2,11].

Land consumption monitoring represents a fundamental activity for effective policies and sustainable planning [6]. Remote sensing has proved over the years to be an extremely powerful tool for land monitoring [7,8,12–14]. In this context, land use (LU) and land cover (LC) classifications play a crucial role in monitoring environment dynamics, studying phenomena such as deforestation, land degradation, or land consumption, and supporting applications such as sustainable spatial planning and resource management policies [15,16].

In detail, the use of remote sensing has contributed to the creation of several cartographic products related to land consumption at the national and European levels. Among European products, there is the imperviousness degree layer (IMD), which is one of the four strata (imperviousness, forest, grassland, water, and wetness) belonging to the Copernicus Pan-European High-Resolution Layers (HRLs). The imperviousness product is a raster map with a spatial resolution of 10 m and with an update frequency of three years, which represents the degree of imperviousness of each pixel ranging from 0 to 100 (https://land.copernicus.eu/pan-european/high-resolution-layers, accessed on 29 April 2022). At the national level, ISPRA has been monitoring land consumption in Italy since 2006 through the publication and the annual update of the National Land Consumption Map (LCM) [2,6,17]. The final product is a 10 m resolution raster for the whole national Italian territory [2], with an overall accuracy of 99.7% [18].

This document presents a first attempt to propose a methodology to automate the monitoring of land consumption, exploiting the innovative technologies of deep learning for the management of big data [19] with a marginal intervention of the photointerpreter. With the development of new technologies, Earth observation environmental data are growing rapidly in complexity, size, and resolution [20], offering an unprecedented monitoring possibility. This information is called "big geospatial data" [21] and it is still underused. The underuse of this technology causes a loss of time and potential in decision-making processes [22].

Recently, the use of the deep learning (DL) has rapidly grown in the data analysis field. The application of this technology was made possible by two main factors: the increased availability of data and the increased computational capability [23]. Due to the progress made by the scientific community, DL has been successfully applied to a variety of fields [24,25] and is now considered a state-of-the-art technology in many areas. Remote sensing image classification is certainly one of these areas [26,27].

DL is a subcategory of machine learning (ML), which corresponds to the concept of artificial neural networks (ANNs). By learning from a set of ground truth data, these algorithms can perform specific tasks without explicit instructions and offer high predictive accuracy even when complex non-linear relationships exist. Hierarchical learning is followed by ANNs, which are made of multiple layers of artificial neurons. In order to produce the outputs, the network combines the inputs and modifies them according to weights, biases, and activation functions [28,29].

A family of DL algorithms known as convolutional neural networks (CNNs) has recently achieved significant results in computer vision [30]. It is particularly effective at detecting lines, angles, curves, etc., in the input image. The convolution operator is what makes them so effective for image classification [31].

CNNs have been used for land cover classification in remote sensing data; as a matter of fact, some studies investigated wetlands [32,33], while others investigated urban areas [34–37]. On the other hand, several studies have primarily focused on natural classes [38,39] or agricultural production areas [40].

These algorithms are nowadays the most popular in remote sensing due to their exceptional performance [29]. Indeed, if compared with other traditional pixel-wise models such as random forest (RF) or support vector machine (SVM), the CNNs have the following advantages: the ability to work directly with raw images, the reduction in input noise, the low overfitting tendency, and the ability to recognize elements even in rotated or distorted images (typical in remote sensing) [23,31,41].

1.2. Operational Requirements

Using traditional methods for cartographic production and updating involves low update frequencies, which compromises the possibility to use them in applications that require rapid and timely intervention [42]. The main LC/LU Copernicus data are updated every 3 or 6 years, while the LCM is the only data available for consumed land mapping at the national scale that has an annual update frequency. LCM updating is extremely time consuming since it requires manual inspection of the entire national territory and photointerpretation of the new land consumption. To this end, ISPRA has launched studies to make this activity less burdensome and less time consuming. Currently, some support products for photointerpretation have been introduced, called "masks of potential changes" [1,17] that narrow the area of investigation (it can be limited to 2.4% of the national territory) and significantly reduce the photointerpretation effort. However, there is still a need for tools that simplify data updating while maintaining high reliability, especially in peri-urban areas, where the greatest density of change occurs.

1.3. Objective

The study aims to show the high potential of CNNs in consumed land mapping as a preliminary step for using these tools in land consumption monitoring. In detail, the activity seeks for the first time to develop tools based on DL techniques that can be integrated into the updating process of the LCM carried out by ISPRA for the Italian territory. As there are few studies on the use of deep learning to assess land consumption, the results of this research offer the opportunity to explore the potential of deep learning to analyze this negative human impact on the landscape.

The objectives of the study are summarized as follows:

1. Testing of the DL in consumed land mapping;
2. Automatic production of thematic maps with the aim of improving accuracy and reducing costs and time compared with traditional techniques;
3. Demonstration of the strategic importance of artificial intelligence for the future of environmental monitoring, with particular reference to land consumption mapping.

The study shows an application of a CNN architecture called ResNet50 to detect consumed land. The training and validation of the model was carried out on a first study area located in the municipality of Rome, which was chosen in order to present a mixed composition in terms of natural and artificial land cover classes. The classification methodology was tested with reference to three different configurations of parameters (called "scenarios"). The most suitable scenario has been applied to new test areas located in the Italian region of Tuscany.

2. Materials and Methods

2.1. Study Area

The study area includes six sites located among the provinces of Florence, Siena, and Arezzo, in the Italian region of Tuscany (Figure 1). The six areas have a total area of 67.6 km^2 and are located at a latitude between 43°29′50.19″ N and 43°36′4.24″ N and a longitude between 11°19′46.39″ E and 11°32′35.33″ E. For the training and validation of the deep learning model, reference was made to an area of 5.7 km^2 located in the municipality of Rome (Figure 2).

(a) (b)

Figure 1. (**a**) Location of the study areas in Italy; (**b**) study areas.

(a) (b)

Figure 2. (**a**) Training area; (**b**) validation area.

2.2. Overview

The proposed methodology applied the deep learning algorithms on orthophoto images to verify the feasibility of creating land cover classification, with particular reference to the phenomenon of land consumption. The methodology follows the Cross-industry standard process for data mining (CRISP-DM) model, which is adapted to the study needs owing to its flexibility of use. The CRISP-DM is already used in the data mining field and describes the main phases of a project and their mutual relationship [43,44]. The classification methodology was tested with reference to three different configurations (called "scenarios"):

1. Scenario 1—binary classification distinguishing buildings from the residual areas;
2. Scenario 2—binary classification distinguishing consumed land and non-consumed land;
3. Scenario 3—identification and distinction of different consumed land classes from natural areas.

For each scenario, different sets of model parameters were tested in order to identify the most promising scenario and its most appropriate set of parameters. The optimal configuration was then used in the deployment phase for the classification of images in areas other than the training ones.

The processes were developed using PyTorch, an open-source machine learning library available for python programming language (https://pytorch.org/, accessed on 29 April 2022). The accuracy of the classifications produced by previously trained models was evaluated with the Semi-automatic Classification Plugin (SCP) for QGIS [45]. The methodology (Figure 3) is described in the following sections.

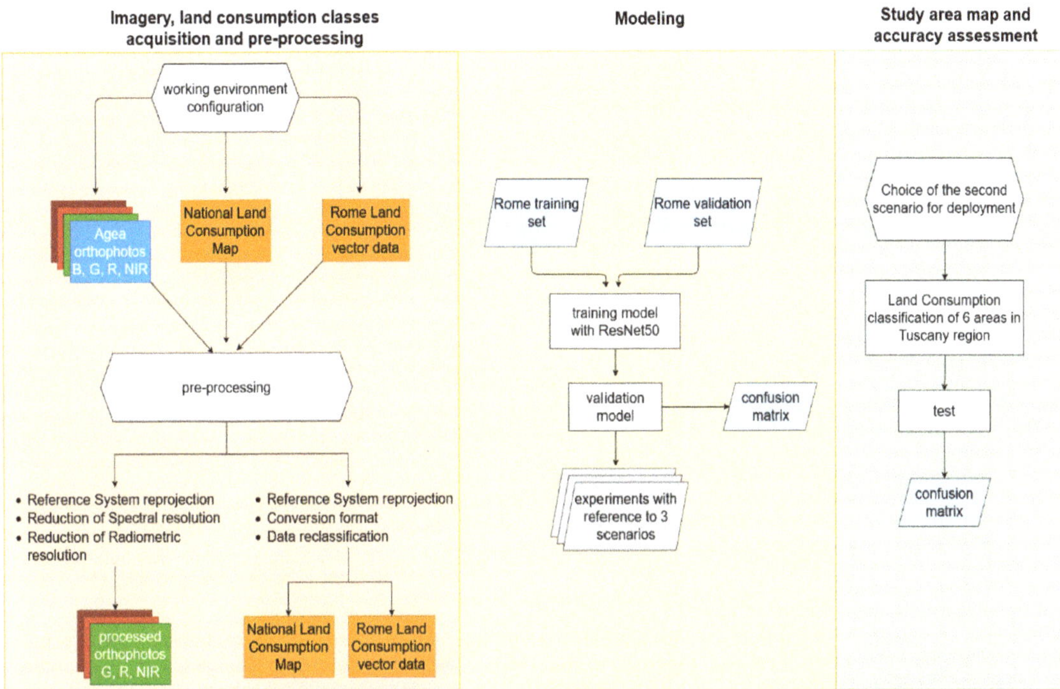

Figure 3. Study workflow.

2.3. Data Pre-Processing and Modeling

The training of the predictive model was carried out starting with two high-resolution images related to a portion of the municipality of Rome. The area was chosen because it is characterized by a fragmented landscape and a heterogeneous composition in terms of land cover, which is suitable for conducting land consumption studies. In detail, AGEA data relating to 2017 with a spatial resolution of 20 cm and four channels (red, green, blue, IR) were used (https://www.agea.gov.it/portal/page/portal/AGEAPageGroup/HomeAGEA/Servizidiutilita/Modulistica/Modulo_richiesta%20_ortofoto, accessed on 29 April 2022). The first orthophoto (a) of Figure 2 is used for the model training, while the second one (b) is used to verify the predictive capabilities of the model and then to validate it. The dataset was labeled using the vector version of the Land Consumption Map (LCM) produced and updated by ISPRA-SNPA. The vector data is available only for the territory of the municipality of Rome and is based on images with the same characteristics as the orthophotos used in this research [2]. The labeling process consisted of combining vector elements with the ground truth provided by the orthophotos, in order to achieve training and validation datasets.

2.3.1. Pre-Processing

The input datasets were preprocessed before the land cover classification based on the DL approach:

1. The AGEA orthophotos (in raster format) were reprojected in the UTM WGS84 reference system and resampled to obtain square pixels. The spectral resolution was reduced to 3 bands: the blue region (500–520 nm) was removed since it is the least significant for artificial surfaces classification, although it led to a reduction in the information content. The radiometric resolution was also reduced: the 16 bit orthophotos were converted into 8 bit ones.
2. The LCM was reprojected in the UTM WGS84 reference system, subjected to topological verification, and converted to Geojson format. The map has a three-level classification system, which distinguishes different consumed land classes. The data was reclassified to be suitable for the analysis of the three scenarios (Table 1).

Table 1. LCM reclassification table with respect to the three scenarios. For the first scenario, the class of "buildings" was highlighted; in the second scenario the consumed land classes were separated from natural surfaces; and in the third one, the different consumed land classes were considered separately.

Class Name	LCM Code	Reclassification Code		
		Scenario 1	Scenario 2	Scenario 3
Buildings	111	1	1	1
Paved roads	112	0	1	2
Other non-built-up sealed areas	116	0	1	3
Unpaved roads	121	0	1	4
Construction sites and other clayey areas	122	0	1	5
Non-consumed land	2	0	0	0
Artificial waterbodies	201	0	0	0
Roundabouts and junctions	202	0	0	0
Unpaved greenhouses	203	0	0	0

2.3.2. Deep Residual Networks

A CNN is a neural network that is an algorithm with a hierarchical structure used to recognize patterns in data. It typically consists of three kinds of layers: convolutional layers, pooling layers, and fully-connected layers. Various CNN models are commonly used in remote sensing investigations such as U-Net [46], DenseNet [47], SegNet [48], and ResNet [49] and the latter model was chosen for the study. It was decided to use ResNet since its structure has shown good capabilities in similar applications and with a wide range of datasets [29,49,50]. ResNet (Residual Network) was first introduced in 2015 by He [51] for image recognition purposes and has had great success in remote sensing applications [49,52,53].

The ResNet model includes several variants which differ in number of neural network layers: ResNet-18, ResNet-34, ResNet-50, ResNet-101, and ResNet-152. These architectures were introduced to solve the vanishing gradient problem in order to improve the accuracy and performance of the model. To this end, ResNet uses a novel component in addition to the traditional ones, called residual block. The new block is a stack of layers that applies the skip connection technique: in other words, the output of a layer is used to feed a layer that is not directly adjacent. Considering the accuracy levels, the available times, and computational resources, ResNet-50 was implemented.

ResNet-50 (Figure 4) consists of 50 layers which include 48 convolutions, 1 max pooling, and 1 average pooling. A final full-connected layer and a softmax function complete the network. Like any ResNet architecture, it starts with a convolution and max-pooling layer using 7×7 and 3×3 kernel sizes, respectively. Afterward, the ResNet-50 proceeds with its 4 stages, which include 3 residual blocks for each [51].

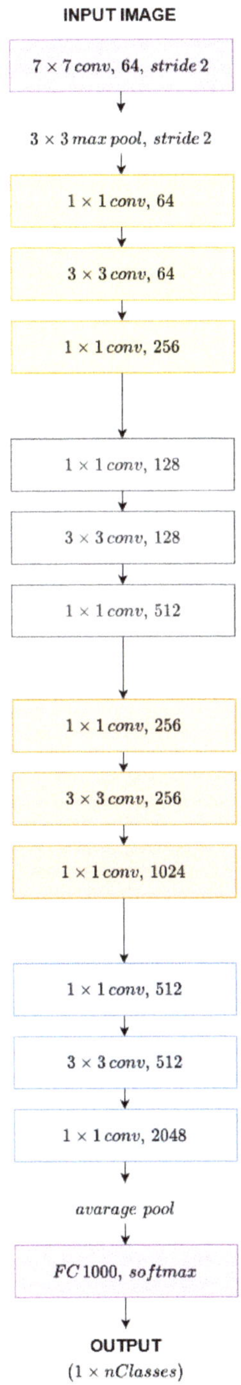

Figure 4. ResNet50 architecture.

2.3.3. Modeling

Modeling is the main phase of a DL experiment. The applied task can be categorized as semantic segmentation or pixel-by-pixel classification. The goal is labeling every pixel of an image with the corresponding semantic class using an automatic procedure. Convolutional neural networks (CNN) are well suited to gain test set labels. For this study, monotemporal input data was used to form a deep learning model, based on Resnet50 architecture [51].

The training phase was carried out considering 4 hyperparameters: chip size, chips per scene, batch size, and number of epochs. A total of 60 sets of values (called "experiments") were defined, each corresponding to a different combination of hyperparameter values; the 60 experiments were applied to the 3 scenarios in order to evaluate which combination of scenario and experiment provides the best results. Hyperparameter tuning is crucial for the quality of the prediction and this depended on the data analyst assessments. Choices were based on computer performance and on "trial and error". Each image was divided into chips (clippings of images) with dimensions ranging from 50 × 50 to 300 × 300 pixels, in relation to the extent of the image. Some of these chips were randomly extracted and used for model training: generally, we opted for a number of chips between 35 and 70 and a batch size smaller than 8. Randomization of image chips is advantageous as it ensures an independent distribution of samples and it is further justified in the case of study areas affected by urban sprawl. To identify the optimal hyperparameter values, several iterations were fitted; the model was trained using a random initialization of 5 epochs, up to the final model with 15 epochs. By increasing the number of epochs to 15, a constant trend was found for each scenario, which made it possible to compare them and identify the most promising one.

The result of the algorithm training was verified graphically by overlapping the produced maps, and through evaluation metrics reading. These metrics were calculated starting from the confusion matrix generated during the validation step.

The main indices computed are shown below:

$$\text{Precision} = \frac{(\text{True Positive})}{(\text{True Positive} + \text{False Positive})}, \quad (1)$$

$$\text{Recall} = \frac{(\text{True Positive})}{(\text{True Positive} + \text{False Negative})}, \quad (2)$$

$$\text{F1-score} = 2 \times \frac{\text{Recall} \times \text{Precision}}{\text{Recall} + \text{Precision}}, \quad (3)$$

2.4. *Deployment and Accuracy Assessment*

To verify the algorithm's ability to identify consumed land and its effectiveness on areas other than training areas, the methodology was applied to 6 AGEA images relating to the Tuscan territory. The areas were chosen in order to test the effectiveness of the algorithm on territories with different degrees of consumed land. Tiles 2 and 3 have a similar presence of consumed land compared with the training areas, while natural areas prevail in tiles 4, 5, and 6, where much of the consumed land is made up of roads. In detail, reference was made to the scenario (and its corresponding set of hyperparameters) that showed the best results. For this application, the chosen scenario was tested on a higher number of epochs, specifically 30. The classifications obtained were then compared with the ISPRA LCM, which is only available in raster format for the study area. This data is a nationwide 10 m resolution raster with a three-level classification system capable of distinguishing different classes of reversible and permanent consumed land. The LCM is obtained starting from the classification of rapid-eye images, integrated with OpenStreetMap data for road mapping and improved and updated through photointerpretation. For a detailed description of the LCM specifications, see [6,17,54]. To compare it with the LCM, the classification was resampled to 10 m of resolution, associating each pixel with a percentage artificial density value. The artificial surface density raster was then reclassified into a binary raster, associating the value 1 (consumed land) to the pixels with over 51% of the territory consumed, and 0 (not consumed land) to the others. These thresholds were chosen to make the two data comparable (in fact they are the same thresholds as the raster version of the LCM). Since the chosen approach uses a binary classification system (consumed land–non consumed land), the binary raster of the LCM was considered for the quantitative verification.

A confusion matrix was generated from the datasets comparison and the accuracy assessment was performed.

Overall accuracy: classification accuracy compared with ground truth:

$$\text{Overall Accuracy } (\%) = \frac{\text{number of correctly classified pixels}}{\text{total number of pixels}} \times 100 \quad (4)$$

User's accuracy: check how many pixels of a class i are correctly classified for the user:

$$\text{User's Accuracy } (\%) = \frac{\text{number of correctly classified pixels in class i}}{\text{number of pixels classified as class i (total row)}} \times 100 \quad (5)$$

Producer's accuracy: check how many pixels of a class i are correctly classified for the producer:

$$\text{Producer's Accuracy } (\%) = \frac{\text{number of correctly classified pixels in class i}}{\text{number of known pixels in class i (total column)}} \times 100 \quad (6)$$

Kappa coefficient: statistical index for assessing the degree of agreement between two assessments:

$$\text{Kappa coefficient} = \frac{\text{observed agreement} - \text{hypothetical probability}}{1 - \text{hypothetical probability}} \quad (7)$$

3. Results

3.1. Modeling

This paragraph shows the results of testing the DL in consumed land mapping using different model parameters and different reference classes. For the training of the algorithm, 60 experiments (distinct configurations of the model parameters) were considered. Figures 5–7 show four representative results for each of the three scenarios.

The results obtained applying the 60 experiments to the three approaches were compared (Figures 5–7) in order to identify the most promising configuration. Scenario 2 shows the best results even by varying the number of epochs and was subjected to evaluation metrics analysis (Table 2) in order to identify the most reliable experiment. This analysis is referred to as a 30-epoch training.

3.2. Deployment

The classification of the six orthophotos (Figure 8) relating to the study area was conducted with reference to experiment B1 of the second scenario. The results were analyzed in order to verify if the methodology allows the achievement of objectives two and three. In detail, two levels of analysis were conducted:

1. First, a visual comparison with orthophotos;
2. Then, calculating the accuracy with the SCP plug-in by comparing predictions with the binary version of the LCM (1 = consumed land, 0 = non-consumed land) (Table 3).

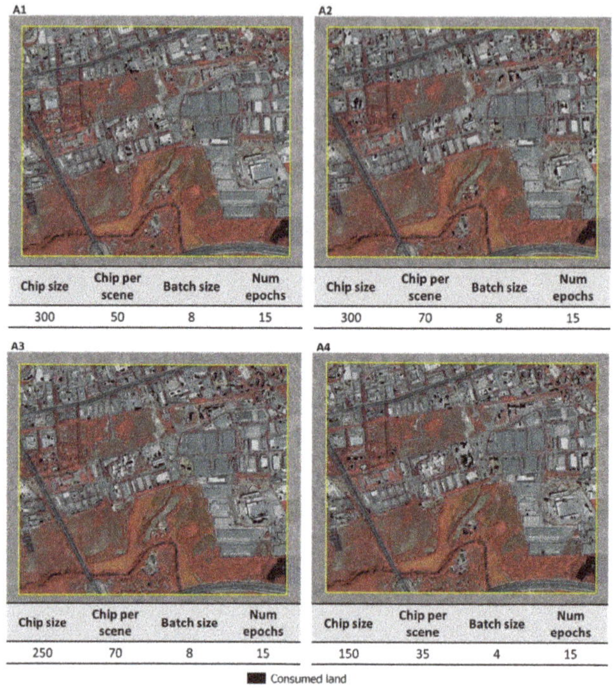

Figure 5. Classification examples relating to the first scenario, with reference to 4 distinct experiments.

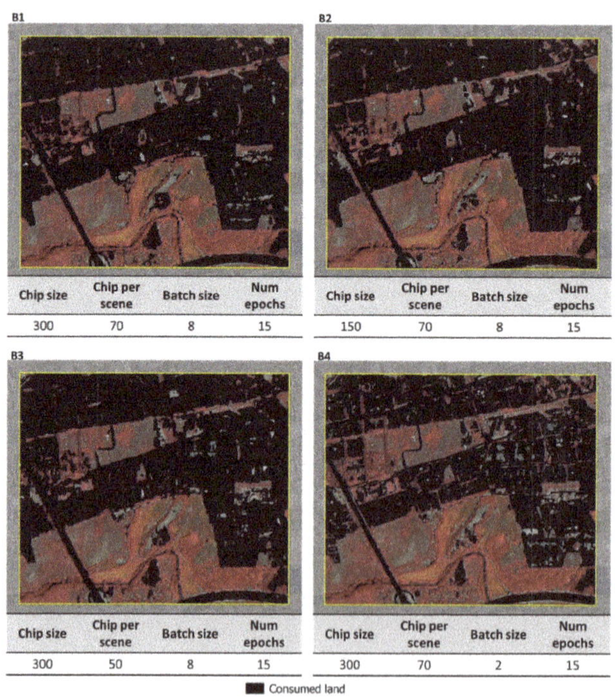

Figure 6. Classification examples relating to the second scenario, with reference to 4 distinct experiments.

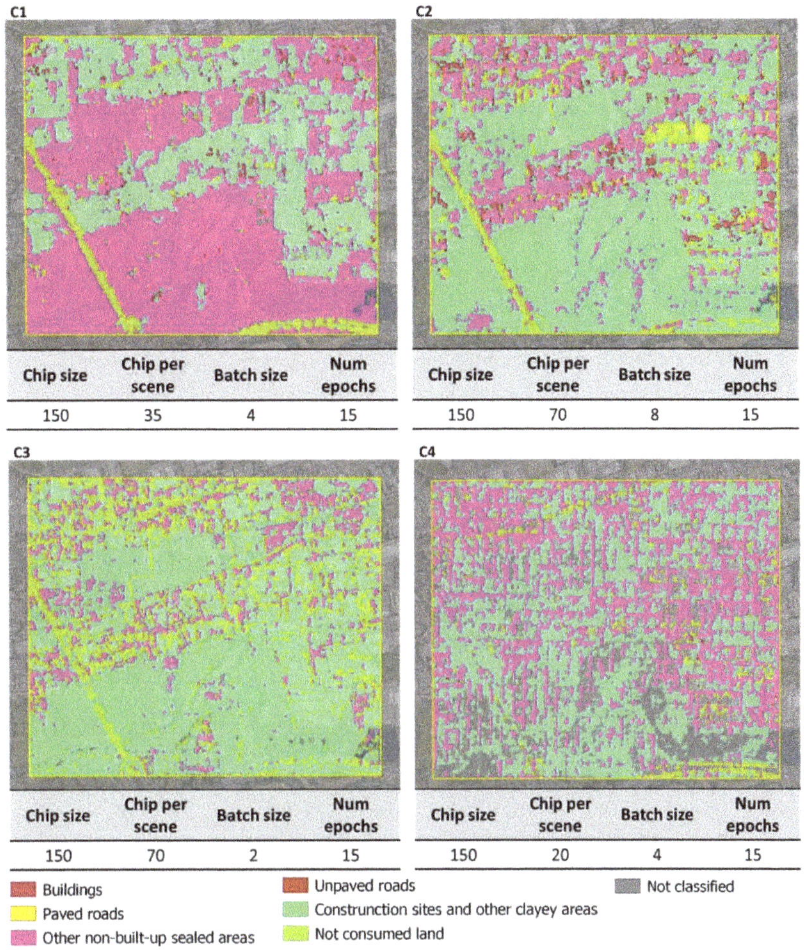

Figure 7. Classification examples relating to the third scenario, with reference to 4 distinct experiments.

Table 2. Scores of the four best experiments related to the second scenario. Class 0 identifies non-consumed land; class 1 identifies consumed land.

Accuracy	Experiment ID							
	B1		B2		B3		B4	
	Class 0	Class 1	Class 0	Class 1	Class 0	Class 1	Class 0	Class 1
Precision	0.931	0.890	0.926	0.863	0.876	0.852	0.881	0.936
Recall	0.897	0.825	0.872	0.822	0.869	0.763	0.944	0.759
F1-score	0.914	0.856	0.898	0.842	0.873	0.805	0.912	0.838

Table 3. Accuracy assessment by comparison with the LCM; 1 = consumed land, 0 = non-consumed land.

Tile 1 **Kappa = 0.32**					
		Land Consumption Map			
		1	**0**	**Tot. surface**	**Commission error**
DL classification	**1**	11	8	19	**41.55**
	0	36	1075	1111	**3.27**
	Tot. surface	48	1083	1131	**-**
	Omission error	**76.34**	**0.74**	-	**96.07**
Tile 2 **Kappa = 0.73**					
		Land Consumption Map			
		1	**0**	**Tot. surface**	**Commission error**
DL classification	**1**	200	32	232	**13.70**
	2	75	824	899	**8.35**
	Tot. surface	275	856	1131	**-**
	Omission error	**27.30**	**3.71**	-	**90.56**
Tile 3 **Kappa = 0.61**					
		Land Consumption Map			
		1	**0**	**Tot. surface**	**Commission error**
DL classification	**1**	95	37	133	**28.13**
	0	61	941	1002	**6.09**
	Tot. surface	156	978	1134	**-**
	Omission error	**39.01**	**3.82**	-	**91.33**
Tile 4 **Kappa = 0.38**					
		Land Consumption Map			
		1	**0**	**Tot. surface**	**Commission error**
DL classification	**1**	4	2	7	**34.49**
	0	11	1120	1131	**0.95**
	Tot. surface	15	1123	1138	**-**
	Omission error	**71.12**	**0.20**	-	**98.86**
Tile 5 **Kappa = 0.37**					

Table 3. *Cont.*

		Land Consumption Map			
		1	0	Tot. surface	Commission error
DL classification	1	6	6	13	**51.03**
	0	14	1168	1183	**1.21**
	Tot. surface	21	1175	1195	-
	Omission error	**69.74**	**0.55**	-	**98.26**
Tile 6 **Kappa = 0.29**					
		Land Consumption Map			
		1	0	Tot. surface	Commission error
DL classification	1	6	3	10	**35.65**
	0	25	1100	1125	**2.25**
	Tot. surface	32	1103	1135	-
	Omission error	**80.32**	**0.31**	-	**97.46**

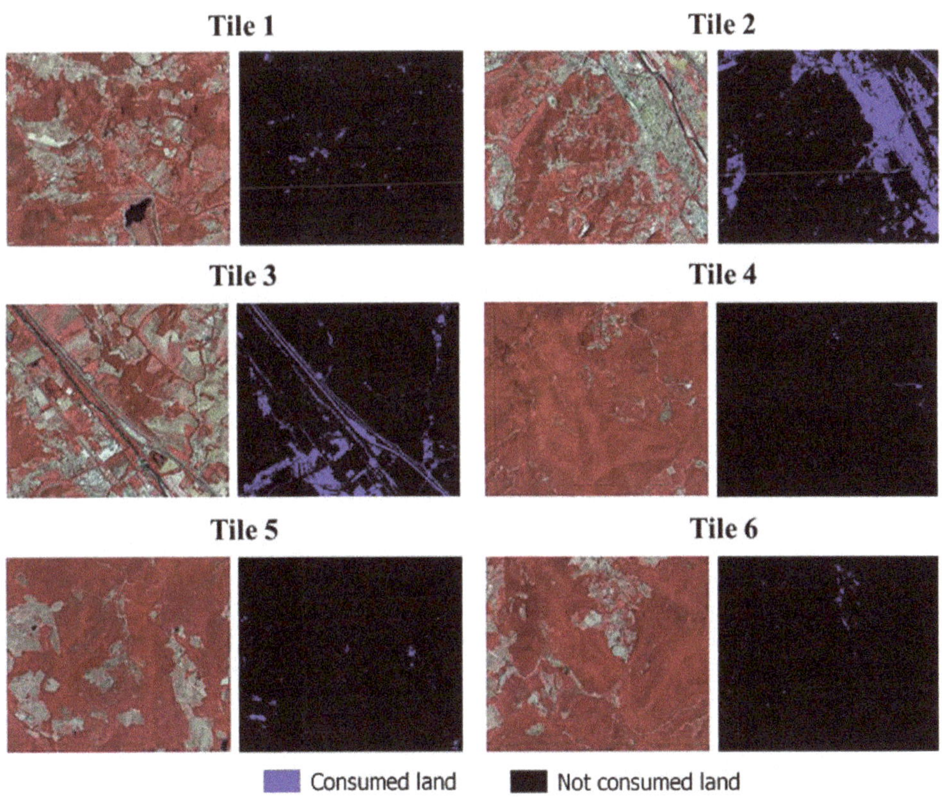

Figure 8. Classification results of the 6 study areas relating to B1 configuration.

4. Discussion

This research represents the first attempt to integrate DL techniques in the process of updating the LCM produced by ISPRA at the national scale and with high frequency. In this sense, the mapping of the consumed land is a preliminary step for the definition of support tools for the photointerpretation of land consumption in the future for the automation of the updating process, limiting the intervention of the operator as much as possible.

The first objective of the research concerned the verification of the applicability of the DL for the mapping of the consumed land, with reference to the classification of very high-resolution images for the Italian territory. Three different scenarios were illustrated: two binary classifications and a multi-category one. The analysis of the three methodologies highlights the second one (binary classification of consumed and non-consumed land) as the most accurate. As a matter of fact, the strategy of merging all the artificial surfaces into a single class produced good results. Slight variations between experiments are dictated by chip and batch size. Decreasing the batch number and using smaller chips, more precision was found in road boundaries and omission errors were reduced.

The second objective of the research concerned the possibility of automatically producing thematic maps with the aim of improving accuracy and reducing costs and time. Overall, the methodology allows to obtain consumed land classifications in a short time with good results especially on artificial areas, while the classification of shaded areas and reversible consumed land is still to be perfected. The first results are promising and show a good ability of the data to identify even small consumed areas. In this sense, we intend to perfect the methodology to produce diachronic classifications of the consumed land to obtain layers of potential changes, suitable in the first instance to support and increase the speed of the photointerpretation process in the most dynamic areas. In detail, the semantic segmentation algorithm was trained and deployed on a personal computer (spec- 12 GB RAM, Intel core i5-6200U CPU @ 2.40 GHz), which took roughly 50 min to complete the most efficient experiment running on CPU. Considering the available computing resources, the entire process required relatively short processing times. The speed of processing confirms the promising capabilities of the methodology in land monitoring. In the accuracy assessment, the classifications were compared with the LCM, which is the national benchmark for this theme. The methodology shows good results in the recognition of the built areas (Figure 9a), even the smallest ones. Looking at the results of Table 3 it is important to take into account that in some cases the algorithm correctly identifies consumed land which is not detected by the LCM (Figure 9d). In other cases, false omission and commission errors on properly mapped land consumed areas are related to the non-alignment caused by the different characteristics of the two data (Figure 9e). In detail, the best results are obtained for tiles 2 and 3 (Kappa = 0.73 and 0.61, respectively) (Table 3), which are characterized by the highest presence of built-up areas, while the lowest values are related to tiles 4, 5, and 6, which are characterized by a low density of buildings and consumed land is primarily determined by the presence of roads. Systematic errors that affect accuracy are due to the still experimental nature of the methodology. Commission errors are mainly located in tiles 2 and 3 and they correspond to a river mapped as consumed land. The error is due to the lack of training areas for this land cover class. Other minor commissions concern uncultivated agricultural areas in the reference period. The methodology shows higher accuracy in built-up areas, while in sparsely urbanized areas the accuracy is strongly affected by the errors associated with the mapping of the roads. In this regard, comparing the results of the accuracy assessment with the LCM at the third classification level helps to analyze the consumed classes that show more classification errors. In detail, omission errors mainly affect the road network, which is not mapped correctly in the presence of buildings or rows of trees near the roads (Figure 9b). Portions of road not covered by shadows or obstacles are correctly mapped. It should be noted that the LCM uses the OpenStreetMap database for mapping roads, which has been updated by photointerpretation; therefore, it is characterized by high accuracy even for hidden roads, which are difficult to map only through spectral information. Other omission errors mainly refer to areas mapped by the LCM as reversible consumed land. This category includes unpaved roads, sports fields, and other areas where soil has been removed (Figure 9c), many of which are partially renaturalized. The complexity in distinguishing bare soil from sealed surfaces is underlined in several experiences [55,56]. In this regard, a bibliographic analysis offers the possibility to introduce and use hyperspectral data [39,57,58], SAR data [59], or consider multitemporal data [60] to improve the recognition of urban or industrial areas and bare soil and to verify their evolution during the year, evaluating the appearance of vegetation or the occurrence of changes.

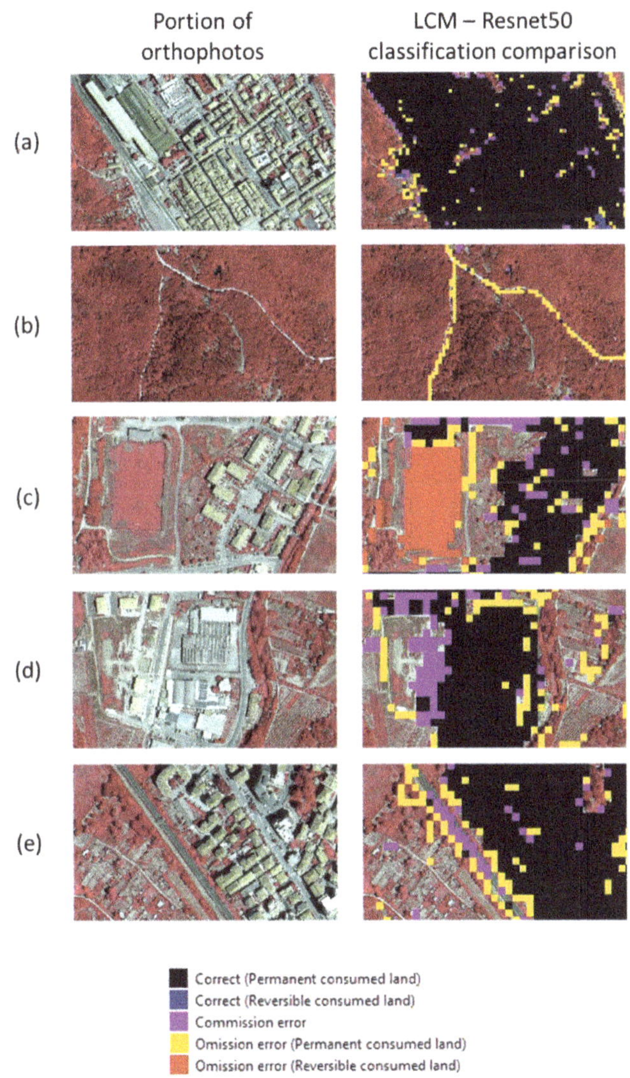

Figure 9. Comparison of the classification results with LCM: (**a**) Example of an excellent recognition of built-up areas; (**b**) omission error of a secondary road in a forest area; (**c**) omission error referred to as a sports field mapped by the LCM as reversible consumed land; (**d**) example of the DL algorithm correctly mapping an area that LCM does not identify as consumed land; and (**e**) errors related to the different resolutions of the two considered data.

The third objective of the research aims to demonstrate the strategic importance of using artificial intelligence for the land consumption mapping. If a first important objective concerns refinement in the identification of reversible consumed land, a further significant step forward will be linked to its use for the detection of land consumption. The study was developed together with other ISPRA activities [1,6,17] for the creation of tools that can facilitate and expedite land consumption monitoring operations. In this sense, the methodology is already a useful tool at present, allowing the rapid production of support layers for photointerpretation and it is useful for identifying the areas of potential change. Compared with other tools already developed by ISPRA [17], the higher spatial resolution of these products makes them particularly suitable for detecting changes in the peri-urban area, where the greater density of small changes is generally concentrated. However, future developments are possible with the methodology. Compared with the traditional algorithms of ML (decision trees, random forest, K-nearest neighbor, etc.), the DL is an adaptive system that is self-regulated according to the input data. The purpose of the DL is to find the most complex and hidden relationships between data, with the aim to reduce the human intervention as much as possible. Furthermore, ANNs allow to analyze raw data without heavy pre-processing and they can recognize distorted and rotated elements. The current state-of-the-art of the DL for land monitoring applications allows achieving good results only by using very high-resolution images, which are not free of charge. An interesting future methodology development could concern the experimentation of the most modern architectures of ANNs with multispectral high-resolution images. For example, the use of Sentinel images of the Copernicus Program would allow the availability of free, multitemporal, and multispectral data, with important advantages from the point of view of classification. The use of multispectral data allows researchers to better distinguish the surfaces based on their spectral characteristics, while analyzing the temporal variation of the land cover surfaces would allow evaluation of their permanence. This, together with the choice of appropriate parameters of the neural network, are potential approaches aimed at improving accuracy. Given the encouraging results in the medium–high-density urban environment, DL algorithms could represent a turning point for land consumption monitoring in peri-urban areas, which are extremely dynamic areas that need special attention and timely interventions.

5. Conclusions

This study introduces a spectral–spatial DL-based model for consumed land mapping. The proposed method experiments with a CNN network called ResNet50 on a reference dataset of high spatial resolution aerial images, considering three scenarios. The performances achieved reveal its promising capabilities in land monitoring if optimal parameters and scenes are used for model training: as a matter of fact, the analysis of the three scenarios shows that the training scene choice has a serious impact on classification accuracy.

Regarding objective two, the comparison with the LCM shows high congruence in permanent consumed land mapping and, in some cases, the CNN model was proven to be more effective than LCM. On the other hand, mapping reversible consumed land is more challenging, due to the need to consider land use characteristics in addition to land cover. In this sense, it is possible to try to improve the accuracy of the methodology by focusing on a more accurate choice of training datasets and integrating different types of data (such as multispectral images or multitemporal datasets) and on the optimization of the parameters. The reliability of the model is strongly influenced by the choice of input data and parameters and is therefore not suitable for application in a vast and heterogeneous territory; however, the results show high reliability in medium–high-density areas. It follows that the methodology would be useful to support land consumption monitoring in peri-urban areas, which are among the areas most affected by the phenomenon [2].

Following the guidelines of the European Commission, in order to achieve the goal "no net land take" by 2050 [61], it is necessary to map and monitor land consumption to avoid its negative consequences, such as loss of high-quality agricultural land, the effects on climate change, biodiversity loss, increased risk of flooding, etc. Overall, regarding the third theme, owing to the high accuracy obtained with the proposed model, improvements in permanent consumed land detection can be achieved. If properly developed, this methodology could provide a valuable aid in change detection applications. Therefore, deep learning techniques could contribute to updating the National Land Consumption Map and to a more efficient production of environmental indicators. Deep learning is a rapidly expanding technology; however, it is still not widespread enough for land monitoring and land consumption mapping, although it would lead to concrete progress in these fields. It is therefore essential to raise research awareness even in the non-informatics field, with the aim to acquire enough knowledge and practice to exploit the most sophisticated cutting-edge techniques.

Author Contributions: Conceptualization, G.C., L.C. and M.M. (Michele Munafò); methodology, G.C. and L.C.; software, G.C., P.D.F. and L.C.; validation, G.C. and P.D.F.; formal analysis, G.C., P.D.F. and L.C.; investigation, G.C. and L.C.; resources, M.M. (Michele Munafò) and M.M. (Marco Marchetti); data curation, G.C., P.D.F. and L.C.; writing—original draft preparation, G.C. and P.D.F.; visualization, G.C. and P.D.F.; supervision, L.C., M.M. (Michele Munafò) and M.M. (Marco Marchetti); and funding acquisition, M.M. (Marco Marchetti). All authors have read and agreed to the published version of the manuscript.

Funding: This research was funded by the University of Molise.

Institutional Review Board Statement: Not applicable.

Informed Consent Statement: Not applicable.

Data Availability Statement: The data are not publicly available because they are part of ongoing research.

Conflicts of Interest: The authors declare no conflict of interest.

References

1. De Fioravante, P.; Luti, T.; Cavalli, A.; Giuliani, C.; Dichicco, P.; Marchetti, M.; Chirici, G.; Congedo, L.; Munafò, M. Multispectral Sentinel-2 and Sar Sentinel-1 Integration for Automatic Land Cover Classification. *Land* **2021**, *10*, 611. [CrossRef]
2. Munafò, M. *Consumo Di Suolo, Dinamiche Territoriali e Servizi Ecosistemici*; Edizione 2021; ISPRA, SNPA: Rome, Italy, 2021; ISBN 9788844810597.
3. Chirici, G.; Giannetti, F.; Mazza, E.; Francini, S.; Travaglini, D.; Pegna, R.; White, J.C. Monitoring Clearcutting and Subsequent Rapid Recovery in Mediterranean Coppice Forests with Landsat Time Series. *Ann. For. Sci.* **2020**, *77*, 1–14. [CrossRef]
4. Marchetti, M.; Vizzarri, M.; Sallustio, L. Towards Countryside Revival: Reducing Impacts of Urban Expansion on Land Benefits. In *Agrourbanism*; Springer: Berlin/Heidelberg, Germany, 2019; pp. 207–222.
5. Costanza, R.; d'Arge, R.; de Groot, R.; Farber, S.; Grasso, M.; Hannon, B.; Limburg, K.; Naeem, S.; O'neill, R.V.; Paruelo, J. The Value of the World's Ecosystem Services and Natural Capital. *Nature* **1997**, *387*, 253–260. [CrossRef]
6. Strollo, A.; Smiraglia, D.; Bruno, R.; Assennato, F.; Congedo, L.; de Fioravante, P.; Giuliani, C.; Marinosci, I.; Riitano, N.; Munafò, M. Land Consumption in Italy. *J. Maps* **2020**, *16*, 113–123. [CrossRef]
7. Shukla, P.R.; Skeg, J.; Buendia, E.C.; Masson-Delmotte, V.; Pörtner, H.-O.; Roberts, D.C.; Zhai, P.; Slade, R.; Connors, S.; van Diemen, S. *Climate Change and Land: An IPCC Special Report on Climate Change, Desertification, Land Degradation, Sustainable Land Management, Food Security, and Greenhouse Gas Fluxes in Terrestrial Ecosystems*; Intergovernmental Panel on Climate Change: Geneva, Switzerland, 2019.
8. Dewan, A.M.; Yamaguchi, Y. Land Use and Land Cover Change in Greater Dhaka, Bangladesh: Using Remote Sensing to Promote Sustainable Urbanization. *Appl. Geogr.* **2009**, *29*, 390–401. [CrossRef]
9. Lal, R. Restoring Soil Quality to Mitigate Soil Degradation. *Sustainability* **2015**, *7*, 5875–5895. [CrossRef]
10. Oldeman, L.R.; Hakkeling, R.T.A.; Sombroek, W.G. *World Map of the Status of Human-Induced Soil Degradation: An Explanatory Note*; International Soil Reference and Information Centre: Wageningen, The Netherlands, 1990; ISBN 9066720425.
11. Cowie, A.L.; Orr, B.J.; Castillo Sanchez, V.M.; Chasek, P.; Crossman, N.D.; Erlewein, A.; Louwagie, G.; Maron, M.; Metternicht, G.I.; Minelli, S.; et al. Land in Balance: The Scientific Conceptual Framework for Land Degradation Neutrality. *Environ. Sci. Policy* **2018**, *79*, 25–35. [CrossRef]
12. Attri, P.; Chaudhry, S.; Sharma, S. Remote Sensing & GIS Based Approaches for LULC Change Detection—A Review. *Int. J. Curr. Eng. Technol.* **2015**, *5*, 3126–3137.
13. Shalaby, A.; Tateishi, R. Remote Sensing and GIS for Mapping and Monitoring Land Cover and Land-Use Changes in the Northwestern Coastal Zone of Egypt. *Appl. Geogr.* **2007**, *27*, 28–41. [CrossRef]

14. Rogan, J.; Chen, D.M. Remote Sensing Technology for Mapping and Monitoring Land-Cover and Land-Use Change. *Prog. Plan.* **2004**, *61*, 301–325. [CrossRef]
15. Winkler, K.; Fuchs, R.; Rounsevell, M.; Herold, M. Global Land Use Changes Are Four Times Greater than Previously Estimated. *Nat. Commun.* **2021**, *12*, 2501. [CrossRef] [PubMed]
16. Solórzano, J.V.; Mas, J.F.; Gao, Y.; Gallardo-Cruz, J.A. Land Use Land Cover Classification with U-Net: Advantages of Combining Sentinel-1 and Sentinel-2 Imagery. *Remote Sens.* **2021**, *13*, 3600. [CrossRef]
17. Luti, T.; de Fioravante, P.; Marinosci, I.; Strollo, A.; Riitano, N.; Falanga, V.; Mariani, L.; Congedo, L.; Munafò, M. Land Consumption Monitoring with Sar Data and Multispectral Indices. *Remote Sens.* **2021**, *13*, 1586. [CrossRef]
18. Munafò, M. *Consumo Di Suolo, Dinamiche Territoriali e Servizi Ecosistemici*; ISPRA, SNPA: Rome, Italy, 2019.
19. Gandomi, A.; Haider, M. Beyond the Hype: Big Data Concepts, Methods, and Analytics. *Int. J. Inf. Manag.* **2015**, *35*, 137–144. [CrossRef]
20. Audebert, N.; Saux, B.L.; Lefèvre, S. Semantic Segmentation of Earth Observation Data Using Multimodal and Multi-Scale Deep Networks. In Proceedings of the Asian Conference on Computer Vision, Taipei, Taiwan, 20–24 November 2016; pp. 180–196.
21. Soille, P.; Burger, A.; de Marchi, D.; Kempeneers, P.; Rodriguez, D.; Syrris, V.; Vasilev, V. A Versatile Data-Intensive Computing Platform for Information Retrieval from Big Geospatial Data. *Future Gener. Comput. Syst.* **2018**, *81*, 30–40. [CrossRef]
22. Gibert, K.; Horsburgh, J.S.; Athanasiadis, I.N.; Holmes, G. Environmental Data Science. *Environ. Model. Softw.* **2018**, *106*, 4–12. [CrossRef]
23. Hoeser, T.; Kuenzer, C. Object Detection and Image Segmentation with Deep Learning on Earth Observation Data: A Review-Part I: Evolution and Recent Trends. *Remote Sens.* **2020**, *12*, 1667. [CrossRef]
24. Barash, Y.; Guralnik, G.; Tau, N.; Soffer, S.; Levy, T.; Shimon, O.; Zimlichman, E.; Konen, E.; Klang, E. Comparison of Deep Learning Models for Natural Language Processing-Based Classification of Non-English Head CT Reports. *Neuroradiology* **2020**, *62*, 1247–1256. [CrossRef] [PubMed]
25. Affonso, C.; Rossi, A.L.D.; Vieira, F.H.A.; de Carvalho, A.C.P.; de Carvalho, L.F. Deep Learning for Biological Image Classification. *Expert Syst. Appl.* **2017**, *85*, 114–122. [CrossRef]
26. Debella-Gilo, M.; Gjertsen, A.K. Mapping Seasonal Agricultural Land Use Types Using Deep Learning on Sentinel-2 Image Time Series. *Remote Sens.* **2021**, *13*, 289. [CrossRef]
27. Makantasis, K.; Karantzalos, K.; Doulamis, A.; Doulamis, N. Deep Supervised Learning for Hyperspectral Data Classification through Convolutional Neural Networks. In Proceedings of the International Geoscience and Remote Sensing Symposium (IGARSS), Milan, Italy, 26–31 July 2015; Volume 2015, pp. 4959–4962.
28. Zhang, X.; Han, L.; Han, L.; Zhu, L. How Well Do Deep Learning-Based Methods for Land Cover Classification and Object Detection Perform on High Resolution Remote Sensing Imagery? *Remote Sens.* **2020**, *12*, 417. [CrossRef]
29. Li, Y.; Zhang, H.; Xue, X.; Jiang, Y.; Shen, Q. Deep Learning for Remote Sensing Image Classification: A Survey. *Wiley Interdiscip. Rev. Data Min. Knowl. Discov.* **2018**, *8*, e1264. [CrossRef]
30. LeCun, Y.; Bottou, L.; Bengio, Y.; Haffner, P. Gradient-Based Learning Applied to Document Recognition. *Proc. IEEE* **1998**, *86*, 2278–2324. [CrossRef]
31. Raschka, S.; Patterson, J.; Nolet, C. Machine Learning in Python: Main Developments and Technology Trends in Data Science, Machine Learning, and Artificial Intelligence. *Information* **2020**, *11*, 193. [CrossRef]
32. Jamali, A.; Mahdianpari, M.; Brisco, B.; Granger, J.; Mohammadimanesh, F.; Salehi, B. Comparing Solo versus Ensemble Convolutional Neural Networks for Wetland Classification Using Multi-Spectral Satellite Imagery. *Remote Sens.* **2021**, *13*, 2046. [CrossRef]
33. DeLancey, E.R.; Simms, J.F.; Mahdianpari, M.; Brisco, B.; Mahoney, C.; Kariyeva, J. Comparing Deep Learning and Shallow Learning for Large-Scalewetland Classification in Alberta, Canada. *Remote Sens.* **2020**, *12*, 2. [CrossRef]
34. Arndt, J.; Lunga, D. Large-Scale Classification of Urban Structural Units from Remote Sensing Imagery. *IEEE J. Sel. Top Appl. Earth Obs. Remote Sens.* **2021**, *14*, 2634–2648. [CrossRef]
35. Reda, K.; Kedzierski, M. Detection, Classification and Boundary Regularization of Buildings in Satellite Imagery Using Faster Edge Region Convolutional Neural Networks. *Remote Sens.* **2020**, *12*, 2240. [CrossRef]
36. Liu, C.; Zeng, D.; Wu, H.; Wang, Y.; Jia, S.; Xin, L. Urban Land Cover Classification of High-Resolution Aerial Imagery Using a Relation-Enhanced Multiscale Convolutional Network. *Remote Sens.* **2020**, *12*, 311. [CrossRef]
37. El Mendili, L.; Puissant, A.; Chougrad, M.; Sebari, I. Towards a Multi-Temporal Deep Learning Approach for Mapping Urban Fabric Using Sentinel 2 Images. *Remote Sens.* **2020**, *12*, 423. [CrossRef]
38. Schmitt, M.; Wu, Y.L. Remote Sensing Image Classification with the SEN12MS Dataset. In Proceedings of the ISPRS Annals of the Photogrammetry, Remote Sensing and Spatial Information Sciences, Göttingen, Germany, 17 June 2021; Volume 5, pp. 101–106.
39. Kumar, V.; Singh, R.S.; Dua, Y. Morphologically Dilated Convolutional Neural Network for Hyperspectral Image Classification. *Signal Process. Image Commun.* **2022**, *101*, 116549. [CrossRef]
40. Saralioglu, E.; Gungor, O. Semantic Segmentation of Land Cover from High Resolution Multispectral Satellite Images by Spectral-Spatial Convolutional Neural Network. *Geocarto Int.* **2022**, *37*, 657–677. [CrossRef]
41. Scherer, D.; Müller, A.; Behnke, S. Evaluation of Pooling Operations in Convolutional Architectures for Object Recognition. In Proceedings of the International Conference on Artificial Neural Networks, Thessaloniki, Greece, 15–18 September 2010; pp. 92–101.

42. Marchetti, M.; Bertani, R.; Corona, P.; Valentini, R. Cambiamenti Di Copertura Forestale e Dell'uso Del Suolo Nell'inventario Dell'uso Delle Terre in Italia. *For. J. Silvic. For. Ecol.* **2012**, *9*, 170.
43. Martinez-Plumed, F.; Contreras-Ochando, L.; Ferri, C.; Hernandez-Orallo, J.; Kull, M.; Lachiche, N.; Ramirez-Quintana, M.J.; Flach, P. CRISP-DM Twenty Years Later: From Data Mining Processes to Data Science Trajectories. *IEEE Trans. Knowl. Data Eng.* **2021**, *33*, 3048–3061. [CrossRef]
44. Wirth, R.; Hipp, J. CRISP-DM: Towards a Standard Process Model for Data Mining. In Proceedings of the 4th International Conference on the Practical Applications of Knowledge Discovery and Data Mining, Manchester, UK, 11–13 April 2000; Volume 1, pp. 29–40.
45. Congedo, L. Semi-Automatic Classification Plugin: A Python Tool for the Download and Processing of Remote Sensing Images in QGIS. *J. Open Source Softw.* **2021**, *6*, 3172. [CrossRef]
46. Wang, S.; Chen, W.; Xie, S.M.; Azzari, G.; Lobell, D.B. Weakly Supervised Deep Learning for Segmentation of Remote Sensing Imagery. *Remote Sens.* **2020**, *12*, 207. [CrossRef]
47. Cai, H.; Chen, T.; Niu, R.; Plaza, A. Landslide Detection Using Densely Connected Convolutional Networks and Environmental Conditions. *IEEE J. Sel. Top. Appl. Earth Obs. Remote Sens.* **2021**, *14*, 5235–5247. [CrossRef]
48. Dong, Y.; Li, F.; Hong, W.; Zhou, X.; Ren, H. Land Cover Semantic Segmentation of Port Area with High Resolution SAR Images Based on SegNet. In Proceedings of the 2021 SAR in Big Data Era, BIGSARDATA 2021—Proceedings, Nanjing, China, 22–24 September 2021.
49. Tong, X.Y.; Xia, G.S.; Lu, Q.; Shen, H.; Li, S.; You, S.; Zhang, L. Land-Cover Classification with High-Resolution Remote Sensing Images Using Transferable Deep Models. *Remote Sens. Environ.* **2020**, *237*, 111322. [CrossRef]
50. Qiu, C.; Mou, L.; Schmitt, M.; Zhu, X.X. Local Climate Zone-Based Urban Land Cover Classification from Multi-Seasonal Sentinel-2 Images with a Recurrent Residual Network. *ISPRS J. Photogramm. Remote Sens.* **2019**, *154*, 151–162. [CrossRef]
51. He, K.; Zhang, X.; Ren, S.; Sun, J. Deep Residual Learning for Image Recognition. *arXiv* **2015**, arXiv:1512.03385.
52. Tao, C.S.; Chen, S.W.; Xiao, S.P. Comparison Study of Multitemporal PolSAR Classification Using Convolutional Neural Networks. In Proceedings of the International Geoscience and Remote Sensing Symposium (IGARSS), Waikoloa, HI, USA, 26 September–2 October 2020; pp. 204–207.
53. Lobry, S.; Marcos, D.; Murray, J.; Tuia, D. RSVQA: Visual Question Answering for Remote Sensing Data. *IEEE Trans. Geosci. Remote Sens.* **2020**, *58*, 8555–8566. [CrossRef]
54. Munafò, M. *Consumo Di Suolo, Dinamiche Territoriali e Servizi Ecosistemici*; Edizione 2022; ISPRA, SNPA: Rome, Italy, 2022; ISBN 9788844811242.
55. Gao, H.; Guo, J.; Guo, P.; Chen, X. Classification of Very-high-spatial-resolution Aerial Images Based on Multiscale Features with Limited Semantic Information. *Remote Sens.* **2021**, *13*, 364. [CrossRef]
56. Luo, X.; Tong, X.; Hu, Z.; Wu, G. Improving Urban Land Cover/Use Mapping by Integrating a Hybrid Convolutional Neural Network and an Automatic Training Sample Expanding Strategy. *Remote Sens.* **2020**, *12*, 2292. [CrossRef]
57. Li, C.; Hang, R.; Rasti, B. EMFNet: Enhanced Multisource Fusion Network for Land Cover Classification. *IEEE J. Sel. Top. Appl. Earth Obs. Remote Sens.* **2021**, *14*, 4381–4389. [CrossRef]
58. Pande, S.; Banerjee, B. Adaptive Hybrid Attention Network for Hyperspectral Image Classification. *Pattern Recognit. Lett.* **2021**, *144*, 6–12. [CrossRef]
59. Avolio, C.; Tricomi, A.; Mammone, C.; Zavagli, M.; Costantini, M. A Deep Learning Architecture for Heterogeneous and Irregularly Sampled Remote Sensing Time Series. In Proceedings of the IGARSS 2019—2019 IEEE International Geoscience and Remote Sensing Symposium, Yokohama, Japan, 28 July–2 August 2019; pp. 9807–9810.
60. Qian, M.; Sun, S.; Li, X. Multimodal Data and Multiscale Kernel-Based Multistream Cnn for Fine Classification of a Complex Surface-Mined Area. *Remote Sens.* **2021**, *13*, 5052. [CrossRef]
61. European Commission. Directorate-General for the Environment. In *FUTURE BRIEF: No Net Land Take by 2050?* Directorate-General for Agriculture and Rural Development: Brussels, Belgium, 2016; ISBN 9789279457395.

land

MDPI

Article

Land Consumption Dynamics and Urban–Rural Continuum Mapping in Italy for SDG 11.3.1 Indicator Assessment

Angela Cimini [1], Paolo De Fioravante [2,*], Nicola Riitano [3], Pasquale Dichicco [2], Annagrazia Calò [2], Giuseppe Scarascia Mugnozza [4], Marco Marchetti [5] and Michele Munafò [2]

[1] Department of Architecture and Project, University of Rome La Sapienza, Piazza Borghese 9, 00186 Roma, Italy
[2] Italian Institute for Environmental Protection and Research (ISPRA), Department of Networks and Environmental Information Systems (SINA), Via Vitaliano Brancati 48, 00144 Rome, Italy
[3] Italian Institute for Environmental Protection and Research (ISPRA), Department for the Geological Survey of Italy, Via Vitaliano Brancati 48, 00144 Rome, Italy
[4] Department for innovation in Biological, Agro-Food and Forest Systems, University of Tuscia, Via S. Camillo de Lellis, 01100 Viterbo, Italy
[5] Dipartimento di Bioscienze e Territorio, Università degli Studi del Molise, C/da Fonte Lappone, 86090 Pesche, Italy
* Correspondence: paolodefioravante@gmail.com

Abstract: For the first time in human history, over half of the world's population lives in urban areas. This rapid growth makes cities more vulnerable, increasing the need to monitor urban dynamics and its sustainability. The aim of this work is to examine the spatial extent of urban areas, to identify the urban–rural continuum, to understand urbanization processes, and to monitor Sustainable Development Goal 11. In this paper, we apply the methodology developed by the European Commission-Joint Research Center for the classification of the degree of urbanization of the Italian territory, using the ISPRA land consumption map and the ISTAT population data. The analysis shows that the availability of detailed and updated spatialized population data is essential to calculate SDG indicator 11.3.1, which assesses the ratio of land consumption rate to population growth rate. Three new indicators are also proposed to describe the main trends in urban sprawl, analyzing the spatial distribution of land consumption in terms of infill and settlement dispersion. The research shows good results in identifying class boundaries and describing the Italian urbanized landscape, highlighting the need for more detailed spatialized demographic data. The classification obtained lends itself to a variety of applications, such as monitoring land consumption, settlement dynamics, or the urban heat islands, and assessing the presence and state of green infrastructures in the urban context, driving the development of policies in urban areas toward sustainable choices focused on urban regeneration.

Keywords: urban–rural continuum; land consumption; land use efficiency; SDG; urban expansions; population dynamics

Citation: Cimini, A.; De Fioravante, P.; Riitano, N.; Dichicco, P.; Calò, A.; Scarascia Mugnozza, G.; Marchetti, M.; Munafò, M. Land Consumption Dynamics and Urban–Rural Continuum Mapping in Italy for SDG 11.3.1 Indicator Assessment. *Land* **2023**, *12*, 155. https://doi.org/10.3390/land12010155

Academic Editor: Hualou Long

Received: 24 November 2022
Revised: 21 December 2022
Accepted: 28 December 2022
Published: 3 January 2023

1. Introduction

1.1. Effects of Urban Growth and Insights on Urbanization Dynamics in Italy

Historically, the main drivers of urban development have been economic and demographic growth but also the effects of local policies and regulations [1,2]. In recent years, the extension of city surface areas has been increasing twice as fast as their population, leading to an estimated increase in global land consumption of 1.2 million square kilometers by 2030 [3]. This expansion will have repercussions on the habitats and ecosystem, influencing the services they provide to humans and other life on Earth [4], e.g., they can lead to hydrological impacts, influencing runoff in urban catchments [5] and crop production [6]; moreover, urban centers are responsible for about 70% of global carbon emissions and more than 60% of resource use, while rapid urbanization is often associated with worsening air pollution [7–9]. These considerations were strengthened after the emergency linked to the COVID-19 pandemic, which highlighted the need for resilient cities that guarantee the highest level of well-being to citizens [10,11]. Human pressure on soil is a key driver of the loss of biodiversity and of ecosystem services [12]. Due to their high density of people and assets, cities must play a central role, which is manifested in one of the sustainable development goals formulated by the United Nations within the 2030 Agenda. The 17 Sustainable Development Goals and the 169 associated targets integrate and balance the three dimensions of sustainable development, covering social, economic, and environmental aspects [13]. The thematic of the SDG 11 is the "Sustainable Cities and Communities" [14]; to monitor progress toward the achievement of target 11.3 ("by 2030 enhance inclusive and sustainable urbanization and capacities for participatory, integrated and sustainable human settlement planning and management in all countries"), the UN established indicator 11.3.1, which measures how efficiently cities use the territory, focusing on the relation between urban expansion and the population growth in urban area [15]. This indicator is categorized under Tier II, meaning the indicator is conceptually clear and an established methodology exists but data on many countries are not yet available [16–18].

In this context, it is essential to understand the state of settlements and the main trends in urbanization and demographic variations. The low-density urban growth known as "sprawl" [19,20] is a common phenomenon in Europe that, in Italy, assumes the characteristics of extreme dispersion, taking the name of "sprinkling" [21]. Compared to sprawl (which is based on urban systems designed through zooning), this settlement dynamic develops spontaneously, following different intensities based on the orography and often in the absence of urban planning. Sprinkling is characterized by aggregates of highly variable dimensions that generate low-density hybrid landscapes in rural areas, whose characteristics mix with those of urban areas [22–24] (Figure 1). This configuration is the result of planning processes of the 1960s and 1970s, which led, on the one hand, to the creation of new zone plans on the peripheral areas and to densification, and on the other, to the construction of new residential complexes in the countryside. This dynamic has led to the proliferation of settlements in a widespread form on the margins of large urban centers, influencing the growth of periurban fringes and the transposition of urban models to the countryside [25]. Therefore, on the one hand, there is the presence of consolidated urban forms with centrality roles, and on the other, discontinuous and low-density suburban or periurban settlements have been formed, with marginal dimensions and roles that limit the triggering of centrality effects [26]. This fraying of the margins of urban areas makes the delimitation of the margins between city and countryside complex and limits the definition of a univocal criterion for circumscribing them [27]. It is, therefore, necessary to analyze the urban landscape on a national scale in terms of the urban–rural continuum, overcoming the urban–rural dichotomy [28–30].

(a) (b)

Figure 1. Example of sprawl in Paris (**a**) and sprinkling in Naples (**b**).

1.2. Existing Data for the Representation of Urban Areas

Several products are available nationally and internationally for the representation of urban areas and the urban–rural continuum, with different geometric and thematic characteristics and different definitions.

1.2.1. Global Data—Global Human Settlement Layer (GHSL) SMOD

Eurostat proposes a harmonized methodology to improve the quality of urban statistics starting from the classification of the entire territory along an urban–rural continuum. The scheme is based on the combination of population data (represented by a grid of 1 km^2) and density of artificial surfaces, with the aim of overcoming the traditional urban–rural dichotomy [28,30]. The Joint Research Center (JRC) has implemented Eurostat's methodology by introducing the Global Human Settlement Layer (GHSL), which offers information on a global scale, based on the combination of total resident population density [31,32].

1.2.2. European Data—Copernicus Land Monitoring Service Data

Copernicus data refer to a definition of urban area that considers permanently populated areas and surfaces dominated by the influence of human activities, excluding agricultural use. These areas include all man-made structures (buildings, road infrastructures, and other paved areas) and the unsealed and vegetated surfaces associated with them.

The CORINE Land Cover (CLC) belongs to the Pan-European component of the CLMS and describes land cover and land use with a minimum mapping unit of 25 hectares for the entire European territory. Class 1 CLC distinguishes "artificial areas" according to the prevalent use into "Urban fabric" (which includes built-up areas), "Areas mainly for commercial and industrial use and infrastructures for transport and logistics" (such as ports, airports, roads, and railways), "Quarries, landfills and construction sites", and "Green urban areas and sport and leisure facilities" [33–35].

Urban Atlas (UA) is another important CLMS data for the description of urban areas. UA belongs to the Local component of the CLMS and provides a high-spatial-resolution land cover and land use mapping of Functional Urban Areas (FUAs), defined as the set of central cities and their commuting areas. The data are available for the years 2006, 2012, and 2018 and have a higher spatial resolution for artificial classes (minimum mappable unit of 0.25 hectares) than for natural areas (1 hectare). There are also two change layers, relating to the periods of 2006–2012 and 2012–2018. Urban Atlas adopts a four-level classification system based on CLC classes, with higher thematic detail for artificial areas [36].

1.2.3. European Data—European Settlement Map

The European Settlement Map (ESM) is a mapping of the human settlements in Europe produced with GHSL technology by the Joint Research Center (JRC) and represents the percentage of built-up area coverage per spatial unit. The ESM for 2016 was obtained by applying machine learning techniques on multispectral and panchromatic bands of SPOT5 and SPOT6 imagery, in order to understand systematic relations between morphological and textural features in the description of human settlement [37,38].

1.2.4. National Data—Degree of Artificialization and Degree of Urbanization

ISPRA has characterized the urban–rural continuum according to two approaches, based on the analysis of the density of artificial surfaces (in the first case) and integrating this information with data on population density (in the second one).

The first approach describes the density of the artificial surfaces starting from the ISPRA National Land Consumption Map (LCM), considering the average density of the artificial surfaces in a radius of 300 m. The LCM is a 10-m-resolution raster, produced and updated annually [39–41]. The data have national coverage and are available for the years 2006, 2012, and all the years between 2015 and 2021. The map adopts a three-level classification system, which distinguishes different classes of permanent and reversible consumed land [42]. The third classification level is adopted for all new changes and is available for most of the built-up areas, for the entire road network, and for all quarries, landfills, and paved and unpaved greenhouses. The national territory was then classified into urban, suburban/periurban, and rural areas, in accordance with the thresholds indicated by sustainable development goal number 11 of the United Nations 2030 Agenda.

The second ISPRA data is based on a procedure similar to the Eurostat/JRC one and considers the consumed land density map together with the population density map. The resulting data have four classes of artificial surface and population density, three of which reflect those of the Eurostat/JRC (High-Density Urban Areas, Medium-Density Urban Areas, and Rural Areas) and the fourth relates to areas with a high density of artificial surfaces and low population density [42].

1.3. Describe Urban–Rural Continuum Using National Data

The study proposes a classification of urban areas for the Italian territory according to the indications of the 2030 Agenda for Sustainable Development and the related Sustainable Development Goals. The research is part of the activities carried out by ISPRA for the characterization of the rural-urban continuum and for the monitoring and analysis of the spatial distribution of land consumption [42]. For this purpose, the methodology adopted by Eurostat and applied by JRC for the realization of the GHS-SMOD was resumed and adapted to the ISPRA LCM and to the ISTAT demographic data. Starting from this characterization of the urban area, an estimate of the SDG 11.3.1 indicator and of the land consumption distribution was carried out for the period 2012–2018.

2. Materials and Methods

2.1. Overview

The study refers to the entire Italian territory (Figure 2) and takes up the methodology proposed by Eurostat for a classification of the urban–rural continuum based on the spatial distribution of the population.

Figure 2. Study area—the analysis refers to the whole Italian territory, which is represented with reference to orography and major road infrastructure.

Three main phases can be identified:

1. Creation of a spatialized population layer based on ISTAT demographic data.

The spatial distribution of the population was analyzed starting from the ISTAT data on total municipal population, considering for each municipality a uniform distribution of the inhabitants among the "residential" pixels. Ancillary data were introduced to identify the residential areas: Starting from the LCM, the buildings were identified and then, using the CLMS and ESM data, the residential ones were selected.

2. Classification of the urban–rural continuum on a national scale starting from the spatialized population data.

Through focal analysis, the population density was evaluated around 1 km^2 of each pixel and population density thresholds were introduced to identify 4 classes of clusters, which were then further divided into 9 subclasses based on the total population and contiguity rules.

3. Application of the classification of the urban–rural continuum for the calculation of SDG indicator 11.3.1.

The metadata of the SDG 11.3.1 indicator suggest calculating the "Ratio of land consumption rate to population growth rate" considering the urban area obtained by applying the Eurostat/JRC methodology as the reference area. However, the most detailed and updated demographic data for Italy refers to the aggregate value of total municipal population. Actually, no further information is available on the spatial distribution of the population and its variations among the different urban–rural classes within the same municipality (for example, how many inhabitants of a municipality live in its urban center or the migrations from urban to periurban/suburban or rural and vice versa). As suggested by Eurostat, each municipality was associated with the prevailing class (the one in which the largest number of inhabitants lived) and the calculation of the indicator was carried out for the entire territory of the urban municipalities, i.e., those where the urban class prevails. To provide information on the evolution of urbanized areas within the municipal territory, three new indicators were introduced that evaluate the distribution of land consumption among the urban suburban/periurban and rural classes.

2.2. Spatialization of the Population

This paragraph describes the procedure for creating a spatialized population layer based on ISTAT demographic data (Figure 3). The analysis refers to 2018 and is conducted for the entire national territory.

ISTAT provides yearly updated population data for the Italian territory, with the highest level of detail at the municipal scale (total municipal population). According to the Eurostat methodology, in the absence of spatialized population data (geocoded census or population register), a disaggregation grid can be created by combining the population of census units with high-resolution land use data from national or global sources [31]. To allocate the population within the territory of the municipalities, ancillary data were introduced for the identification of residential areas. First, the built-up areas were identified, starting from the ISPRA LCM, where areas classified as 111 (buildings) and as 1 (consumed land not yet reclassified to the third thematic level of detail) were selected. Starting from these built-up areas, residential buildings were identified using the Copernicus CLMS data (CORINE Land Cover, Urban Atlas, Coastal Zones, Natura 2000 and Riparian Zones) and the ESM data. CLMS class 1 distinguishes different artificial surfaces, making it possible to separate the "urban fabric" from "industrial, commercial, transport, and logistics areas", "extraction sites", "green urban areas", and "sport and leisure facilities". In areas not covered by CLMS class 1, ESM was used, which provides thematic information on residential and non-residential buildings derived by remote sensing imagery and training data. For each municipality, the total population of ISTAT was spatialized considering a uniform distribution among the "residential building" pixels, as no supporting data are available (such as the height of the buildings or updated census sections) to distribute the population on the municipality in a more detailed way. In the following, we refer to the spatialized population data as "population raster".

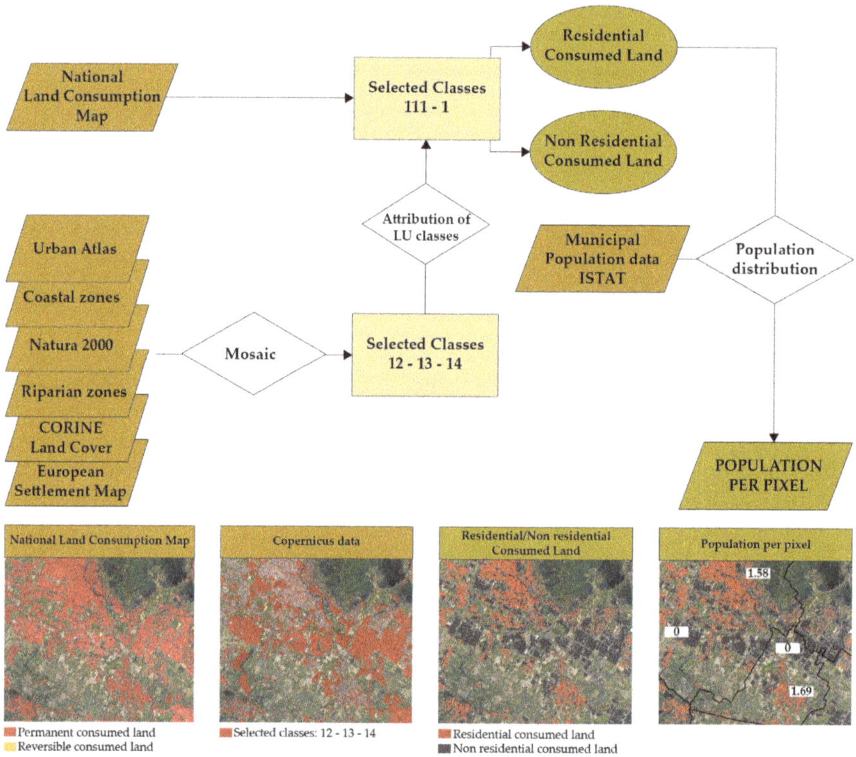

Figure 3. Workflow of the spatialization of population. The built-up areas were identified with the ISPRA LCM; among these, the residential buildings were selected thanks to the CLMS data and the ESM. The population was then distributed uniformly among all residential pixels.

2.3. *Classification of the Urban–Rural Continuum*

This paragraph describes the methodology for the classification of the urban–rural continuum on a national scale starting from the population raster. The analysis refers to 2018 and is conducted for the entire national territory.

The classification adopted by Eurostat and JRC requires information on population density per square kilometer. These data were obtained by calculating the focal statistics on the population raster and associating with each pixel its own population density value around 1 km^2. The data were, therefore, reclassified into four classes, according to the following thresholds (Table 1):

- Class 1, with over 2000 inhabitants per km^2;
- Class 2, for areas with between 500 and 2000 inhabitants per km^2;
- Class 3, for areas with between 100 and 500 inhabitants per km^2;
- Class 4, with less than 100 inhabitants per km^2.

Table 1. Classes defined by Eurostat/JRC for the classification of settlements and their adaptation to the Italian context. For the "Water" class, there must be no population and at least half a pixel must be covered with water. The revision of the thresholds takes into account the differences between the GHS-POP data (to which the Eurostat/JRC thresholds refer) and the proposed data, in terms of spatial resolution and a different criterion for the spatialization of the population.

Code	Definition	Eurostat/JRC Thresholds		Proposed Thresholds	
		Pop. Density	Pop. Size	Pop. Density	Pop. Size
30	Urban Centers	≥1500	≥50,000	≥2000	>10,000
23	Dense urban cluster	≥1500	5000–49,999	≥2000	≤10,000
22	Semi-dense urban cluster	300–1500	5000–49,999	500–2000	>5000
21	Suburban cells	300–1500	-	500–2000	-
13	Rural cluster	300–1500	500–4999	500–2000	≤5000
12	Rural low density	50–300	-	100–500	-
11	Rural very low-density	<50	-	<100	-
50	Industrial, commercial and services	n.a.	n.a.	-	-
60	Water	0	0	-	-

Class 1 was further subdivided based on the total population, making it possible to identify:

- Urban Centers with over 2000 inhabitants per km^2 and a population of over 10,000 inhabitants;
- Dense Urban Clusters with over 2000 inhabitants per km^2 and a population of less than 10,000 inhabitants.

Class 2 was divided into two subclasses using a contiguousness criterion:

- Suburban cells, for areas contiguous to class 1 and with a population density between 500 and 2000 inhabitants per km^2;
- Semi-dense Urban Clusters, for areas not contiguous to class 1 and with a population density between 500 and 2000 inhabitants per km^2 and a total population of over 5000 inhabitants;
- Rural Clusters with population density between 500 and 2000 inhabitants per km^2 and a population of less than 5000 inhabitants.

Class 4 was defined "Low-Density Rural Cells", while the areas that do not fall into any of the above classes were classified as "Very Low-Density Urban Cells". An additional class was also introduced with respect to the Eurostat and JRC classification, relating to areas with high built-up density and low population density, which are mostly associated with production, commercial, or logistics areas. This class was identified with a focal analysis on non-residential buildings and includes areas that have a built density higher than 10% in an area of 1 km^2.

The Eurostat/JRC thresholds were revised through trial and error in order to take into account the characteristics of the population raster, maintaining a result as comparable as possible to the GHS-SMOD. In detail, the population raster compared to the GHS-SMOD has a higher spatial resolution (10 m against 1 km) and, for the spatialization of the population, it does not consider the census sections, whose data are updated to 2011.

2.4. Calculation of the SDG 11.3.1 Indicator

The paragraph describes the application of the classification of the urban–rural continuum for the calculation of SDG indicator 11.3.1. The calculation was conducted with respect to the period of 2012–2018 on urban municipalities, i.e., those in which the population is mainly concentrated in the urban classes. Here, a set of three unpublished indicators was also calculated to analyze the spatial distribution of land consumption among the urban, suburban/periurban, and rural classes.

2.4.1. SDG 11.3.1 Indicator

The indicator SDG 11.3.1 "Ratio of land consumption rate to population growth rate" (RLCRPGR) relates the variation in land consumption with the variation in population in a reference period. In detail, it is defined as the ratio of land consumption rate (LCR) to population growth rate (PGR) [16]:

$$\text{RLCRPGR} = \frac{\text{LCR}}{\text{PGR}} = \frac{\left[\frac{\left(\frac{CL_{t+n}-CL_t}{CL_t}\right)}{T}\right]}{\left[\frac{ln\left(\frac{Pop_{t+n}}{Pop_t}\right)}{y}\right]} \tag{1}$$

where:

CL_t = Total areal extent of the consumed land for the first year.

CL_{t+n} = Total areal extent of the consumed land for the last year.

T = Number of years between CL_{t+n} and CL_t.

Pop_t = Total population in the first year.

Pop_{t+n} = Total population in the last year.

y = Number of years between the two measurement periods.

The indicator metadata suggest calculating the RLCRPGR assuming the territory classified as an urban area as the reference area, according to the Eurostat/JRC methodology. The assessment requires deep knowledge of the spatial distribution of land consumption and population variation, in order to be able to accurately evaluate the variation in the two quantities within the reference urban area.

With regard to the LCR, the LCM provides spatialized and yearly updated information on land consumption, allowing for the identification of the changes occurring in the reference urban area.

Regarding the PGR, ISTAT produces and annually updates the official demographic data for Italy, providing the total municipal population as more detailed data. These data do not allow for spatially characterizing in which areas of the municipality population variations occur. Actually, it is not possible to evaluate the population variation with respect to urban areas, as, in some cases, they only occupy a portion of the municipal territory, and in others, they include more municipalities. In accordance with the Eurostat/JRC methodology, the calculation was, therefore, carried out on a municipal scale, associating the prevailing urban–rural class with each administrative unit (i.e., the one in which the majority of the population is concentrated) [31]. The indicator was calculated for the period 2012–2018 on the municipalities classified as "Urban Center", "Dense Urban Cluster", and "Semi-dense Urban Cluster".

The value of the indicator depends on the two terms LCR and PGR. If they are both positive (increase in land consumption and population) or negative (decrease in land consumption and population), the indicator assumes positive values; otherwise, it is negative. For the interpretation of the RLCRPGR indicator, it is important to observe the sign of LCR and PGR, as the same RLCRPGR value can represent different dynamics (for example, a negative RLCRPGR value can derive from an increase in PGR and a decrease in LCR or from an increase in LCR and a decrease in PGR) (Table 2).

Table 2. Interpretation of the RLCRPGR based on the values of the two terms LCR (Land Consumption Rate) and PGR (Population Growth Rate).

LCR	PGR	RLCRPGR	Description
>0	>0	>1	Land consumption grows much more than population
		0–1	Population grows much more than land consumption
	<0	0–−1	Population decreases significantly and land consumption increases
		<−1	Population decreases and land consumption increases significantly
<0	>0	<−1	Population grows while land consumption decreases significantly
		−1–0	Population grows significantly while land consumption decreases
	<0	0–1	Population decreases much more than land consumption
		<−1	Land consumption decrease much more than population
0	>0, <0	0	-

2.4.2. Normalized Difference of Consumed Land

In addition to the calculation of SDG 11.3.1, the auxiliary indicator "Normalized Difference of Consumed Land" (NDCL) was also introduced, with reference to urban areas (2), suburban/periurban areas (3), and rural areas (4):

$$NDCL_u = \frac{\text{LC in urban area} - \text{LC in rural and suburban area}}{\text{Total LC}} \quad (2)$$

$$NDCL_s = \frac{\text{LC in suburban area} - \text{LC in rural and urban area}}{\text{Total LC}} \quad (3)$$

$$NDCL_r = \frac{\text{LC in rural area} - \text{LC in urban and suburban area}}{\text{Total LC}} \quad (4)$$

The indicators were calculated for the same urban municipalities considered for the RLCRPGR indicator. The calculation was based on the ISPRA LCM for the two reference years and the interpretation of the three indicators allows for describing the spatial distribution of land consumption in the urban municipalities. Each of the three indicators evaluates the portion of land consumption that affects each of the three classes (urban, suburban/periurban, and rural) and the one that assumes the highest value represents the prevailing trend of that municipality, in terms of:

- Urban Infill, where land consumption mainly affects urban areas, thus filling the gaps between the existing urban fabric.
- Periurban Infill, when land consumption is concentrated in periurban areas on the edge of the dense urban fabric.
- Dispersion, when low-density land consumption occurs in rural areas.

3. Results

3.1. Spatialization of the Population

The spatialization of the population refers to the residential areas identified on the ISPRA LCM. Starting from the total consumed land of the LCM (2,130,088 hectares in 2018), buildings (class 111) and consumed land at the first classification level (class 1) were selected, for a total of 536,579 and 689,645 hectares, respectively. The introduction of CLMS and ESM data made it possible to distinguish the residential component (71.42% of the built-up area) from the non-residential component (28.57% of the built-up area) (Table 3), where the population was evenly distributed.

Table 3. Extension of total consumed land and of classes 1 and 111 of the LCM and extension of residential and non-residential areas within classes 1 and 111 of the LCM.

	Area (ha)	Percentage (%)		
		On Total National Area	On LCMC onsumed Land	On LCM Built-Up Area
Consumed land	**2,130,088**	**7.06**	-	-
Built-up-class 1	689,645	-	2.37	-
Built-up-class 111	536,579	-	25.19	-
Residential	875,794	-	-	71.42
Non residential	350,430	-	-	28.57

3.2. Classification of the Urban–Rural Continuum

The classification of the settlements of the national territory according to the eight Eurostat/JRC classes provided the results shown in Figures 4 and 5 and Table 4, expressed in terms of hectares and percentages on the national surface. The data were also compared with the GHS-SMOD data.

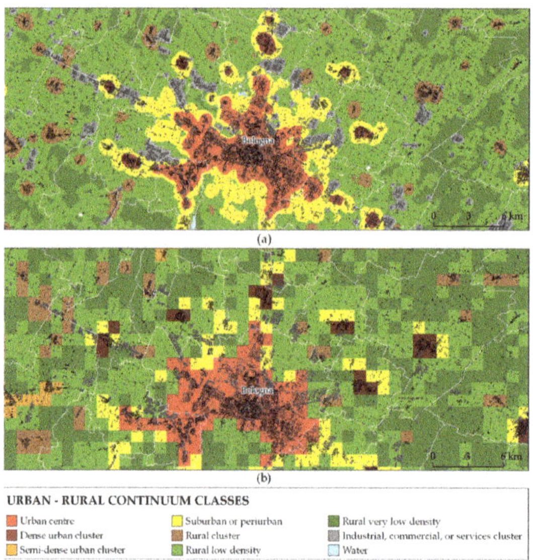

Figure 4. Urban area of Bologna: comparison between (**a**) classification of the urban–rural continuum from the proposed data (2018) and (**b**) GHS-SMOD (2020). The two data are consistent with regard to the distribution of the different classes; however, the proposed data represent with higher geometric detail the edges of dense urban centers and the spatial configuration of small and medium-sized clusters that have developed around the main metropolitan areas.

The Italian territory shows a prevalence of rural classes with low and very low population density. The areas with the highest population density (Urban Centers and Dense Urban Clusters) occupy just over 2% of the national territory, while the surrounding suburban areas occupy under 5%. Compared to the GHS-SMOD, there is a similar distribution of the national territory between urban, suburban/periurban, and rural classes, with the main differences concerning the classes of Dense Urban Cluster and Semi-Dense Urban Cluster, which are larger in the GHS-SMOD than in the proposed data.

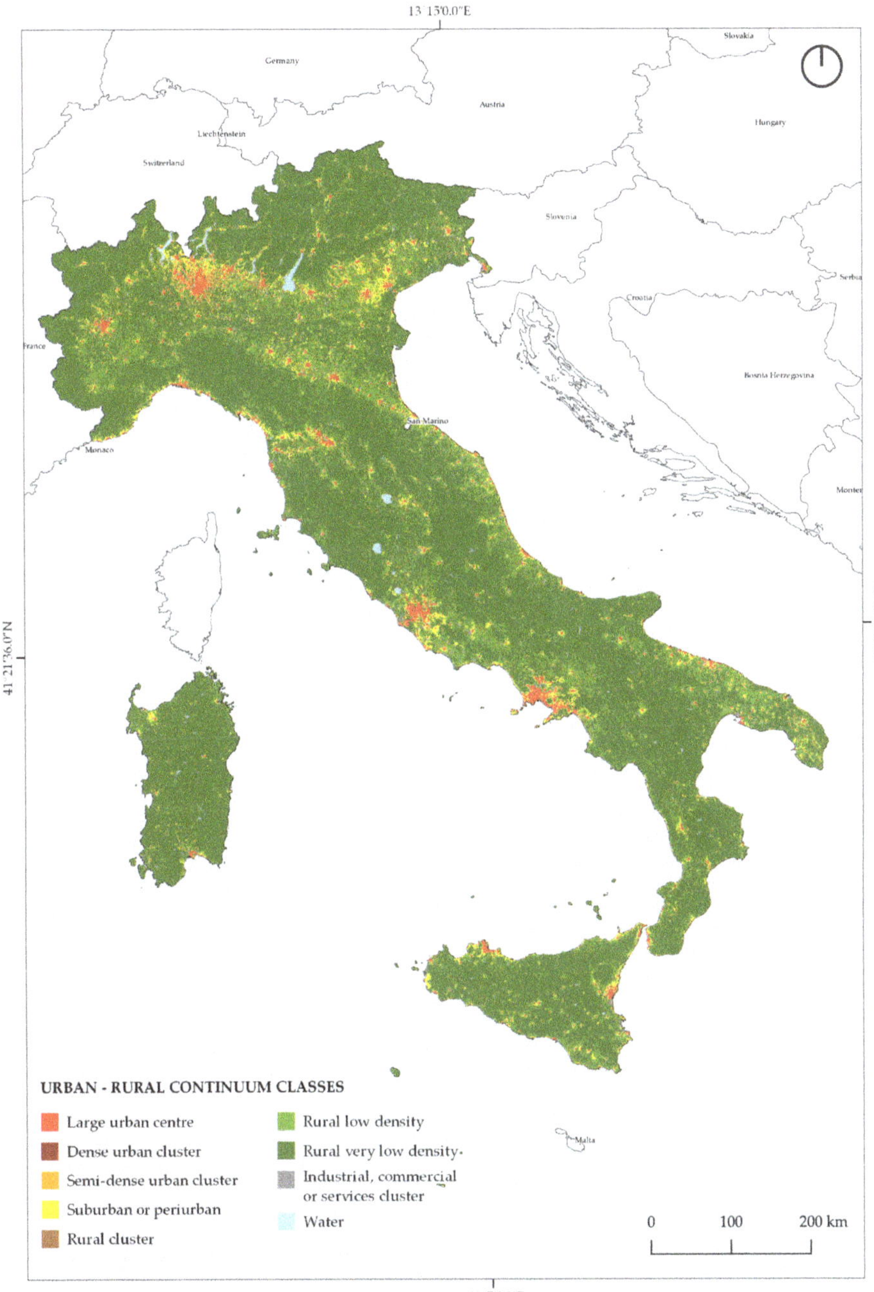

Figure 5. Urban–rural continuum in Italy (2018). The map shows the location and spatial configuration of the urban clusters and the extension of the related expansion areas. Here, the rural spaces close to the dense urban cores have undergone a conversion into urban uses with a tendency to connect with each other.

Table 4. Extension of the urban–rural continuum classes on the national territory, expressed in hectares and percentage and compared with the GHS-SMOD data.

Classes	URCC		GHS-SMOD	
	ha	%	ha	%
Urban center	478,965	1.59	421,293	1.39
Dense urban cluster	161,683	0.54	412,635	1.36
Semi-dense urban cluster	44,219	0.15	253,155	0.83
Suburban cells	1,400,857	4.65	1,315,428	4.33
Rural cluster	660,323	2.19	800,973	2.64
Rural low density	5,921,639	19.65	4,555,108	15.00
Rural very low-density	20,855,863	69.20	22,381,960	73.70
Industrial, commercial, and services	441,594	1.47	-	-
Water	173,558	0.58	227,783	0.75
Total	30,138,702	100.00	30,368,335	100.00

A total of 2825 urban clusters are identified on the national territory, of which 719 are classified as Urban centers, 2046 as Dense Urban Clusters, and the remaining 60 as Semi-Dense Urban Clusters (Table 5). Overall, the Urban Centers occupy about three times the surface of the Dense Urban Clusters and over 10 times that of the Semi-Dense Urban Clusters. About half of the Urban Centers are in Lombardy (which accounts for a quarter of the class), Campania, and Lazio. Lombardy is also the region where the class of Dense Urban Clusters is most extended, while about one third of the Semi-Dense Urban Clusters fall in Veneto, Latium, and Sicily.

Table 5. Surface and number of clusters present on the Italian territory for the three classes of Urban Centers, Dense Urban Clusters, and Semi-Dense Urban Clusters.

	Large Urban Center		Dense Urban Cluster		Semi-Dense Urban Cluster		Largest Urban Center	LUCPI
	Count	ha	Count	ha	Count	ha	ha	%
Piedmont	32	29,926	138	10,646	3	2515	14,809	34.37
Aosta Valley	1	730	3	290	0	0	730	71.56
Lombardy	66	108,648	452	34,143	4	2817	48,016	32.98
Trentino A.A.	6	5291	48	3273	0	0	1509	17.62
Veneto	23	27,278	197	17,286	12	6780	5299	10.32
Friuli V.G.	28	8674	33	2265	6	4162	2903	19.22
Liguria	119	14,626	44	3524	1	601	7623	40.66
Emilia-Romagna	27	27,843	154	9975	2	1263	6737	17.24
Tuscany	33	29,906	109	7720	5	3472	8833	21.49
Umbria	4	3007	26	1628	4	3124	1289	16.61
Marche	18	8193	50	4258	2	1333	1438	10.43
Latium	37	52,989	142	13,173	5	5140	39,061	54.78
Abruzzo	14	8277	33	2735	0	0	4226	38.38
Molise	2	1393	9	685	0	0	836	40.24
Campania	125	74,469	103	7859	3	2176	49,065	58.06
Apulia	80	24,305	157	13,710	6	4530	4280	10.06
Basilicata	2	1381	21	2115	0	0	770	22.02
Calabria	15	7553	79	6260	1	812	2253	15.41
Sicily	74	36,482	177	14,231	5	4898	11,089	19.94
Sardinia	13	8289	71	5987	1	598	4181	28.11
ITALY	**719**	476,837	**2046**	161,762	**60**	44,219	-	-

Regarding the extension of the larger urban centers, in Lombardy and Campania, the centers of Milan and Naples each occupy almost 50,000 hectares, while, in Latium, the urban area of Rome reaches just under 40,000 hectares. In the other regions, the main centers do not reach 10,000 hectares, except for Piedmont and Sicily, which stop just above this threshold. The "Largest Urban Patch Index" (LUPI) expresses the extent of the largest patch as a percentage on the total regional extent of the three urban classes. In Campania and Latium, the two main clusters are among the largest in Italy and occupy between 50 and 60% of the regional urban class, while Lombardy's Milan cluster, despite being the largest in Italy, includes less than a third of regional urban areas, reflecting the high level of urbanization of this region. The highest LUPI value is in the Valle d'Aosta, where the Aosta cluster (despite being the smallest of the largest regional urban centers) includes over 70% of the urbanized area of the region, confirming the low level of artificialization of that territory.

3.3. Calculation of the SDG 11.3.1 Indicator

The attribution of the prevailing urban–rural class to the municipalities provided the results of Table 6 and Figures 6 and 7.

Table 6. Classification of municipalities among the urban–rural classes based on the prevailing class.

	Municipality	
Classes	**n.**	**%**
Urban center	605	7.65
Dense urban cluster	470	5.94
Semi-dense urban cluster	56	0.71
Suburban cells	1216	15.38
Rural cluster	1640	20.74
Rural low density	2777	35.13
Rural very low-density	1142	14.44

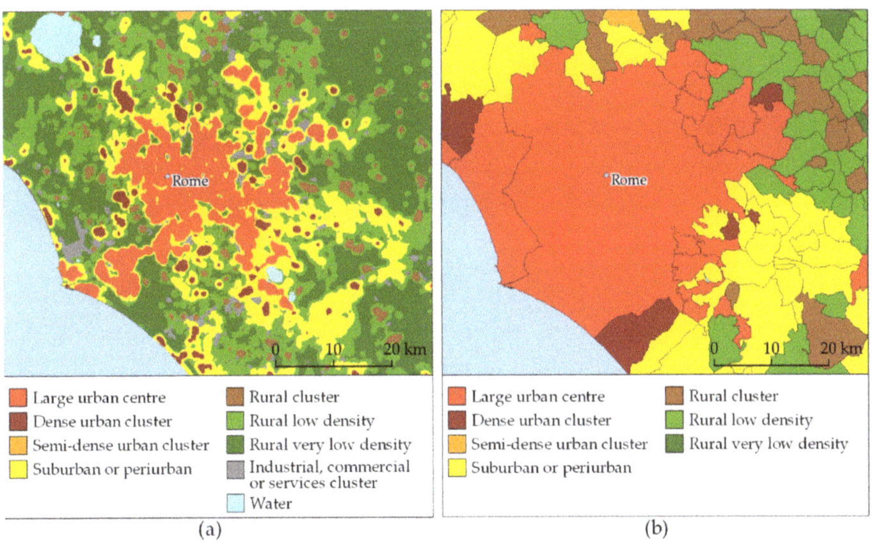

Figure 6. The metropolitan area of Rome. (**a**) Urban–rural continuum classes. (**b**) Attribution of the prevailing urban–rural class to the municipalities. This generalization involves a loss of detail, which precludes the possibility of using indicator 11.3.1 for intra-municipal analyses, but it is the only way to exploit ISTAT demographic data, which have annual updates at the most at the municipal level.

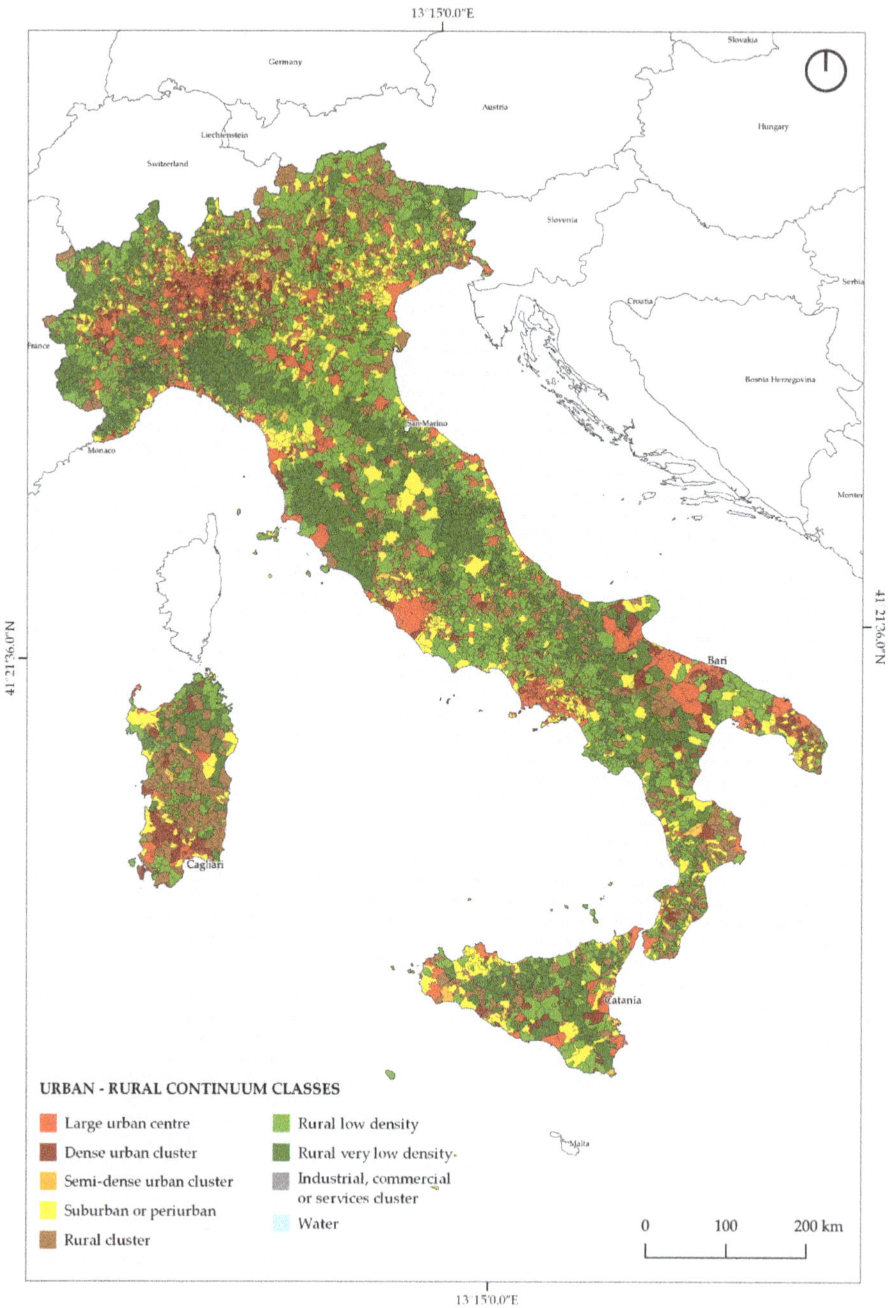

Figure 7. Urban–rural continuum classes in Italy (2018)—Attribution of the prevailing urban–rural class to the municipalities. Compared to Figure 5, this generalization leads to a loss of detail, which limits intra-municipal analyses, but still allows evaluations and comparisons to be made between different municipalities attributable to the same classes.

For the calculation of the RLCRPGR indicator, municipalities defined as "Urban Center", "Dense Urban Cluster", and "Semi-dense Urban Cluster" were taken into consideration (Figure 7). Overall, 1131 municipalities belong to the three classes (14% of the total), of which 605 classified as Urban Center, 470 as Dense Urban Cluster, and 56 as Semi-dense Urban Cluster.

The calculation of the RLCRPGR indicator provided the values shown in Figure 8, relating to municipalities with an increase in land consumption and, therefore, positive LCR values. In detail, 13 of the 1131 municipalities show negative LCR values; in 6, there are no changes in LCR, and 1112 show an increase in land consumption. For all urban municipalities with a decrease in land consumption (LCR < 0), the whole territory was inspected using the ISPRA LCM. In all cases, the restorations take place in areas classified as reversible land consumption, in correspondence with areas of maneuvering and storage surrounding the construction sites of new buildings, roads, and infrastructures, as in the example of Figure 9. The following analysis, therefore, focused only on municipalities with LCR > 0.

With regard to the municipalities with increasing land consumption, the following results were obtained (Figure 10):

- In 571 municipalities (51.3% of those with LCR > 0), the indicator has values less than 0. In these areas, the increase in land consumption corresponds to a decrease in population. In detail, in the 397 municipalities with SDG values between 0 and -1, the rate of population decrease is significantly higher than the rate of increase in land consumption, while in the 174 municipalities with SDG < -1, the increase in land consumption is higher than the population decrease.
- The SDG has positive values in 541 municipalities (48.6% of urban municipalities with increasing land consumption). About two thirds of them have values between 0 and 1, i.e., a rate of increase in the population higher than the rate of increase in land consumption, while in the remaining 164 municipalities, the population increases less than the increase in land consumption.

The SDG indicator was supported by an analysis on the territorial distribution of land consumption between urban, periurban/suburban, and rural areas relating to the period of 2012–2018 (Table 7).

Table 7. Distribution of land consumption (2012–2018) among the different urban–rural classes.

CLASSES	ha	%	ha	%
Urban center	2172	19.45		
Dense urban cluster	347	3.11	2628	23.53
Semi-dense urban cluster	108	0.97		
Periurban/suburban area	2819	25.24	2819	25.24
Rural cluster	278	2.49		
Rural low density	2170	19.43	3749	33.57
Rural very low-density	1301	11.65		
Industrial, commercial, and services	1972	17.66	1972	17.66
Water	-	-	-	-
Total	11,169	100.00	11,169	100.00

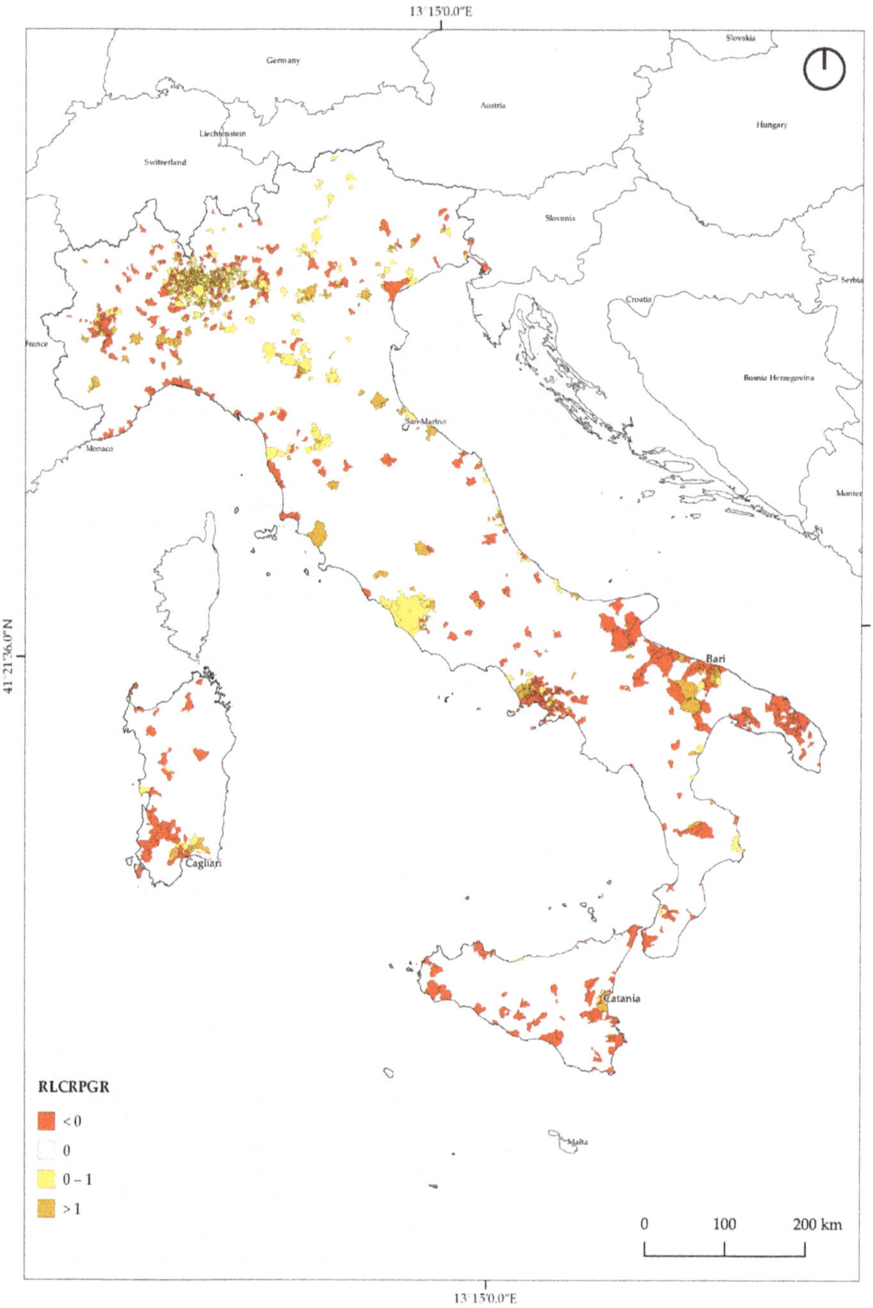

Figure 8. RLCRPGR values in the municipalities classified as "Urban Center", "Dense Urban Cluster", and "Semi-dense Urban Cluster" with an increase in land consumption (LCR > 0). In 51.3% of these municipalities, the population decreases, while in the remaining ones, the population increases together with land consumption.

Figure 9. The image shows a road construction site area (reversible consumed land) in Novara between 2012 (**a**) and 2018 (**b**). The central area has been transformed into a paved road (permanent consumed land), while the purple areas have been restored after the completion of the road. In all urban municipalities with a decrease in land consumption, the changes are associated with the restoration of the maneuvering and storage areas surrounding the construction sites of new buildings, roads, and infrastructures.

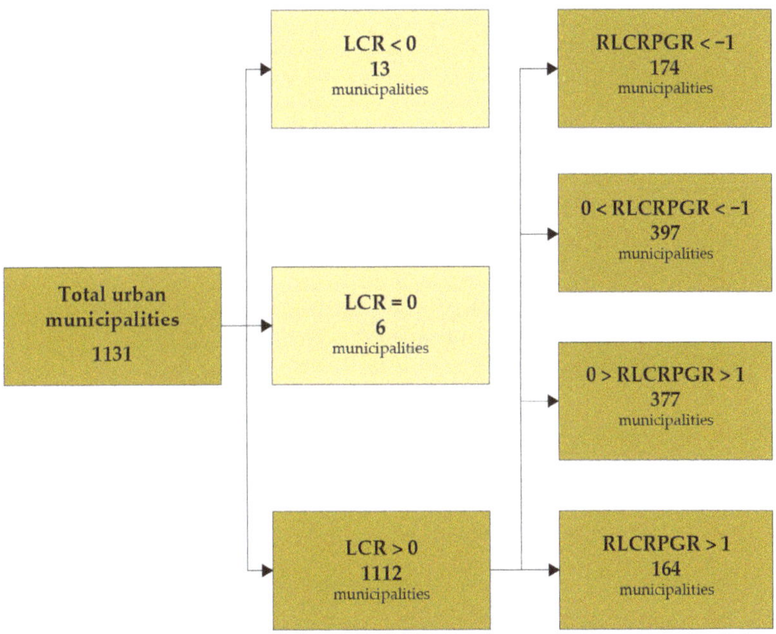

Figure 10. Value of the RLCRPGR indicator in the urban municipalities. In almost all urban municipalities, there is an increase in land consumption. Among these municipalities, just over half also show an increase in population, while in the remaining municipalities, the variations in land consumption and population are uncorrelated.

Land consumption between 2012 and 2018 prevails in rural areas; in fact, over one third of the changes fall into the "Rural Cluster", "Rural Low Density", and "Rural Very Low Density" classes. Urban classes and periurban/suburban areas each host about a quarter of the total changes, while about 17% of land consumption is in non-residential areas for productive and commercial use. The trend toward land consumption outside the denser urban classes is also recognizable from the analysis of the three Normalized

Difference of Consumed Land (NDCL) indicators, which describe the prevailing trend in each of the analyzed municipalities. The three indicators $NDCL_u$, $NDCL_s$, and $NDCL_r$ assume values between -1 and 1 and, for each municipality, they have a sum equal to -1; the one with the highest value indicates the prevailing tendency of land consumption to be concentrated in urban areas (prevalence of $NDCL_u$), periurban/suburban areas (prevalence of $NDCL_s$), or rural areas (prevalence of $NDCL_r$).

The calculation of the three indicators for the 1131 urban municipalities shows the results of Figures 11 and 12, with a tendency to infill in the main provincial municipalities and a concentration of land consumption in the rural area in the other municipalities, with a tendency toward dispersion in areas with low settlement density.

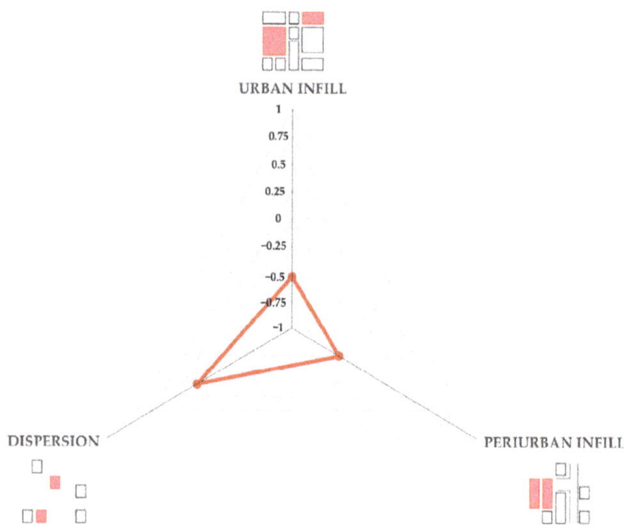

Figure 11. Value of the three indicators $NDCL_u$, $NDCL_s$, and $NDCL_r$ for the Italian urban municipalities. Overall, a tendency toward dispersion emerges, caused by a concentration of new buildings in rural areas.

In detail (Table 8), in 493 municipalities, land consumption is concentrated in rural areas (settlement dispersion). In 317 municipalities, land consumption has the characteristics of the completion of periurban/suburban areas, and in the remaining 321 municipalities, it shows a tendency to infill the voids between the dense and semi-dense consolidated urban fabric.

Table 8. Land consumption distribution between urban, periurban/suburban, and rural municipalities according to NDCL values.

CLASSES	Count	%
Urban Infill	321	28.38
Periurban Infill	317	28.03
Dispersion	493	43.59
Total	1131	100.00

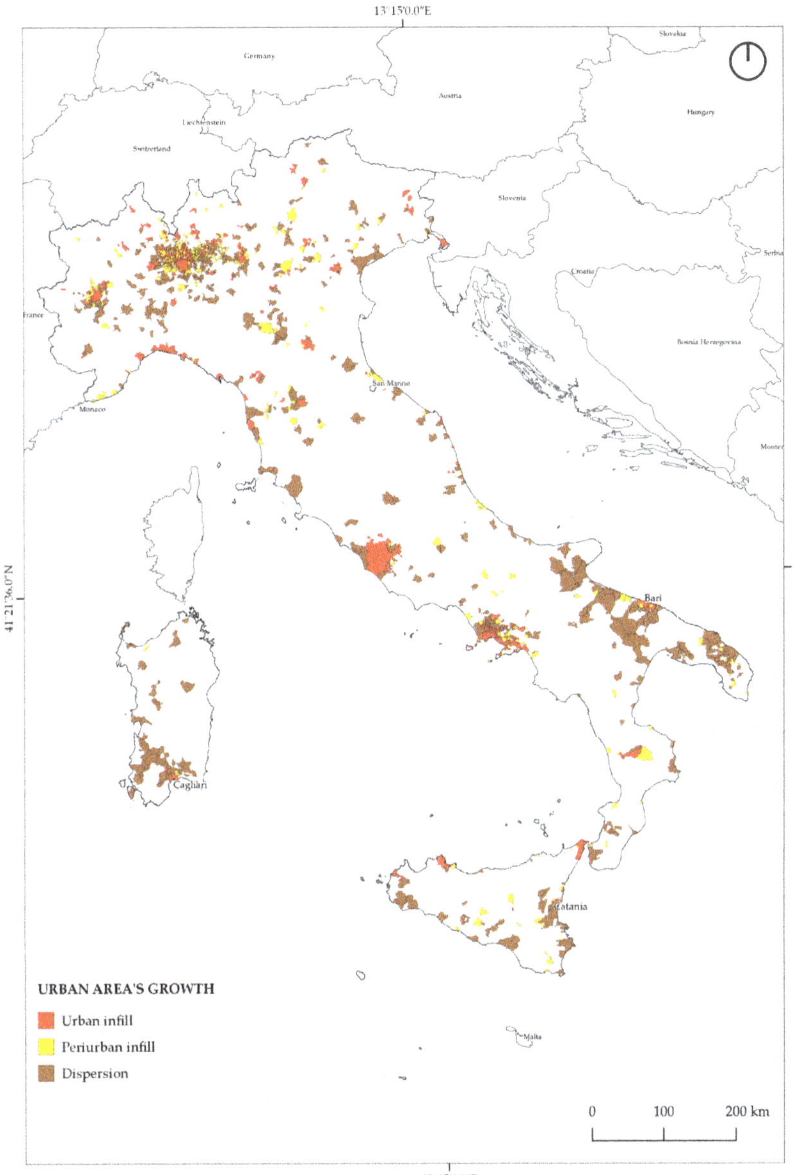

Figure 12. $NDCL_u$, $NDCL_s$, and $NDCL_r$ values in the municipalities classified as "Urban Center", "Dense Urban Cluster", and "Semi-dense Urban Cluster". With the exception of the main capitals, in most urban municipalities, land consumption takes place in rural areas, assuming the characteristics of settlement dispersion.

4. Discussion

The availability of adequate input data for the description of the spatial distribution of built-up areas and population is one of the main factors influencing the representation of settlements in compliance with the Eurostat methodology [17,43]. ISPRA LCM represents the built-up areas on the Italian territory with high thematic and spatial detail and yearly updates [40,42], while, at the European level, the CLMS data allow a description of the artificial areas LU with good coverage and high thematic detail while maintaining an adequate update frequency (indicator SDG 11.3.1 recommends conducting the calculation over a reference period of 5–10 years) [16]. Finally, on a global scale, it is possible to study urban areas starting from the classification of artificial surfaces obtained from free satellite data, such as Sentinel-2 multispectral images or Sentinel-1 SAR data [41]. Currently, the main limitation is linked to the availability and homogeneity of demographic data [44]. ISTAT data are the national reference for demographic analysis in Italy and they are updated yearly for the national scale and at NUTS 2, NUTS 3, and municipal levels [45]. In this study, the availability of high-resolution LU data has made it possible to uniformly spatialize the population on the residential areas of each municipality; however, the absence of an updated spatialized population data limits the possibility of carrying out analysis on a local scale, for example, linked to population flows between different urban clusters of the same municipality. A version of the data with higher spatial detail refers to the census sections but is subject to ten-year updates, since 1991, 2001, and 2011 [46], and the data as of 2021 have not yet been published.

The proposed representation of the urban–rural continuum is in line with Eurostat indications, adapting to the specificities of the Italian territory and bringing the description of these areas to a higher level than the ISPRA "degree of urbanization" [42]. The data show a concentration of urban areas along the coastline, in lowland areas, and in the valley floors, particularly in the Po valley. At the regional level, Lombardy has the largest extension and the highest number of urban clusters (about 18% of large urban centers, dense urban clusters, and semi-dense urban clusters are in this region). The yellow (suburban/periurban areas) and light green (rural low-density areas) areas show that the expansion of the urban clusters is concentrated along the main road infrastructures (Figure 5, Tables 4 and 5). Compared to the GHS-SMOD, the data allow for yearly updates and the use of LCM offers a better description of the low-density built-up areas typical of the sprinkling of Italian rural and periurban landscapes. Moreover, the introduction of the class of "Industrial, commercial, and services" areas, which occupy 1.47% of the total national surface, makes it possible to understand their position and role with respect to the urban–rural continuum, which constitutes a useful support tool for orienting effective governance policies for the sustainable management of productive areas [47].

The classification of the urban–rural continuum was used to calculate the SDG 11.3.1 indicator on Italian urban areas. The LCM adequately identifies land consumption within urban areas, while the population data do not allow for intra-municipal analysis. The calculation of the RLCRPGR was, therefore, carried out on a municipal scale, associating with each municipality the class of the urban–rural continuum where most of the population is concentrated; the RLCRPGR was then calculated in municipalities classified as "Urban Center", "Dense Urban Cluster", and "Semi-dense Urban Cluster". The calculation of RLCRPGR conducted on urban municipalities highlighted an increase in land consumption in 1112 of the 1131 urban municipalities considered (over 98% of the total), and half of them show a decrease in the population (571 of the 1131 considered municipalities), indicating a disconnection between land consumption and population variation; however, the lack of spatialized population data prevents an exact assessment of the migration flows within the administrative boundaries of the municipality, limiting the effectiveness of the RLCRPGR indicator. To better understand the dynamics occurring within the territory of the urban municipalities, a set of new indicators was also introduced for the assessment of the spatial distribution of the land consumption that occurred between 2012 and 2018, evaluated on the ISPRA LCM. The analysis of the distribution of changes between the different classes of the urban–rural continuum shows a tendency to fill urban voids in the consolidated built-up area (infill), which affects 28.38% of the 1131 urban municipalities considered. A similar percentage (28.03%) of the municipalities shows a tendency toward the expansion of the cities in the periurban fringes (periurban infill), while, in the remaining 43.59% of the urban municipalities, the land consumption is concentrated in the low-density rural areas, confirming the consolidated trend toward sprinkling. This trend affects the rural areas located on the edge of the dense urban core, which are subject to a conversion and a fusion in periurban fringes.

The application of the Eurostat/JRC methodology to the Italian context shows encouraging results, improving the representation of the territory compared to other current available national and international data. The distinction between low-density clusters and surrounding periurban areas can be further improved, but more detailed information on the spatial distribution of the population would be needed. Furthermore, the future introduction of LU information on the LCM will also allow a more accurate discrimination of residential areas from non-residential ones; the integration with an auxiliary database (i.e., satellite images and socioeconomic data) could support the creation of dasymetric maps useful for improvement of the limitation related to the availability of disaggregated census data [48].

5. Conclusions

Historically, city size and economic growth have been described as strictly correlated [49] and, in combination with population growth, have generally driven land consumption dynamics. This paradigm has lost its validity in recent years; indeed, the significant urban growth detected in recent decades is completely unrelated to the demographic dynamics both in Italy and at a global scale [3]. Furthermore, the intense migratory flow to urban areas has led to an unplanned growth of cities, which usually expand their geographic boundaries [50]. In this context, monitoring the state and evolution of urban areas is essential for defining strategies aimed at sustainable management and development of settlements [51].

Regarding spatial expansion, the historical evolution of the Italian landscape has led to an increase in land consumption in rural areas, and, in particular, in the urban fringe, assuming the characteristics of sprinkling and causing a low-density expansion of the urban fabric toward the surrounding areas [52]. This trend has led to a shift from a landscape in which a dichotomy between urban and rural areas was clearly distinguishable, to a continuum where dense urban fabric expands and merges with the surrounding rural areas, generating large and dilated suburban and periurban bands [29]. In this context, it is necessary to define and orient the planning tools in such a way that they capture these specificities of the territory and adequately orient the decision-making processes, while remaining connected with the European framework in terms of achieving the sustainable development goals.

This study falls within the framework of the research activities carried out by ISPRA for the definition of tools for the monitoring of land consumption and its influence on urban areas and for the characterization of settlement dynamics [39–42,53–55]. These topics are also of interest at the legislative level; the proposed law on urban regeneration [56] and the future national law on land consumption (envisaged among the reforms of the PNRR) [57] underline the need of tools for monitoring land consumption and the effectiveness of urban regeneration interventions, functional to the conservation of natural and semi-natural areas within and outside the urban area. The monitoring activity should be ensured at different scales of investigation and be based on scientific data collected in representative contexts of the urban area, in order to provide data and information to support decision-making and land-government activities.

The presented work fits within this request by providing a geographic interpretation of the urban–rural continuum of the whole national territory with high geometric detail, allowing for the analysis of settlement dynamics within urban areas and in the surrounding landscape, to make comparisons between different contexts and to stand as a supporting basis for further multi-scalar analysis [58,59]. Along these lines, future insights will involve analysis of urban morphology even at the neighborhood scale with the aim of identifying recurring and recognizable settlement types that can support strategies to restore the urbanized landscape in accordance with the physical elements that compose it [60–65].

Actually, this representation of the urban–rural continuum has the characteristics to be a reference for a wider range of further analysis at the regional scale, such as the management and conservation of urban green infrastructures [66,67] or the monitoring of the Urban Heat Island along the urban–rural continuum [68,69], and can support activities related to the introduction of nature-based solutions [67] or with the study of the fragmentation of urban forms [70].

This study shows that the availability of appropriate input data significantly affects the quality of the resulting monitoring tools. The input data introduced for the description of land consumption and the population have yearly updates, allowing the data to be updated with a high frequency. In addition, the introduction of high-spatial-resolution data for the representation of the consumed land has made it possible to obtain a classification of the urban–rural continuum consistent with the GHS-SMOD, but with a higher spatial detail in the description of the composition of the territory. On the other hand, the lack of information on the spatial distribution of the population was an important limitation in conducting analysis related to demographic dynamics, as demonstrated in the calculation of Indicator SDG 11.3.1, where the reduction in detail of the analysis has limited the possibility of observing intra-municipal dynamics.

Author Contributions: Conceptualization, A.C. (Angela Cimini), N.R. and P.D.F.; methodology, A.C. (Angela Cimini), P.D. and P.D.F.; software, A.C. (Angela Cimini); validation, A.C. (Angela Cimini), A.C. (Annagrazia Calò) and P.D.F.; formal analysis, A.C. (Angela Cimini); investigation, A.C. (Angela Cimini) and P.D.F.; resources, M.M. (Michele Munafò); data curation, A.C. (Angela Cimini); writing—original draft preparation, A.C. (Angela Cimini) and P.D.F.; writing—review and editing, A.C. (Angela Cimini) and P.D.F.; visualization, A.C. (Angela Cimini); supervision, M.M. (Michele Munafò), M.M. (Marco Marchetti) and G.S.M.; project administration, M.M. (Michele Munafò), M.M. (Marco Marchetti) and G.S.M.; funding acquisition, M.M. (Michele Munafò). All authors have read and agreed to the published version of the manuscript.

Funding: This research was funded by the Italian Institute for Environmental Protection and Research (ISPRA) structural funds.

Data Availability Statement: Data presented in this study are available on request from the corresponding author. The data are not publicly available because they are part of ongoing research.

Conflicts of Interest: The authors declare no conflict of interest.

References

1. Mahtta, R.; Fragkias, M.; Güneralp, B.; Mahendra, A.; Reba, M.; Wentz, E.A.; Seto, K.C. Urban Land Expansion: The Role of Population and Economic Growth for 300+ Cities. *Npj Urban Sustain.* **2022**, *2*, 5. [CrossRef]
2. Mahendra, A.; Seto, K.C. *Upward and Outward Growth: Managing Urban Expansion for More Equitable Cities in the Global South*; WRI: World Resources Institute: Washington, DC, USA, 2019.
3. United Nations. *Department of Economic and Social Affairs. Population Division World Urbanization Prospects The 2018 Revision*; United Nations: New York, NY, USA, 2019.
4. Alberti, M. The Effects of Urban Patterns on Ecosystem Function. *Int. Reg. Sci. Rev.* **2005**, *28*, 168–192. [CrossRef]
5. Jacobson, C.R. Identification and Quantification of the Hydrological Impacts of Imperviousness in Urban Catchments: A Review. *J. Environ. Manag.* **2011**, *92*, 1438–1448. [CrossRef] [PubMed]
6. Ceccarelli, T.; Bajocco, S.; LUIGI, P.L.; Luca, S.L. Urbanisation and Land Take of High Quality Agricultural Soils-Exploring Long-Term Land Use Changes and Land Capability in Northern Italy. *Int. J. Environ. Res.* **2014**, *8*, 181–192.
7. Churkina, G. The Role of Urbanization in the Global Carbon Cycle. *Front. Ecol. Evol.* **2016**, *3*, 144. [CrossRef]
8. Crippa, M.; Guizzardi, D.; Pisoni, E.; Solazzo, E.; Guion, A.; Muntean, M.; Florczyk, A.; Schiavina, M.; Melchiorri, M.; Hutfilter, A.F. Global Anthropogenic Emissions in Urban Areas: Patterns, Trends, and Challenges. *Environ. Res. Lett.* **2021**, *16*, 074033. [CrossRef]
9. Heinonen, J.; Junnila, S. Case Study on the Carbon Consumption of Two Metropolitan Cities. *Int. J. Life Cycle Assess.* **2011**, *16*, 569–579. [CrossRef]
10. UN-Habitat. *Cities and Pandemics: Towards a More Just, Green and Healthy Future*; UN-Habitat: Nairobi, Kenya, 2021; ISBN 9789211328776.
11. EEA. *Ensuring Quality of Life in Europe's Cities and Towns Tackling the Environmental Challenges Driven by European and Global Change*; EEA Report 05/2009; EEA: Copenhagen, Denmark, 2009.
12. Haines-Young, R. Land Use and Biodiversity Relationships. *Land Use Policy* **2009**, *26*, S178–S186. [CrossRef]

13. Ferri, N. United Nations General Assembly. *Int. J. Mar. Coast. Law* **2010**, *25*, 271–287. [CrossRef]
14. Akuraju, V.; Pradhan, P.; Haase, D.; Kropp, J.P.; Rybski, D. Relating SDG11 Indicators and Urban Scaling—An Exploratory Study. *Sustain. Cities Soc.* **2020**, *52*, 101853. [CrossRef]
15. UN-Habitat. *Utilisation Efficace Des Terres*; UN-Habitat: Nairobi, Kenya, 2022.
16. United Nations. Department of Economic and Social Affairs.SDG Indicators Metadata Repository. Available online: https://unstats.un.org/sdgs/metadata/files/Metadata-11-03-01.pdf (accessed on 11 May 2022).
17. Laituri, M.; Davis, D.; Sternlieb, F.; Galvin, K. Sdg Indicator 11.3.1 and Secondary Cities: An Analysis and Assessment. *ISPRS Int. J. Geo-Inf.* **2021**, *10*, 713. [CrossRef]
18. Paganini, M.; Petiteville, I.; Ward, S.; Dyke, G.; Steventon, M.; Harry, J. Satellite earth observations in support of the sustainable development goals. In *The CEOS Earth Observation Handbook*, 2018th ed.; Earth Observation Graphic Bureau (ESA): Venice, Italy, 2018.
19. European Environment Agency. *Urban Sprawl in Europe*; EEA Report 11/2016; EEA: Luxembourg, 2016.
20. Ewing, R.H. Characteristics, Causes, and Effects of Sprawl: A Literature Review. *Environ. Urban Stud.* **1994**, *21*, 519–535.
21. Romano, B.; Zullo, F.; Ciabò, S.; Fiorini, L.; Marucci, A. Il Modello Italiano Di Dispersione Urbana: La Sfida Dello "Sprinkling". *Sentieri Urbani* **2016**, *19*, 15–22.
22. Romano, B.; Zullo, F.; Fiorini, L.; Ciabò, S.; Marucci, A. Sprinkling: An Approach to Describe Urbanization Dynamics in Italy. *Sustainability* **2017**, *9*, 2. [CrossRef]
23. Romano, B.; Fiorini, L.; Marucci, A. Italy without Urban "Sprinkling". A Uchronia for a Country That Needs a Retrofit of Its Urban and Landscape Planning. *Sustainability* **2019**, *11*, 3469. [CrossRef]
24. Manganelli, B.; Murgante, B.; Saganeiti, L. The Social Cost of Urban Sprinkling. *Sustainability* **2020**, *12*, 2236. [CrossRef]
25. Silvia, V.; Andrea, A.; Carlo Alberto, B.; Francesco Domenico, M. *Sul Consumo Di Suolo*, Roma, Italy, 2018.
26. Palazzo, A.L.; Aristone, O. Né Città Né Campagna. La Nuova "Forma Città". *Agriregionieuropa* **2016**, *12*, 7–9.
27. Pagliacci, F. Measuring EU Urban-Rural Continuum Through Fuzzy Logic. *Tijdschr. Voor Econ. En Soc. Geogr.* **2017**, *108*, 157–174. [CrossRef]
28. Urso, G. Metropolisation and the Challenge of Rural-Urban Dichotomies. *Urban Geogr.* **2021**, *42*, 37–57. [CrossRef]
29. Cattivelli, V. Planning Peri-Urban Areas at Regional Level: The Experience of Lombardy and Emilia-Romagna (Italy). *Land Use Policy* **2021**, *103*, 105282. [CrossRef]
30. Van Vliet, J.; Birch-Thomsen, T.; Gallardo, M.; Hemerijckx, L.M.; Hersperger, A.M.; Li, M.; Tumwesigye, S.; Twongyirwe, R.; van Rompaey, A. Bridging the Rural-Urban Dichotomy in Land Use Science. *J. Land Use Sci.* **2020**, *15*, 585–591. [CrossRef]
31. Florczyk, A.; Corbane, C.; Ehrlich, D.; Carneiro Freire, S.M.; Kemper, T.; Maffenini, L.; Melchiorri, M.; Pesaresi, M.; Politis, P.; Schiavina, M.; et al. *GHSL Data Package 2019: Public Release GHS P2019*; Publications Office of the European Union: Luxembourg, 2019; Volume JRC117104, ISBN 978-92-76-13186-1.
32. European Commission; Statistical Office of the European Union; Organisation for Economic Co-Operation and Development; Food and Agriculture Organization of the United Nations; United Nations Human Settlements Programme; World Bank. *Applying the Degree of Urbanisation: A Methodological Manual to Define Cities, Towns and Rural Areas for International Comparisons: 2021 Edition*; Eurostat: Luxembourg, 2021; ISBN 9789276203063.
33. Büttner, G.; Kosztra, B.; Maucha, G.; Pataki, R.; Kleeschulte, S.; Hazeu, G.; Vittek, M.; Littkopf, A. *Copernicus Land Monitoring Service CORINE Land Cover Product User Manual (Version 1.0)*; Copernicus Publications: Göttingen, Germany, 2021.
34. Diaz-Pacheco, J.; Gutiérrez, J. Exploring the Limitations of CORINE Land Cover for Monitoring Urban Land-Use Dynamics in Metropolitan Areas. *J. Land Use Sci.* **2014**, *9*, 243–259. [CrossRef]
35. Bielecka, E.; Jenerowicz, A. Intellectual Structure of CORINE Land Cover Research Applications in Web of Science: A Europe-Wide Review. *Remote Sens.* **2019**, *11*, 2017. [CrossRef]
36. EC. *Mapping Guide v6.2 for a European Urban Atlas Regional Policy*; European Commission: Brussels, Belgium, 2020.
37. EEA. *GMES Initial Operations/Copernicus Land Monitoring Services-Validation of Products European Settlement Map 2016 VALIDATION REPORT*; EEA: Copenhagen, Denmark, 2017.
38. Sabo, F.; Corbane, C.; Politis, P.; Kemper, T. *The European Settlement Map 2019 Release Application of the Symbolic Machine Learning to Copernicus VHR imagery*; Publications Office: Luxembourg, 2019. [CrossRef]
39. Luti, T.; De Fioravante, P.; Marinosci, I.; Strollo, A.; Riitano, N.; Falanga, V.; Mariani, L.; Congedo, L.; Munafò, M. Land Consumption Monitoring with SAR Data and Multispectral Indices. *Remote Sens.* **2021**, *13*, 1586. [CrossRef]
40. Strollo, A.; Smiraglia, D.; Bruno, R.; Assennato, F.; Congedo, L.; De Fioravante, P.; Giuliani, C.; Marinosci, I.; Riitano, N.; Munafò, M. Land Consumption in Italy. *J. Maps* **2020**, *16*, 113–123. [CrossRef]
41. De Fioravante, P.; Luti, T.; Cavalli, A.; Giuliani, C.; Dichicco, P.; Marchetti, M.; Chirici, G.; Congedo, L.; Munafò, M. Multispectral Sentinel-2 and Sar Sentinel-1 Integration for Automatic Land Cover Classification. *Land* **2021**, *10*, 611. [CrossRef]
42. Munafò, M. *Consumo Di Suolo, Dinamiche Territoriali e Servizi Ecosistemici*, 2022nd ed. Report SNPA, 32/22. 2022. Available online: https://www.snpambiente.it/wp-content/uploads/2022/07/Rapporto_consumo_di_suolo_2022.pdf (accessed on 24 October 2022).
43. Wang, Y.; Huang, C.; Feng, Y.; Zhao, M.; Gu, J. Using Earth Observation for Monitoring SDG 11.3.1-Ratio of Land Consumption Rate to Population Growth Rate in Mainland China. *Remote Sens.* **2020**, *12*, 357. [CrossRef]
44. Bai, Z.; Wang, J.; Yang, F. Research Progress in Spatialization of Population Data. *Prog. Geogr.* **2013**, *32*, 1692–1702.

45. ISTAT. *Ricostruzione Della Popolazione Residente per Sesso, Età e Comune*; ISTAT: Rome, Italy, 2021.
46. ISTAT. *Descrizione Dei Dati Geografici e Delle Variabili Censuarie Delle Basi Territoriali per i Censimenti: Anni 1991, 2001, 2011*; ISTAT: Rome, Italy, 2016.
47. Guerri, G.; Crisci, A.; Congedo, L.; Munafò, M.; Morabito, M. A Functional Seasonal Thermal Hot-Spot Classification: Focus on Industrial Sites. *Sci. Total Environ.* **2022**, *806*, 151383. [CrossRef]
48. Branislav, B.; Nikola, K.; Mileva, S.-P.; Milan, K. Dasymetric Modelling of Population Dynamics in Urban Areas. *Geod. Vestn.* **2013**, *57*, 777–792.
49. Frick, S.A.; Rodríguez-Pose, A. Big or Small Cities? On City Size and Economic Growth. *Growth Change* **2018**, *49*, 4–32. [CrossRef]
50. Seto, K.C.; Güneralp, B.; Hutyra, L.R. Global Forecasts of Urban Expansion to 2030 and Direct Impacts on Biodiversity and Carbon Pools. *Proc. Natl. Acad. Sci. USA* **2012**, *109*, 16083–16088. [CrossRef] [PubMed]
51. Zitti, M.; Ferrara, C.; Perini, L.; Carlucci, M.; Salvati, L. Long-Term Urban Growth and Land Use Efficiency in Southern Europe: Implications for Sustainable Land Management. *Sustainability* **2015**, *7*, 3359–3385. [CrossRef]
52. Fiorini, L.; Zullo, F.; Marucci, A.; Romano, B. Land Take and Landscape Loss: Effect of Uncontrolled Urbanization in Southern Italy. *J. Urban Manag.* **2019**, *8*, 42–56. [CrossRef]
53. Cecili, G.; De Fioravante, P.; Congedo, L.; Marchetti, M.; Munafò, M. Land Consumption Mapping with Convolutional Neural Network: Case Study in Italy. *Land* **2022**, *11*, 1919. [CrossRef]
54. De Fioravante, P.; Strollo, A.; Assennato, F.; Marinosci, I.; Congedo, L.; Munafò, M. High Resolution Land Cover Integrating Copernicus Products: A 2012–2020 Map of Italy. *Land* **2022**, *11*, 35. [CrossRef]
55. Munafò, M.; Marinosci, I. *Territorio Processi e Trasformazioni in Italia*; ISPRA: Roma, Italy, 2018.
56. AS Ddl n.1131 Misure per La Rigenerazione Urbana. XVIII Legislatura. 2019. Available online: https://upel.va.it/wp-content/uploads/2021/04/51435.pdf (accessed on 24 October 2022).
57. Ministero della Transizione Ecologica. Strategia Nazionale Biodiversità 2030. 2022. Available online: https://www.isprambiente.gov.it/it/archivio/notizie-e-novita-normative/notizie-ispra/2022/04/consultazione-pubblica-della-strategia-nazionale-biodiversita-2030 (accessed on 24 October 2022).
58. Clifton, K.; Ewing, R.; Knaap, G.J.; Song, Y. Quantitative Analysis of Urban Form: A Multidisciplinary Review. *J. Urban.* **2008**, *1*, 17–45. [CrossRef]
59. Schirmer, P.M.; Axhausen, K.W. A Multiscale Classification of the Urban Morphology. *J. Transp. Land Use* **2015**, *9*, 101–130. [CrossRef]
60. Ronchi, S.; Salata, S.; Arcidiacono, A. *An Indicator of Urban Morphology for Landscape Planning in Lombardy (Italy)*; Emerald Publishing Limited: Bingley, UK, 2018.
61. Duany, A.; Talen, E. *Transect Planning*; American Planning Association: Chicago, IL, USA, 2002; Volume 68.
62. Southworth, M.; Owens, P.M. The Evolving Metropolis Studies of Community, Neighborhood, and Street Form at the Urban Edge. *J. Am. Plan. Assoc.* **1993**, *59*, 271–287. [CrossRef]
63. Wheeler, S.M. Built Landscapes of Metropolitan Regions: An International Typology. *J. Am. Plan. Assoc.* **2015**, *81*, 167–190. [CrossRef]
64. Rashid, M. The Evolving Metropolis after Three Decades: A Study of Community, Neighbourhood and Street Form at the Urban Edge. *J. Urban Des.* **2018**, *23*, 624–653. [CrossRef]
65. Case Scheer, B.; Petkov, M. Edge City Morphology. A Comparision of Commercial Centers. *J. Am. Plan.* **1998**, *64*, 298–310.
66. Comitato per lo Sviluppo del Verde. *Strategia Nazionale Del Verde Urbano*; Ministero dell'Ambiente e della Tutela del Territorio e del Mare (MATTM): Roma, Italy, 2018.
67. Sallustio, L.; Lasserre, B.; Blasi, C.; Marchetti, M. Infrastrutture Verdi Contro Il Consumo Di Suolo. *Reticula* **2020**, *25*, 21–31.
68. Marando, F.; Salvatori, E.; Sebastiani, A.; Fusaro, L.; Manes, F. Regulating Ecosystem Services and Green Infrastructure: Assessment of Urban Heat Island Effect Mitigation in the Municipality of Rome, Italy. *Ecol. Modell.* **2019**, *392*, 92–102. [CrossRef]
69. Guerri, G.; Crisci, A.; Messeri, A.; Congedo, L.; Munafò, M.; Morabito, M. Thermal Summer Diurnal Hot-Spot Analysis: The Role of Local Urban Features Layers. *Remote Sens.* **2021**, *13*, 538. [CrossRef]
70. Zambrano, L.; Aronson, M.F.J.; Fernandez, T. The Consequences of Landscape Fragmentation on Socio-Ecological Patterns in a Rapidly Developing Urban Area: A Case Study of the National Autonomous University of Mexico. *Front. Environ. Sci.* **2019**, *7*, 152. [CrossRef]

land

Article

Ecosystem Mapping and Accounting in Italy Based on Copernicus and National Data through Integration of EAGLE and SEEA-EA Frameworks

Paolo De Fioravante [1], Andrea Strollo [1], Alice Cavalli [2], Angela Cimini [3,*], Daniela Smiraglia [1], Francesca Assennato [1] and Michele Munafò [1]

1 Italian Institute for Environmental Protection and Research (ISPRA), Via Vitaliano Brancati 48, 00144 Rome, Italy
2 Department for Innovation in Biological, Agro-Food and Forest Systems, University of Tuscia, Via S. Camillo de Lellis, 01100 Viterbo, Italy
3 Department of Architecture and Project, University of Rome La Sapienza, Piazza Borghese 9, 00186 Rome, Italy
* Correspondence: angela.cimini@uniroma1.it

Abstract: Developing appropriate tools to understand and protect ecosystems and the services they provide is of unprecedented importance. This work describes the activity performed by ISPRA for the mapping of the types of ecosystems and the evaluation of their related ecosystem services, to meet the needs of the "ecosystem extent account" and "ecosystem services physical account" activities envisaged by the SEEA-EA framework. A map of the types of ecosystems is proposed, obtained by integrating the main Copernicus data with the ISPRA National Land Consumption Map, according to the MAES (Mapping and Assessment of Ecosystems and their Services) classification system. The crop production and carbon stock values for 2018 were then calculated and aggregated with respect to each ecosystem. The ecosystem accounting was based on the land cover map produced by ISPRA integrating, according to an EAGLE compliant classification system, the same Copernicus and National input data used for mapping the types of ecosystems. The analysis shows the importance of an integrated reading of the main monitoring tools and the advantages in terms of compatibility and comparability, with a view to enhancing the potential of Copernicus land monitoring instruments also in the context of ecosystem accounting activities.

Keywords: ecosystem accounting; Copernicus; land cover; SEEA-EA; MAES; EAGLE; ecosystem services; carbon stock; crop production

Citation: De Fioravante, P.; Strollo, A.; Cavalli, A.; Cimini, A.; Smiraglia, D.; Assennato, F.; Munafò, M. Ecosystem Mapping and Accounting in Italy Based on Copernicus and National Data through Integration of EAGLE and SEEA-EA Frameworks. *Land* **2023**, *12*, 286. https://doi.org/10.3390/land12020286

Academic Editor: Richard Smardon

Received: 24 November 2022
Revised: 13 January 2023
Accepted: 15 January 2023
Published: 19 January 2023

1. Introduction

Ecosystems represent the base units to detect land changes and to assess environmental conditions, which enables the recognition of past and current ecological processes and the related services supplied and the analysis of future scenarios [1–3]. The ecosystem's biotic and abiotic characteristics and state affect the energy flow, the nutrient cycle, and the availability of resources, species, and habitat [4–6]. In this sense, an exhaustive, effective, and operational ecosystems distribution knowledge is a fundamental aspect in land monitoring activities [7].

The whole legislative framework settled for by the EU in recent years refers to ecosystem accounting, e.g., the Biodiversity Strategy for 2030 [8] and the related European Soil Strategy [9], the Nature Restoration Law [10] adopted by the Commission in June 2022, and the Healthy Soils Law that is under preparation. The United Nations Statistical Commission, in its 52nd session of March 2021, attributed to the National Accounting System and to the System of Environmental Economic Accounting (SEEA) the assessment of the contribution of the environment to the economy and the impact of the economy on the environment. This is in order to provide data, indicators, and statistics to stakeholders, to monitor these interactions, and to identify more sustainable development strategies. SEEA Ecosystem Accounting (SEEA-EA) is now the reference framework under the proposal for the amendment of Regulation (EU) No. 691/2011 on European environmental economic accounts to include a new module on ecosystem accounts [11]. The proposed legal module on ecosystem accounts has been adopted by the Commission in July 2022 and proposed to the Council and Parliament for final approval. Relevant policies consider the SEEA-EA framework, including the Nature Directives, the Marine Strategy Framework Directive, the Common Agriculture Policy, the EU Forest Strategy, and the EU Pollinators Initiative [12–15]. The SEEA-EA may also support the reporting under the 8th Environmental Action Programme or the revised Regulation on Land Use Land-Use Change, and Forestry (LULUCF) (EU) 2018/841 [11]; it can also be an example for further analysis models to implement in international projects such as the Horizon 2020 project SERENA (Soil Ecosystem seRvices and soil threats modElling aNd mApping), which is conducting an in-depth analysis of models about soil threats, soil ecosystem services, and their bundles at the European level [16]. The Natural Capital Accounting and Ecosystem Service Valuation (NCAVES) project highlighted the potential of the SEEA-EA in supporting the calculation and mainstreaming of many Aichi target indicators and Sustainable Development Goal (SDG) indicators [17] and the link with other key international environmental conventions and platforms, including the UNCCD, Ramsar, and IPBES [18]. The UN Committee of Experts on Environmental-Economic Accounting (UNCEEA) conducted a "Broad Brush Analysis of SDG Indicators" identifying 40 SEEA-relevant SDG indicators that are or can be aligned with the SEEA [19]. In detail, the usefulness of the SEEA as a tool to mainstream the environment and biodiversity into national planning processes is explicitly recognised via SDG Indicator 15.9.1, "Progress towards national targets established in accordance with Aichi Biodiversity Target 2 of the Strategic Plan for Biodiversity 2011–2020", which in part B requires the "integration of biodiversity values into national accounting and reporting systems, defined as implementation of the SEEA" [20]. The need to produce systematic physical and monetary measurements of the flow of ecosystem services on a national scale is also expressed at the national level by law 221/2015, which requires the development of consolidated statistical and accounting systems for natural capital and the introduction of indicators for assessing the impact of policies on it.

The SEEA-EA is a spatially based integrated framework conceived to create a coherent and comprehensive view of ecosystems, allowing the organisation of biophysical data, the measurement of ecosystem services, the tracking of changes in ecosystem assets, and the linking of this information to economic aspects and other human activity. The SEEA-EA aims to become the tool for ecosystem accounting globally by standardising data production, thus making it accessible and comparable [21–23]. In detail, the following set of accounts is defined [22]:

- Ecosystem extent account, which organises information on the extent of different ecosystem types (e.g., forests, wetlands, agricultural areas, marine areas) as a starting point for ecosystem accounting.
- Ecosystem condition account, which considers the ecological integrity of ecosystems, evaluating the distance from a reference condition with respect to different biophysical characteristics.
- Ecosystem services physical and monetary flow account, which measures the supply and use of ecosystem services by economic units.
- Monetary ecosystem asset account, which evaluates stocks and changes in stocks of ecosystem assets, based on the monetary valuation of ecosystem services at the beginning and end of each accounting period.
- Thematic accounts, which organise data on themes of specific policy relevance, such as biodiversity, climate change, urban areas, and oceans.

In order to provide a comprehensive measurement of Natural Capital and Ecosystem Services, the European Commission initiative MAES (Mapping and Assessment of Ecosystems and their Services) is a strong tool to support the effort to map and classify the extent of ecosystems, providing the conceptual framework, methodologies, and indicators to collect information on ecosystems and their services in Europe, in order to address policy decisions. MAES responded to Action 5 of the European Biodiversity Strategy to 2020, which lays the foundation for the development of a methodology for understanding the condition of biodiversity and ecosystems and the pressures they are subject to. In response to this task, since 2012, the MAES working group has developed a classification system for ecosystems [24], and then deepened the issues related to monitoring and analysis of ecosystem services [2]. The ecosystem classification method proposed in MAES is based on the use of CLC classes aggregated on the basis of the relationships between land cover (LC) and land use (LU) classes and EUNIS habitats in order to represent and collect information on large-scale ecosystems [24]. The result is a 12-class classification system, which gathers 7 terrestrial ecosystems, a freshwater one, and 4 related to marine areas. The ecosystem types proposed in MAES were used for the EU ecosystem extent accounts within the INCA project (The official name of the INCA project is knowledge innovation project on an Integrated system of Natural Capital and ecosystem services Accounting for the European Union), which was used to test the System of Environmental-Economic Accounting-Experimental Ecosystem Accounting (SEEA-EEA). The results were included in the most recent version of the SEEA-EA handbook, adopted in 2021 [25]. In this context, the INCA project has shown that by following the SEEA-EA guidelines, it is possible to produce large amounts of data on ecosystem accounting, thus enabling consistent and comparable information on ecosystems and ecosystem services at the European scale.

The MAES classification system has also been adopted by the Copernicus Land Monitoring Service (CLMS) Local Component (details are available at https://land.copernicus.eu/local, accessed on 11 January 2023), which includes high spatial and thematic resolution vector data relating to three main categories of areas that require specific and detailed monitoring: the Riparian Zones data offer a mapping of riparian areas [26], the Coastal Zones data map a buffer zone of 10 km from the coast line [27], and the Natura 2000 data map protected areas [28]. The Copernicus CLMS also includes other relevant LC/LU products, such as the CORINE Land Cover [29], Urban Atlas [30], and the new CLC Plus Backbone data [31], and also numerous other geographic information on soils and related variables (e.g., the state of the vegetation or the water cycle). One of the objectives of Copernicus is to provide data organised according to criteria that guarantee comparability and interchange between different EU countries. This need is particularly important in the context of LC/LU monitoring, where products born in different contexts and for different needs have required the definition of specific classification systems, which are often difficult to compare. In order to coordinate data flows from a thematic point of view, within CLMS, the EAGLE concept (EIONET Action Group on Land monitoring in Europe) was introduced. EAGLE aims to define a conceptual methodology to describe land cover and land use information from different classification systems by tracing them to the three categories: Land cover components (LCC), Land use attributes (LUA), and Landscape characteristics (CH) [32]. This allows the understanding of overlaps and the possible conversions between different classification systems, but also to define new ones. The EAGLE model aims to separate the LC and LU components through data modelling systems applicable at different scales and in different contexts, while maintaining compatibility with existing datasets.

Over the years, the Italian Institute for Environmental Protection and Research (ISPRA) has introduced many different LC and LU products on a national scale for the Italian territory, as much as possible in line with the EAGLE model, both through the classification of Sentinel-1 and -2 data [33–35], and through the integration of existing data. In this sense, a methodology was developed for the integration of Copernicus and national data, which made it possible to produce a national scale EAGLE compliant land cover map starting from the integration of CLMS Local (Coastal Zones, Riparian Zones, Urban Atlas, Natura

2000) and Pan European (CORINE Land Cover, CLC Plus Backbone) LC/LU data with the ISPRA National Land Consumption Map (LCM) [36]. This product overcomes some of the main limitations of the CLC and MAES classification systems, such as the widespread use of mixed LC/LU classes, and makes it possible to meet the needs of SEEA-EA Ecosystem services' physical accounting, becoming the basis for the assessment of ecosystem service proposed by the annual ISPRA report on land consumption [37].

This research is a first attempt to map the MAES ecosystem types at a national scale, in order to provide support data to the ecosystem extent accounting activity with respect to the classes recognised internationally by SEEA-EA and considered at a national level for the future accounting of ecosystem services. In this sense, a map of the types of ecosystems was produced by integrating the main CLMS data and the LCM, while the ecosystem services associated with crop production and carbon stocks in the soil were calculated and expressed as a function of these types of ecosystems.

2. Materials and Methods

2.1. Overview

The following methodology describes the input data and the procedures adopted to produce a map of the MAES types of ecosystems, useful for the activities of ecosystem extent accounting in compliance with the SEEA-EA approach. The ecosystems thus identified were associated with the values of ecosystem services relating to crop production and carbon storage. The two ecosystem services are estimated with reference to the total stock, starting from the LC map produced that integrated the main CLMS data with the LCM [37] according to an EAGLE compliant classification system. The analysis was conducted for the entire national territory, with reference to 2018 (Figure 1).

2.2. SEEA-EA Compliant Types of Ecosystems Map

To produce a map of the ecosystem typologies in compliance with the United Nations SEEA-EA approach [22], the classification system of Table 1 was defined.

The proposed classification system of the types of ecosystems refers to the MAES classes, which are divided at the first classification level into terrestrial ecosystems, freshwater ecosystems, and marine ecosystems, and then further characterised at the second classification level.

Terrestrial ecosystems are delineated starting from CORINE Land Cover classes and are subdivided into:

- Settlements and other artificial areas, i.e., urban areas where most of the human population live and which also include significant areas for synanthropic species associated with urban habitats. This class significantly affects other ecosystem types and includes urban, industrial, commercial, and transport areas, green urban areas, and mines, dumping, and construction sites.
- Cropland, i.e., areas mainly dedicated to agricultural production, even with the presence of important natural areas.
- Grassland, i.e., areas with a prevalence of herbaceous vegetation, which can include managed pastures and natural and semi-natural pastures.
- Forest and woodland include areas dominated by woody vegetation; they are very important from the point of view of the provision of ecosystem services.
- Heathland and shrub are dominated by moors, heathland, and sclerophyllous vegetation.
- Sparsely vegetated ecosystems are all naturally unvegetated or sparsely vegetated habitats, usually with extreme climatic conditions, such as bare rocks, glaciers, dunes, beaches, and sand plains.
- Inland wetlands include natural or modified mires, bogs, and fens, as well as peat extraction sites. In these areas, water regulation and peat-related processes are associated with specific species of animals and plants.

Figure 1. Study area. The analysis was conducted for the entire Italian national territory.

Table 1. MAES ecosystem types adopted for the realisation of the ecosystem type map by integrating Copernicus data with the LCM.

Ecosystem Types	
I Classification Level	**II Classification Level**
Terrestrial ecosystems	Settlements and other artificial areas Cropland Grassland (pastures, semi-natural and natural grasslands) Forest and woodland Heathland and shrub Sparsely vegetated ecosystems Inland wetlands
Freshwater ecosystems	Rivers, canals, lakes, and reservoirs
Marine ecosystems	Marine inlets and transitional waters

Freshwater ecosystems include permanent freshwater, inland water courses, and water bodies.

For marine ecosystems, the only class considered was "Marine inlets and transitional water", which includes coastal wetlands, lagoons, estuaries, i.e., areas on the land–water interface under the influence of tides and with salinity levels greater than 0.5 ‰. The other types of marine ecosystems were not considered as they relate to areas outside the study area; furthermore, they are not considered in CLMS data.

In detail, the map integrates the main local and Pan-European CLMS data and the ISPRA LCM for the year 2018 (Table 2).

Table 2. Input data used for the mapping of ecosystem types and to produce the land cover map used for the ecosystem services assessment; LC= Land Cover classes, LU= Land Use classes, CLMS= Copernicus Land Monitoring Service.

	Name	Data Type	Classes	MMU
National data	Land Consumption map (ISPRA)	Raster	17 (LC)	Pixel 10 × 10 m
CLMS Pan-European Component	CLC Plus Backbone	Raster	12 (LC)	Pixel 10 × 10 m
	CORINE Land Cover	Vector	44 (LC, LU)	25 ha (status) 5 ha (changes)
CLMS Local Component	Coastal Zones Natura 2000 Riparian Zones	Vector	55 (LC, LU)	0.5 ha
	Urban Atlas		27 (LC, LU)	0.25 ha (class 1) 1 ha (class 2–5)

The input data were reclassified according to ecosystems of Table 1 and merged into a single 10 × 10 m raster mosaic. The proposed typologies represent the basic units for ecosystems state and services assessments from the local to national scale. This map aims to provide a representation of the Italian territory with respect to the MAES classes in order to be a useful support for the ecosystem extent accounting activity.

These ecosystem types have been associated with the values of the services relating to agricultural productivity and organic carbon storage, calculated using the procedure described in the following paragraphs.

2.3. EAGLE Compliant Land Cover for the Assessment of Ecosystem Services

The same input data described before (Table 2) were combined to produce a land cover map for 2018 according to the methodology described in De Fioravante et al. [33].

The input data were converted to a 10 × 10 m resolution raster and reclassified according to the classification system of Table 3, which is based on previous activities of the ISPRA working group [33,36] and adopts a combination of land cover classes directly attributable to the EAGLE model Land Cover Components (LCC), integrated with appropriate Land Characteristics (LCH) in order to increase the thematic detail of the classification system and preserve the information content of the input data.

Table 3. Land cover map classification system.

Land Cover									
I Level		**II Level**		**III Level**		**IV Level**		**V Level**	
1	Abiotic non-vegetated surfaces	11	Artificial abiotic						
		12	Natural abiotic	121	Consolidated (bare rocks, cliffs)				
				122	Unconsolidated (beaches, dunes, sands)				
2	Biotic vegetated surfaces	21	Woody vegetation	211	Trees	2111	Broad-leaved	21111	Preval. of oaks and other evergreen broad-leaved
								21112	Preval. of deciduous oaks
								21113	Preval. of other native broad-leaved
								21114	Preval. of chestnut
								21115	Preval. of beech
								21116	Preval. of hygrophytes
								21117	Preval. of exotic broad-leaved
								21118	Preval. of olive trees
								21119	Preval. of orchards
						2112	Needle-leaved	21121	Preval. of Mediterranean pines and cypresses
								21122	Preval. of gold-Mediterranean and mountain pines
								21123	Preval. of spruce
								21124	Preval. of larch and/or Swiss pine
								21125	Preval. of needle-leaved exotics
				212	Shrubs	2121	Vineyards		
						2122	Shrubland		
		22	Herbaceous vegetation	221	Periodically	2211	Pastures		
						2212	Arable land		
				222	Permanent				
3	Water surfaces	31	Water bodies						
		32	Permanent snow and ice						
4	Wetlands								

At the first classification level, four macro-classes are defined (Abiotic non-vegetated areas, Biotic vegetated areas, Water surfaces, and Wetlands). Abiotic non-vegetated areas include any unvegetated surfaces and are subdivided into man-made artificial structures (artificial abiotic surfaces) and natural material surfaces (natural abiotic surfaces, both consolidated and unconsolidated). Biotic vegetated areas include any vegetated surfaces, with or without anthropogenic influence. At the second classification level, woody and herbaceous vegetation are distinguished. Woody vegetation is further subdivided at the third, fourth, and fifth classification levels in different classes of broad-leaved trees, needle-leaved trees, and shrubs, while for herbaceous vegetation, the classes of natural unmanaged grassland, pastures, and arable land are distinguished. Water surfaces include natural or artificial solid water (permanent ice) and liquid water (regardless of shape, position, salinity, and origin). Wetlands are defined according to the definition provided by CORINE Land Cover, and include inland wetlands (inland marshes and peat bogs) and coastal wetlands (salt marshes, salines, and intertidal flats), while lagoons and estuaries are associated with water bodies. The wetlands class was introduced at the first classification level to preserve the information content of the input data, but it is not directly compatible with the EAGLE model and will be better integrated in the classification system in future studies. For a more detailed description of the classes, reference can be made to [36] and to the official ISPRA [37] and EAGLE group [32] documentation.

The new Copernicus CLC Plus Backbone has allowed the distinguishing of different typologies of LC and LU in the mixed classes. In detail, the woody component from the shrub and herbaceous vegetation and the agricultural use from natural areas have been identified. The reclassified data were then mosaicised, giving priority to the CLMS Local data, which have higher geometric detail than the CLC. The latter was included in the areas not covered by Local data, while the fourth CLC classification level (available for Italy) made it possible to detail the different broad-leaved and needle-leaved classes. The map was used for the evaluation of ecosystem services described in the following two paragraphs.

2.4. Crop Production

Crop production, understood as an ecosystem service, is the ecological contribution to the growth of cultivated crops that can be harvested and used as raw material [38]. Crop production was calculated using the methodology defined by ISPRA [37,39], based on data from the Italian National Institute of Statistics (ISTAT) agricultural census of 2013 [40]. These data are available at the municipality scale and provide information on the area occupied by the different types of crops (expressed in hectares), and on their total production (expressed in quintals (1 Quintal (q): 1 q = 100 kg; the unit is officially used for data published by ISTAT, although it is not part of the International System of Units)), with reference to both herbaceous and woody crops. Five classes of crops have been identified (arable land, pastures, olive groves, vineyards, and orchards), and for each, the productivity in terms of quintals produced per hectare occupied by the class has been assessed. The productivity value was then traced back to the area of the single pixel (equal to 10×10 m) for each province and attributed to all the pixels of that class on the 2018 land cover map. The ecosystem service was calculated starting from the classes in Table 3, obtaining a map with agricultural productivity values per pixel, which were then aggregated with respect to the types of ecosystems. In this way, the agricultural production in 2018 associated with each of the five crops classes for each ecosystem type was evaluated.

2.5. Carbon Storage

Carbon storage is a regulation service provided by terrestrial and marine ecosystems thanks to their ability to fix greenhouse gases [41]. This service contributes to the regulation of the climate at a global level and plays a fundamental role in the context of climate change mitigation and adaptation strategies. The analysis of the carbon storage capacity referred to the entire national territory, starting from the methodology reported by ISPRA [37] and De Fioravante et al. [36], applied to the 2018 land cover map described above. The study provided estimates of the stored carbon for each portion of the territory and each type of land cover with reference to four main carbon pools [42], recognised and classified by the Intergovernmental Panel on Climate Change [43]:

- Above-Ground Biomass (AGB) includes all the tissues of plant organisms outside the soil (such as stems, branches, leaves, seeds, etc.). The fraction of stored carbon is calculated starting from the growing stock volume multiplied by specific multiplicative coefficients.
- Below-Ground Biomass (BGB) includes the root system of plants. The volume is calculated according to [44], considering the growing stock volume, the wood basic density, the crown/roots ratio [44,45], and a biomass expansion factor.
- The carbon content in the Dead Organic Substance (DOS) includes the necromass, the woody plant residues, the litter, and the residues not yet decomposed.
- The soil carbon considers organic and mineral layers up to a thickness of 30 cm. The calculation is based on the 1 km resolution raster produced by CREA-ABP, CNR-Ibimet as part of the Global Soil Partnership/FAO initiative [46], the data of the National Inventory of Forests and Forest Carbon Tanks (INFC) [47], and other data from the literature [33], assuming zero carbon stored by artificial areas.

The service was calculated starting from the classes of Table 3, obtaining a map of carbon stock values per pixel, which were then aggregated with respect to the types of ecosystems. In this way, it was possible to evaluate the carbon stock in 2018 for each ecosystem type.

3. Results

3.1. SEEA-EA Compliant Types of Ecosystems Map

Figure 2 shows the map of the types of ecosystems for 2018, obtained from the reclassification of the CLMS and LCM data according to the MAES classes of Table 1.

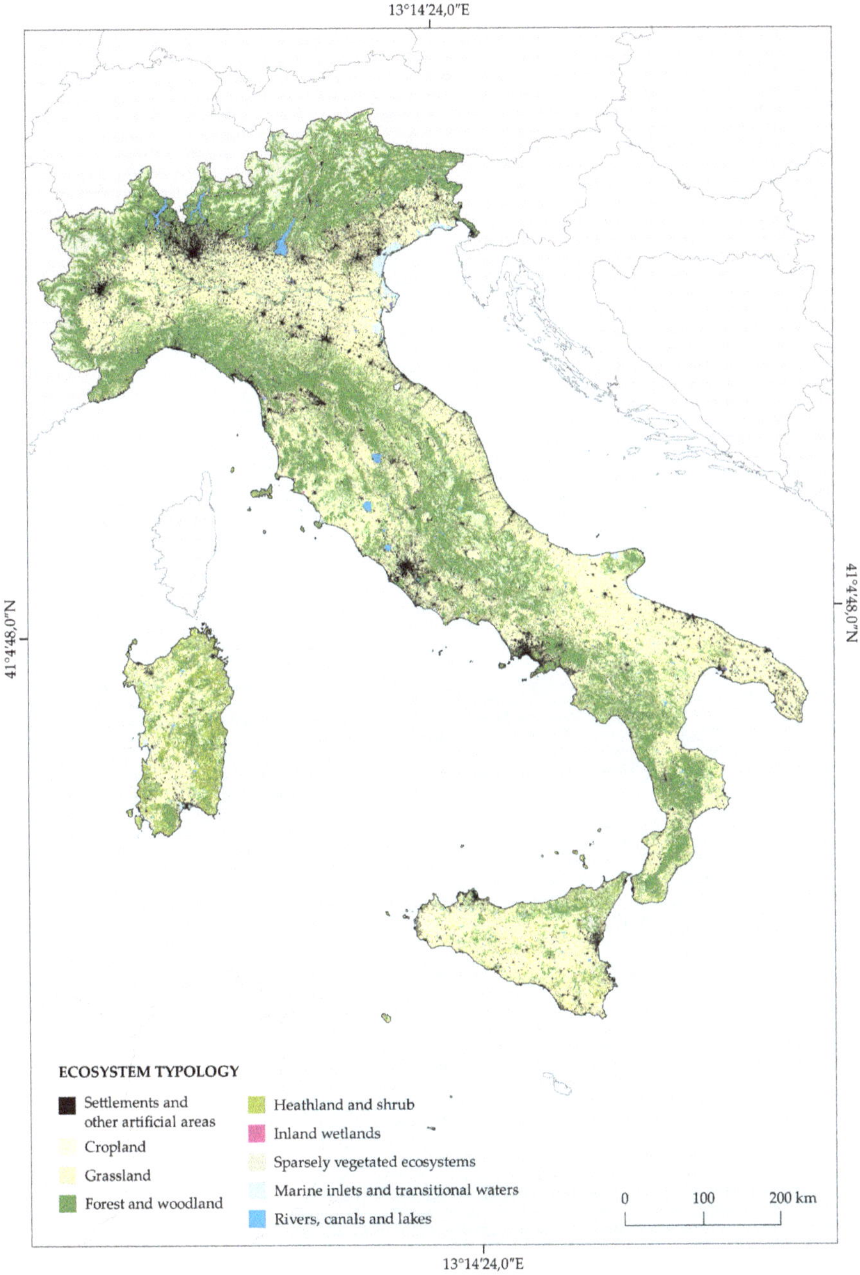

Figure 2. Ecosystem type map with MAES compliant classification system (2018).

The spatial distribution of the ecosystem typologies of Figure 2 shows the following results (Table 4).

Table 4. Surface statistics relating to the ecosystem typologies map for the Italian territory (2018).

Region	Ecosystem Typologies																	
	Settlements and Other Artificial Areas		Cropland		Grassland		Forest and Woodland		Heathland and Shrub		Sparsely Vegetated Ecosystems		Inland Wetlands		Rivers, Canals, and Lakes		Marine Inlets and Transitional Waters	
	km^2	%	km^2	%	km^2	%	km^2	%	km^2	%	km^2	%	km^2	%	km^2	%	km^2	%
Piedmont	2376	7.7	8597	7.1	2469	8.9	9045	9.8	407	3.4	2231	18.0	1	0.7	275	9.2	0	0.0
Aosta Valley	96	0.3	99	0.1	342	1.2	1135	1.2	101	0.8	1476	11.9	1	0.7	13	0.4	0	0.0
Lombardy	3989	13.0	8569	7.1	2445	8.9	6375	6.9	253	2.1	1443	11.7	28	15.6	776	25.9	0	0.0
Trentino–Alto Adige	589	1.9	747	0.6	2120	7.7	7138	7.7	377	3.1	2530	20.4	1	0.7	102	3.4	0	0.0
Veneto	3005	9.8	8038	6.6	1276	4.6	4101	4.4	268	2.2	561	4.5	9	5.1	397	13.2	682	44.6
Friuli–Venezia Giulia	902	2.9	2367	2.0	392	1.4	3371	3.6	212	1.7	457	3.7	2	1.3	67	2.3	148	9.7
Liguria	600	2.0	436	0.4	385	1.4	3767	4.1	143	1.2	67	0.5	0	0.2	20	0.7	1	0.1
Emilia–Romagna	2782	9.1	12,125	10.0	1085	3.9	5668	6.1	155	1.3	251	2.0	45	25.4	219	7.3	171	11.2
Tuscany	2024	6.6	7983	6.6	1410	5.1	10,868	11.8	345	2.8	163	1.3	35	19.9	118	3.9	42	2.7
Umbria	664	2.2	3396	2.8	614	2.2	3544	3.8	32	0.3	47	0.4	6	3.1	152	5.1	0	0.0
Marche	896	2.9	4937	4.1	617	2.2	2702	2.9	40	0.3	106	0.9	0	0.1	26	0.9	1	0.1
Latium	2426	7.9	6852	5.6	1520	5.5	5498	5.9	239	2.0	384	3.1	6	3.1	261	8.7	16	1.1
Abruzzo	782	2.5	3585	3.0	1872	6.8	3844	4.2	258	2.1	411	3.3	3	2.0	39	1.3	1	0.1
Molise	266	0.9	2103	1.7	486	1.8	1452	1.6	84	0.7	29	0.2	1	0.5	20	0.7	1	0.1
Campania	2001	6.5	5599	4.6	920	3.3	4490	4.9	352	2.9	172	1.4	5	2.6	56	1.9	5	0.3
Apulia	2120	6.9	13,720	11.3	1409	5.1	1401	1.5	381	3.1	56	0.5	7	3.7	38	1.3	224	14.6
Basilicata	477	1.6	4372	3.6	1346	4.9	3273	3.5	260	2.1	202	1.6	1	0.6	59	2.0	2	0.1
Calabria	1084	3.5	5359	4.4	1091	3.9	6414	6.9	732	6.0	316	2.6	2	1.0	74	2.5	11	0.7
Sicily	2368	7.7	14,261	11.8	3069	11.1	3064	3.3	2007	16.5	802	6.5	13	7.5	104	3.5	31	2.0
Sardinia	1265	4.1	8214	6.8	2755	10.0	5326	5.8	5505	45.3	669	5.4	11	6.2	179	6.0	194	12.7
Italy	30,712	100.0	121,360	100.0	27,623	100.0	92,479	100.0	12,151	100.0	12,372	100.0	177	100.0	2996	100.0	1530	100.0

From the analysis of Figure 2 and Table 4, a prevalence of the "Cropland" class emerges; it occupies more than 40% of the national territory, concentrating in Sicily, Apulia, and Emilia–Romagna, where one third of the class falls. Additionally important is the presence of the "Forest and woodland" class, which occupies approximately 30% of the national territory, with a prevalence in Tuscany, Piedmont, and Trentino–Alto Adige (more than a quarter of the surface occupied by this class falls in these three regions). The "Sparsely vegetated ecosystems" are mainly present in the alpine regions, while the "Heathland and shrub" typologies fall in the two major islands. Almost half of the coastal ecosystems are concentrated in Veneto and Friuli–Venezia Giulia, while more than 60% of those of the "Inland Wetlands" are in Tuscany, Emilia–Romagna, and Lombardy.

3.2. *Assessment of Ecosystem Services through an EAGLE Compliant Land Cover Map*

The 2018 land cover map used for the assessment of the two ecosystem services is shown in Figure 3.

The map refers to the EAGLE compliant classification system of Table 3, while the composition of the land cover (Figure 4) shows a prevalence of forest areas, which occupy approximately one third of the national territory (of which more than 80% consists of broad-leaved trees) and arable land, which covers 30% of the national territory. Permanent herbaceous vegetation and areas with sparse or no vegetation prevail in the high-altitude alpine areas, while olive groves and orchards occupy approximately 5% of the national territory and are particularly present in the south.

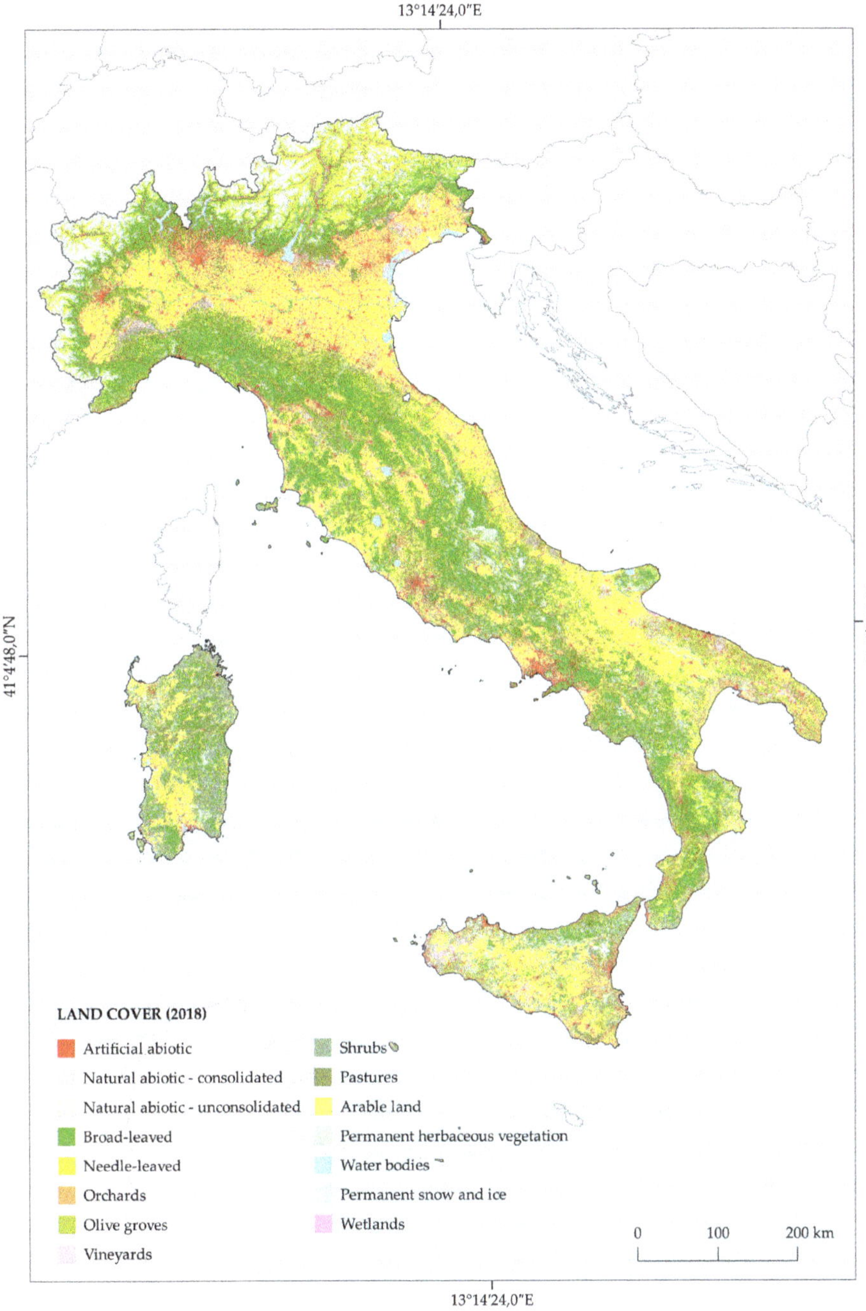

Figure 3. Land cover map with EAGLE compliant classification system (2018).

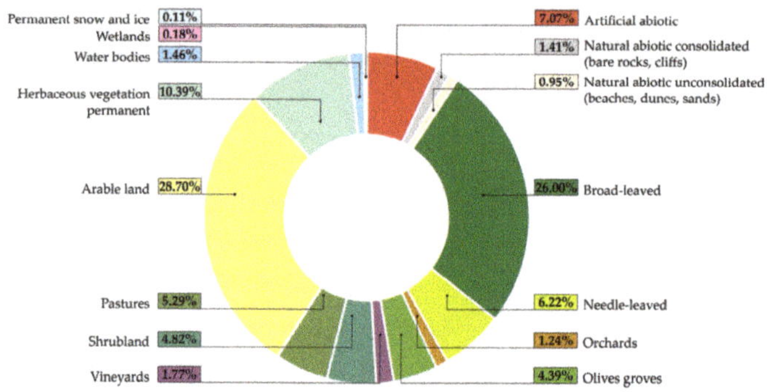

Figure 4. Surface statistics relating to the classes of the 2018 land cover map for the Italian territory.

The assessment of ecosystem services provided the results displayed in Figure 5 for organic carbon stocks and Figure 6 for agricultural productivity, with both referring to 2018.

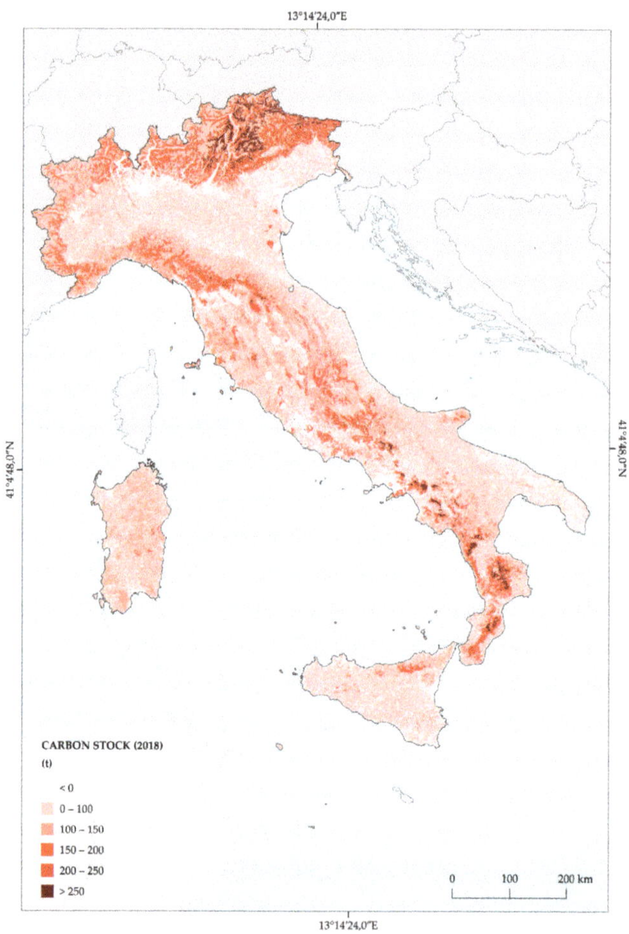

Figure 5. Carbon stock (2018).

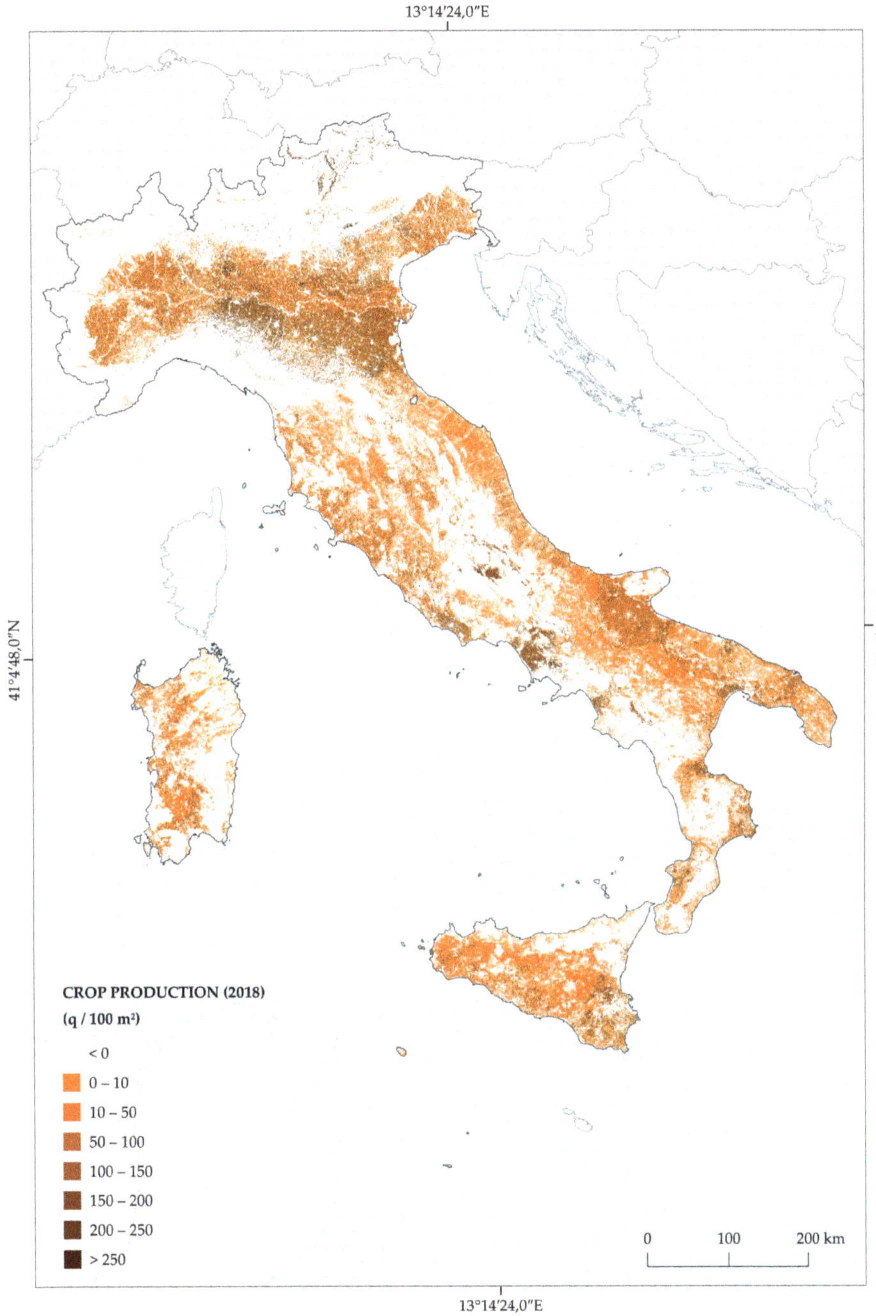

Figure 6. Crop production (2018).

The highest carbon stock values are recorded in mountainous areas with a high presence of forest cover, in particular in the Central and Eastern Alps and in the southern Apennines. In detail, the maximum value is recorded in Trentino–Alto Adige (Figure 7), due to the significant presence of needle-leaved trees, and in Piedmont, followed by Tuscany, which is the region with the greatest extension of forest cover.

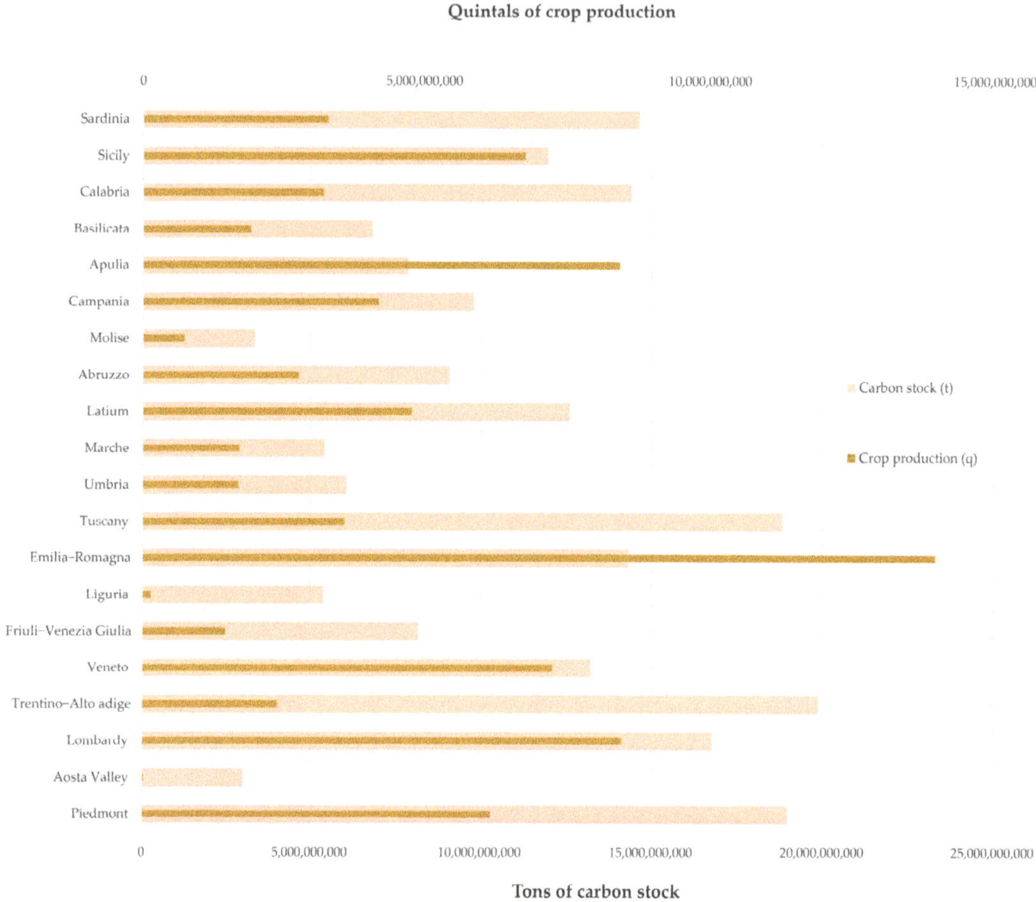

Figure 7. National and regional values of crop production and carbon stock (2018).

Agricultural productivity varies as a function of the productivity values considered for the different territories; it shows a maximum in the Po Valley and in the hinterland of Naples (Figure 6). The maximum values are relative to Emilia–Romagna, followed by Apulia and Lombardy. Overall, productivity exceeds 10 billion quintals in 6 of the 20 regions (Figure 7).

3.3. Ecosystem Services Provided by the Types of Ecosystems

The results of the ecosystem services assessment refer to the single pixel of 10 × 10 m (Figures 5 and 6). Starting with the spatialised data, the values were aggregated with respect to the types of ecosystems, allowing the results that are shown in Figures 8 and 9, which respectively refer to carbon stocks and crop production by ecosystem, both for 2018.

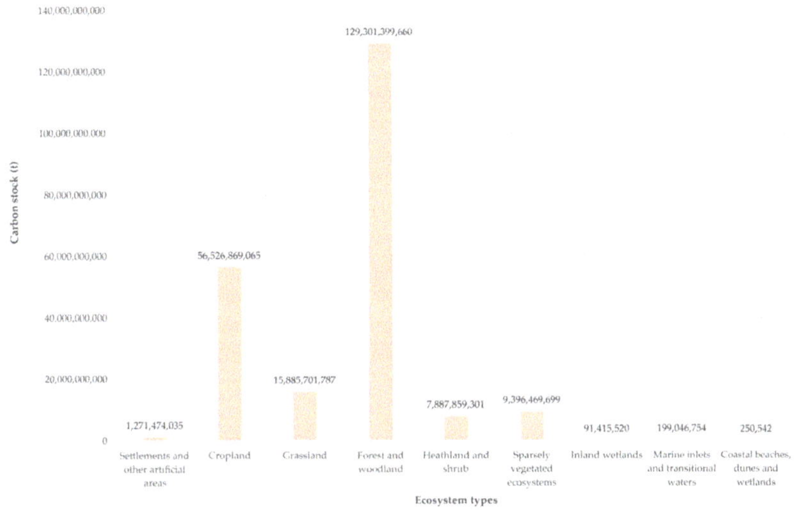

Figure 8. Carbon stock with respect to the types of ecosystems (2018).

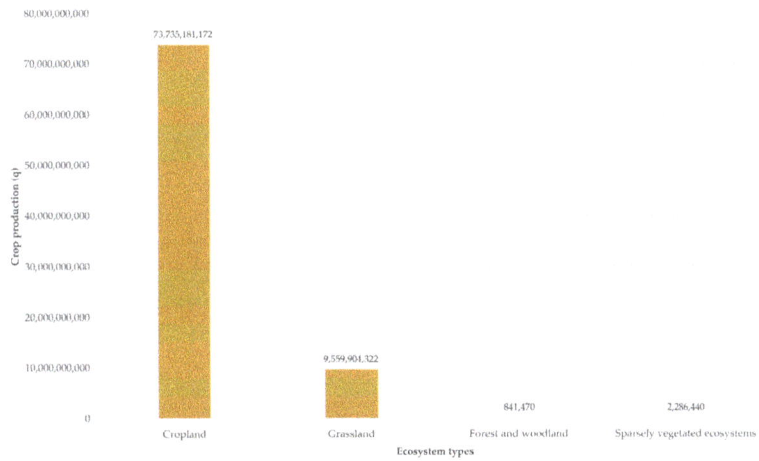

Figure 9. Crop production with respect to the types of ecosystems (2018).

Figure 8 shows a concentration of carbon stocks in the ecosystem type of "Forest and woodland", where 58% of the total carbon stock is concentrated, followed by "Croplands" and "Grasslands". The remaining classes have negligible values, due to the scarce presence of vegetation.

More than 99% of the crop production is concentrated in the "Cropland" and "Grassland" classes, while marginal values can be found in "Forest and woodland" and "Sparsely vegetated ecosystems".

4. Discussion

The approach adopted by SEEA-EA is increasingly a reference tool in the management and production of environmental data connected with the ecosystem assets and the services they provide, also allowing the analysis of the connections with economic aspects and with other human activities. In Europe, this process is rapidly evolving. Indeed, many countries have started to identify ecosystems' extent and condition, and assess biodiversity, ecosystems, and ecosystem services at the national scale [48,49]. In some cases, these activities are aimed at improving the degree of thematic detail in ecosystem mapping, supporting the achievement of EU legislative frameworks [50,51]. The assessments require spatially explicit data and information to identify ecosystems and to delimit the LC/LU classes to be considered for the calculation of ecosystem services. Delimiting the ecosystem extent is the first of the five core accounts of SEEA-EA, with the approach defined by MAES and adopted by the UN to perform it based on a classification of the types of ecosystems according to classes derived from CLC. In recent years, several LC/LU CLMS data have been introduced based on the MAES classification system and referring to critical areas for ecosystem conservation and protection, such as Riparian Zones, Coastal Zones, and Natura 2000. This research analyses and elaborates on these data, synthesising them with the ISPRA LCM in a semantically consistent representation of the territory in terms of types of ecosystems, which can be a useful basis for conducting further studies, primarily the activity of the ecosystem extent account. The map constitutes a reference for the aggregation of the results of the physical or economic assessment of ecosystem services, informing about the relevance of each ecosystem in the provision of a given service. It also lends itself to further refinements, for example, through the introduction of geological or climatic data, or additional information for the ecosystem condition accounts, which record the condition of specific characteristics of ecosystems at specific points in time. In this sense, the map of the types of ecosystems is also suitable for conducting diachronic analysis for evaluating the changes that occurred in a given period within an ecosystem, and in its ability to provide ecosystem services or assessing trade-offs and synergies between ecosystem services [44–46,52–54]. Actually, one of the first future research developments will concern the calculation of the variation in the provision of ecosystem services associated with the LC/LU changes that occurred between 2012 and 2018, to be evaluated starting from the revised version of the ISPRA LC/LU map for 2012.

However, the map of the ecosystem typologies needs to be accompanied by more detailed spatial data for the application of estimation models for the ecosystem services physical flow account. In fact, the methodologies for calculating agricultural productivity and carbon stock described above require a detailed knowledge of the shape and composition of agricultural and natural areas, allowing also to distinguish different types of tree cover or periodic and permanent crops. The ISPRA EAGLE compliant land cover map obtained integrating CLMS data with the LCM is conceived with the aim of maximising the description of the territory from a thematic point of view, distinguishing trees from herbaceous vegetation and shrubs in mixed areas classified as "Heterogeneous agricultural areas" in the CLC and MAES input data, also considering the presence of agricultural activities or natural areas; this allows for the estimation of ecosystem services over a large area with a higher geometric detail and a better thematic characterisation of the territory compared to CLC.

The fact that the land cover map derives from the same input data used for the production of the map of the ecosystem typologies constitutes an added value for the ecosystem services accounting, since the two products are coherent from a geometric point of view and their integrated reading makes it possible to understand the composition of the land cover within a certain ecosystem or, vice versa, to deepen the ecosystem functions of the different portions of the territory pertaining to a given land cover class. In fact, comparing the two products, the map of the ecosystem typologies shows a composition of the territory in line with that found by observing the ISPRA land cover map, with a prevalence of "Cropland" in the lowland areas and of "Forest and woodland" in the alpine and Apennine areas. The shrubs are mainly on the islands, while sparse or herbaceous vegetation prevails at high altitudes. Most of the carbon stocks fall in areas classified as "Forest and woodland", while almost all agricultural productivity falls in the "Croplands"; however, in the two ecosystems, there are not only the LC/LU classes of (respectively) trees and agricultural areas. The example of Figure 10 shows an agricultural area with a patch of natural vegetation in the central part (Figure 10a), which CLC classifies as 243 "Land principally occupied by agriculture, with significant areas of natural vegetation" (Figure 10b). By tracing the CLC class to the corresponding MAES ecosystem typology, the area falls into the "Cropland" typology (Figure 10d), while in the LC/LU map, the use of the CLC Backbone data allowed the disambiguation of the natural woody vegetation component from the agricultural component.

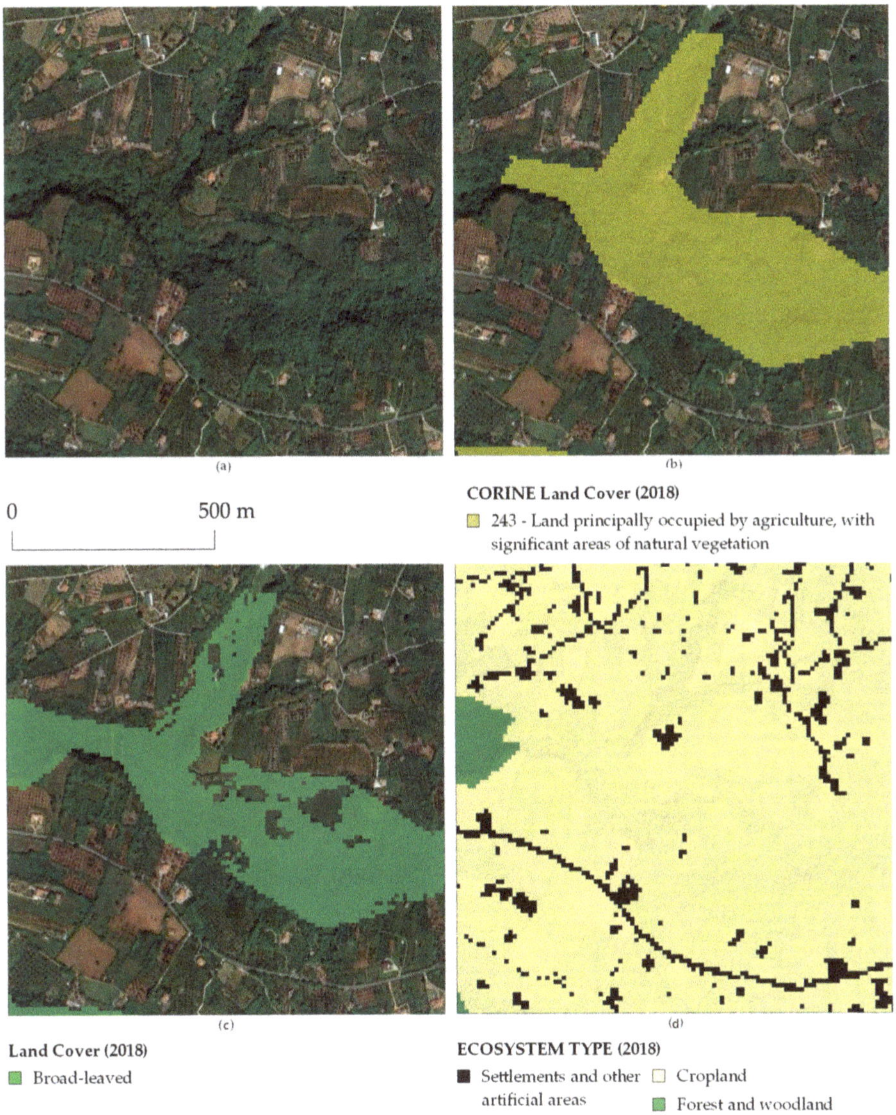

Figure 10. Examples of land cover and land use classes within the types of ecosystems. In the example there is a patch of natural vegetation surrounded by croplands (**a**). The area is classified as 243 "Land mainly occupied by agriculture, with significant areas of natural vegetation" by CLC (**b**) and corresponds to the "Cropland" MAES ecosystem typology (**d**), while in the land cover map the natural woody vegetation component the from the agricultural component (c) has been disambiguated.

In this sense, the overall values of ecosystem services obtained for each ecosystem (in this case, cropland) should be understood as the synthesis of several different contributions, deriving from the presence of different classes of land cover and land use in each type of ecosystem. Figure 11 shows that the "Forest and woodland" ecosystem type also includes areas with permanent herbaceous land cover. In other cases, areas classified in the LC map as arable land or permanent crops, but also areas covered by arboreal vegetation, both broad-leaved and needle-leaved, fall into the "Cropland" typology.

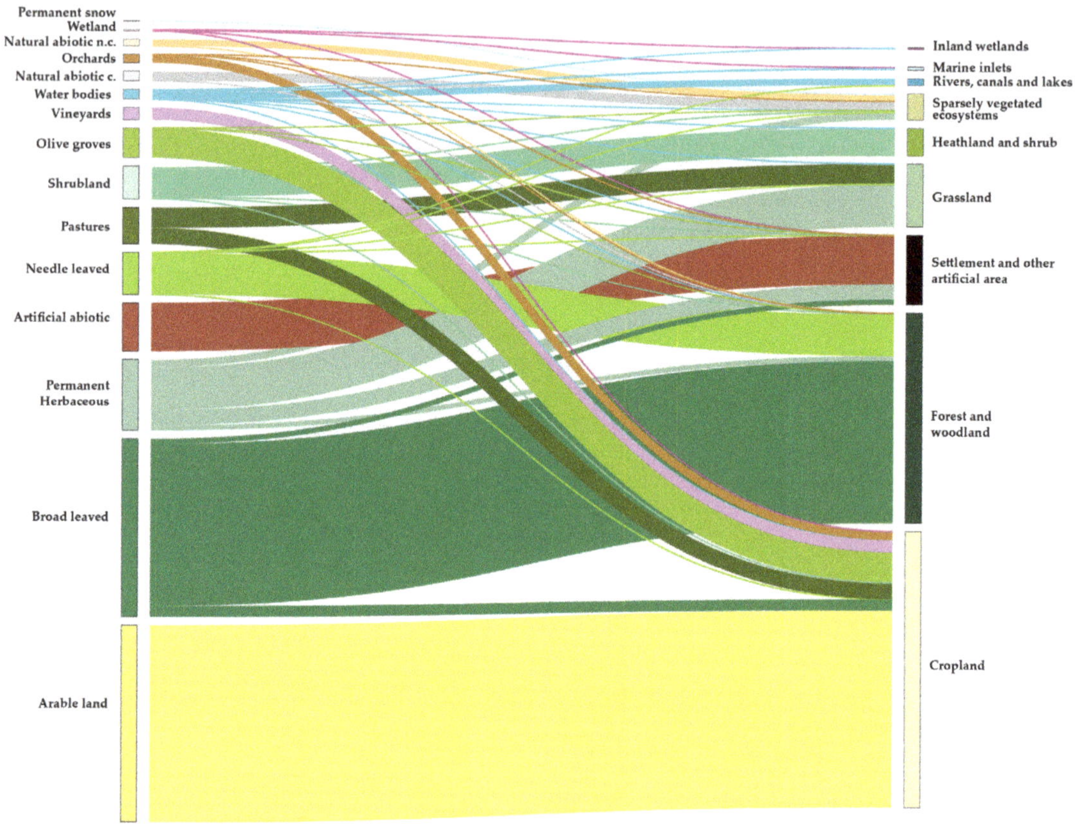

Figure 11. Flows of land cover and land use classes within the types of ecosystems.

As in many other studies on the mapping of ecosystem types, the proposed methodology uses the CLC as reference data, trying to increase its geometric and thematic detail. Some mapping experiences conducted on a national [55] and regional [47,56] scale exploit classifications of satellite data to increase the detail of the CLC, while in other cases, the improvement is conducted on a national scale by photointerpretation with the help of ancillary data [50] or using national datasets [57,58].

In this research, to increase the detail in the description of the ecosystems, data available for the entire European territory were used (the CLMS Local data), making the methodology also applicable to medium-scale studies in other contexts, generating homogeneous and comparable data (the LCM can be replaced by Copernicus HRL Imperviousness). Furthermore, the application of the methodology is not time consuming, as it does not require significant pre-processing of the input data or intense photo-interpretation activities. The areas not covered by CLMS Local data have a lower spatial detail; however, the use of

CLC Plus Backbone data still allows improvement of the description of the territory in the mixed CLC classes, disambiguating the LC classes.

5. Conclusions

This study provides insights on the integration between CLMS-derived EAGLE compliant LC/LU data and the UN approach to ecosystems classification, to supply the SEEA-EA framework for the ecosystem accounting. The five SEEA-EA core accounts integrate the main available and forthcoming data about the ecosystem assets extent, condition, and value at multiple spatial scales into a standardised, robust, and modular framework, also indicating data and knowledge gaps to be filled for a more comprehensive assessment; actually, available data often do not ensure adequate spatial and/or temporal consistency, conditioning the effectiveness of the assessment. This work focuses on ecosystem extent accounting and ecosystem services physical accounting, which require spatial LC/LU data with good thematic detail and broad coverage. The Copernicus Land Monitoring Service framework plays a fundamental role in this area, e.g., CORINE Land Cover (CLC) is one of the most suitable datasets currently used for Ecosystem accounting [24], acting as a proxy of ecosystem types for reporting purposes, although for a detailed assessment of ecosystem condition and ecosystem services physical accounting, more accurate data are needed. In this research, the CLC was integrated with the CLMS Local data and the ISPRA LCM, providing a land cover map and a map of ecosystem typologies that represent the territory in more detail and which satisfy the following requirements:

- They are in line with the EAGLE (the land cover map) and MAES (the ecosystem typology map) standards in terms of classification systems;
- They are comparable and compatible with each other from a geometric and thematic point of view;
- They are suitable for conducting ecosystem extent accounting (the land cover map) and ecosystem services physical accounting (the type of ecosystem map).

The approach adopted by SEEA-EA is increasingly a reference tool in the management and production of environmental data connected with the ecosystem assets and the services they provide, assuming an important role in directives, reporting activities, international projects, international conventions, and platforms such as the UNCCD, Ramsar, and IPBES, and supporting the calculation and mainstreaming of Aichi indicators and SDG indicators. On the other hand, the fact that there are currently few applications of the SEEA-EA methodology [11] makes this field challenging and open to developments, especially for the definition and identification of suitable input data. This research is a first attempt to apply the SEEA-EA methodology to the Italian territory, starting from the currently available data and proposing an approach in line with the UN and EAGLE standards. On the one hand, this can be replicated in other EU territories with CLMS data availability and on the other hand, it has the potential for development, both in the integration of other SEEA-EA account activities (ecosystem condition account, monetary flow account, monetary ecosystem asset account, thematic accounts) and in the refinement of those already conducted, e.g., increasing the information content by integrating future CLC Plus products.

Author Contributions: Conceptualization, P.D.F., F.A., and M.M.; methodology, P.D.F., A.S. and A.C. (Alice Cavalli); software, P.D.F., A.S. and A.C. (Alice Cavalli); validation, P.D.F., A.S., A.C. (Alice Cavalli) and A.C. (Angela Cimini); formal analysis, P.D.F. and A.C. (Angela Cimini); investigation, P.D.F., A.S. and A.C. (Alice Cavalli); resources, M.M.; data curation, P.D.F., A.S., A.C. (Alice Cavalli) and A.C. (Angela Cimini).; writing—original draft preparation, P.D.F. and A.C. (Angela Cimini); writing—review and editing, P.D.F., A.S., A.C. (Alice Cavalli), A.C. (Angela Cimini), F.A. and D.S.; visualization, P.D.F. and A.C. (Angela Cimini); supervision, M.M. and F.A.; project administration, M.M. and F.A.; funding acquisition, M.M. and F.A. All authors have read and agreed to the published version of the manuscript.

Funding: This research was funded by the Italian Institute for Environmental Protection and Research (ISPRA) structural funds.

Data Availability Statement: Data presented in this study are available on request from the corresponding author. The data are not publicly available because they are part of ongoing research.

Conflicts of Interest: The authors declare no conflict of interest.

References

1. Simon, F.; Karachepone, N.N.; Paul, L.; Rob, A. *The Methodological Assessment Report on Scenarios and Models of Biodiversity and Ecosystem Services*; Secretariat of the Intergovernmental Science-Policy Platform on Biodiversity and Ecosystem Services: Bonn, Germany, 2016; pp. 1–348.
2. Joachim, M.; Anne, T.; Markus, E.; Bruna, G.; José, I.B.; Maria Luisa, P.; Sophie, C.; Francesca, S.; Alberto, O.; Arwyn, J.; et al. *Mapping and Assessment of Ecosystems and Their Services. An Analytical Framework for Mapping and Assessment of Ecosystem Condition in EU*; European Union: Luxembourg, 2018.
3. Vargas, L.; Willemen, L.; Hein, L. Assessing the Capacity of Ecosystems to Supply Ecosystem Services Using Remote Sensing and An Ecosystem Accounting Approach. *Environ. Manage.* **2019**, *63*, 1–15. [CrossRef] [PubMed]
4. Bailey, R.G. Ecoregions of the United States. In *Ecosystem Geography*; Springer: Berlin/Heidelberg, Germany, 1996; pp. 83–104.
5. Dale, V.H.; Brown, S.; Haeuber, R.A.; Hobbs, N.T.; Huntly, N.; Naiman, R.J.; Riebsame, W.E.; Turner, M.G.; Valone, T.J. Ecological Principles and Guidelines for Managing the Use of Land Sup> 1. *Ecol. Appl.* **2000**, *10*, 639–670.
6. Keith, D.A.; Rodríguez, J.P.; Brooks, T.M.; Burgman, M.A.; Barrow, E.G.; Bland, L.; Comer, P.J.; Franklin, J.; Link, J.; McCarthy, M.A. The IUCN Red List of Ecosystems: Motivations, Challenges, and Applications. *Conserv. Lett.* **2015**, *8*, 214–226. [CrossRef]
7. Keith, D.A.; Ferrer-Paris, J.R.; Nicholson, E.; Bishop, M.J.; Polidoro, B.A.; Ramirez-Llodra, E.; Tozer, M.G.; Nel, J.L.; Mac Nally, R.; Gregr, E.J.; et al. A Function-Based Typology for Earth's Ecosystems. *Nature* **2022**, *610*, 513–518. [CrossRef]
8. European Commission. *EU Biodiversity Strategy for 2030. Bringing Nature Back into Our Lives*; European Commission: Brussels, Belgium, 2020.
9. European Commission. *EU Soil Strategy for 2030. Reaping the Benefits of Healthy Soils for People, Food, Nature and Climate*; European Commission: Brussels, Belgium, 2021.
10. European Commission. *Regulation of the European Parliament and of the Council on Nature Restoration*; European Commission: Brussels, Belgium, 2022.
11. Vallecillo, S.; Maes, J.; Teller, A.; Babí Almenar, J.; Barredo, J.I.; Trombetti, M.; Malak, A. *EU-Wide Methodology to Map and Assess Ecosystem Condition. Towards a Common Approach Consistent with a Global Statistical Standard*; European Commission: Brussels, Belgium, 2022; ISBN 9789276570295.
12. European Union. *Establishing a Framework for Community Action in the Field of Marine Environmental Policy (Marine Strategy Framework Directive)*; European Commission: Brussels, Belgium, 2008.
13. European Commission. *Establishing Rules on Support for Strategic Plans to Be Drawn up by Member States under the Common Agricultural Policy (CAP Strategic Plans) and Financed by the European Agricultural Guarantee Fund (EAGF) and by the European Agricultural Fund for Rural Development (EAFRD) and Repealing Regulation (EU) No1305/2013 of the European Parliament and of the Council and Regulation (EU) No 1307/2013 of the European Parliament and of the Council*; European Commission: Brussels, Belgium, 2018.
14. European Commission. *A New EU Forest Strategy: For Forests and the Forest-Based Sector*; European Commission: Brussels, Belgium, 2013.
15. European Commission. *EU Pollinators Initiative*; European Commission: Brussels, Belgium, 2018.
16. van Egmond, F.M. Towards Climate-Smart Sustainable Management of Agricultural Soils. Deliverable 6.1. Report on Harmonized Procedures for Creation of Databases and Maps. Available online: https://ejpsoil.eu/fileadmin/projects/ejpsoil/WP6/EJP_SOIL_D6.1_Report_on_harmonized_procedures_for_creation_of_databases_and_maps_revised_vf__1_.pdf (accessed on 10 October 2022).
17. United Nations. *Transforming Our World: The 2030 Agenda for Sustainable Development*; United Nations: New York, NY, USA, 2015.
18. UNSD. *Using the SEEA EA for Calculating Selected SDG Indicators. Report of the NCAVES Project*; UNSD: New York, NY, USA, 2020.
19. UN Cover Note for Area A: Coordination, Discussion on UN Sustainable Development Goals (SDGs). Available online: https://seea.un.org/sites/seea.un.org/files/sdg_cover_note_broadbrush.pdf (accessed on 3 November 2022).
20. Secretariat of the Convention on Biological Diversity; UN Environment; UN Statistics Division. SDG Indicator 15.9.1 Metadata. Available online: https://unstats.un.org/sdgs/metadata/files/Metadata-15-09-01.pdf (accessed on 3 November 2022).
21. United Nations Committee of Experts on Environmental-Economic Accounting. *System of Environmental Economic Accounting 2012: Experimental Ecosystem Accounting 2012*; UNCEEA: New York, NY, USA, 2014; ISBN 9789211615753.
22. *UN-Department of economic and social affairs System of Environmental-Economic Accounting-Ecosystem Accounting: Final Draft*; UN DESA: New York, NY, USA, 2021.
23. Edens, B.; Maes, J.; Hein, L.; Obst, C.; Siikamaki, J.; Schenau, S.; Javorsek, M.; Chow, J.; Chan, J.Y.; Steurer, A.; et al. Establishing the SEEA Ecosystem Accounting as a Global Standard. *Ecosyst. Serv.* **2022**, *54*, 101413. [CrossRef]
24. European Commission. *Directorate-General for the Environment. Mapping and Assessment of Ecosystems and Their Services: An Analytical Framework for Ecosystem Assessments under Action 5 of the EU Biodiversity Strategy to 2020: Discussion Paper—Final, April 2013*; European Commission: Brussels, Belgium, 2013; ISBN 9789279293696.

25. Veronika, V.; Joachim, M.; Jan-Erik, P.; Alessandra, L.N.; Sara, V.; Nerea, A.; Eva, I.; Anne, T. *Accounting for Ecosystems and Their Services in the European Union (INCA). Final Report from Phase II of the INCA Project Aiming to Develop a Pilot for an Integrated System of Ecosystem Accounts for the EU*; Publications office of the European Union: Luxembourg, 2021.
26. European Environmental Agency. *Riparian Zones Nomenclature Guideline 2021*; European Environmental Agency: Copenhagen, Denmark, 2021.
27. European Environmental Agency. *Copernicus Land Monitoring Service-Local Component: Coastal Zones Monitoring Nomenclature Guideline 2021*; European Environmental Agency: Copenhagen, Denmark, 2021.
28. Buck, O.; Sousa, A. *Copernicus Land Monitoring Service N2K User Manual (Version 1.0)*; European Environmental Agency: Copenhagen, Denmark, 2021.
29. Büttner, G.; Kosztra, B.; Maucha, G.; Pataki, R.; Kleeschulte, S.; Hazeu, G.; Vittek, M.; Littkopf, A. *Copernicus Land Monitoring Service CORINE Land Cover Product User Manual (Version 1.0)*; Environmental Agency: Copenhagen, Denmark, 2021.
30. European Commission. *Mapping Guide v6.2 for a European Urban Atlas Regional Policy*; European Commission: Brussels, Belgium, 2020.
31. Kleeschulte, S.; Banko, G.; Smith, G.; Arnold, S.; Scholz, J.; Kosztra, B.; Maucha, G. Refined Details on CLC+ Backbone Specifications, Criteria for CLC+. Available online: https://land.copernicus.eu/user-corner/technical-library/clc-core-consultations-for-the-technical-specifications (accessed on 3 November 2022).
32. Arnold, S.; Kosztra, B.; Banko, G.; Milenov, P.; Smith, G.; Hazeu, G.; Bock, M.; Perger, C.; Caetano, M. *Explanatory Documentation of the EAGLE Concept-Version 3.1.2*; European Environmental Agency: Copenhagen, Denmark, 2021.
33. De Fioravante, P.; Luti, T.; Cavalli, A.; Giuliani, C.; Dichicco, P.; Marchetti, M.; Chirici, G.; Congedo, L.; Munafò, M. Multispectral Sentinel-2 and Sar Sentinel-1 Integration for Automatic Land Cover Classification. *Land* **2021**, *10*, 611. [CrossRef]
34. Spadoni, G.L.; Cavalli, A.; Congedo, L.; Munafò, M. Analysis of Normalized Difference Vegetation Index (NDVI) Multi-Temporal Series for the Production of Forest Cartography. *Remote Sens. Appl. Soc. Environ.* **2020**, *20*, 100419. [CrossRef]
35. Luti, T.; De Fioravante, P.; Marinosci, I.; Strollo, A.; Riitano, N.; Falanga, V.; Mariani, L.; Congedo, L.; Munafò, M. Land Consumption Monitoring with SAR Data and Multispectral Indices. *Remote Sens.* **2021**, *13*, 1586. [CrossRef]
36. De Fioravante, P.; Strollo, A.; Assennato, F.; Marinosci, I.; Congedo, L.; Munafò, M. High Resolution Land Cover Integrating Copernicus Products: A 2012–2020 Map of Italy. *Land* **2022**, *11*, 35. [CrossRef]
37. Michele, M. *Consumo Di Suolo, Dinamiche Territoriali e Servizi Ecosistemici*; Report SNPA, 32/22; Istituto superiore per la protezione e la ricerca ambientale: Roma, Italy, 2022.
38. Haines-Young, R.; Potschin, M. *Common International Classification of Ecosystem Services (CICES) V5.1 Guidance on the Application of the Revised Structure*; Fabis Consulting Ltd.: Nottingham, UK, 2018.
39. Assennato, F.; Smiraglia, D.; Cavalli, A.; Congedo, L.; Giuliani, C.; Riitano, N.; Strollo, A.; Munafò, M. The Impact of Urbanization on Land: A Biophysical-Based Assessment of Ecosystem Services Loss Supported by Remote Sensed Indicators. *Land* **2022**, *11*, 236. [CrossRef]
40. ISTAT 6° Censimento Generale Dell'Agricoltura. Available online: https://www.istat.it/it/files//2014/03/Atlante-dellagricoltura-italiana.-6°-Censimento-generale-dellagricoltura.pdf (accessed on 20 October 2022).
41. Hutyra, L.R.; Yoon, B.; Alberti, M. Terrestrial Carbon Stocks across a Gradient of Urbanization: A Study of the Seattle, WA Region. *Glob. Chang. Biol.* **2011**, *17*, 783–797. [CrossRef]
42. Li, S.; Liu, Y.; Yang, H.; Yu, X.; Zhang, Y.; Wang, C. Integrating Ecosystem Services Modeling into Effectiveness Assessment of National Protected Areas in a Typical Arid Region in China. *J. Environ. Manage.* **2021**, *297*, 113408. [CrossRef] [PubMed]
43. Penman, J.; Gytarsky, M.; Hiraishi, T.; Krug, T.; Kruger, D.; Pipatti, R.; Buendia, L.; Miwa, K.; Ngara, T.; Tanabe, K. *Good Practice Guidance for Land Use, Land-Use Change and Forestry*; Institute for Global Environmental Strategies: Kanagawa, Japan, 2003.
44. Obiang Ndong, G.; Therond, O.; Cousin, I. Analysis of Relationships between Ecosystem Services: A Generic Classification and Review of the Literature. *Ecosyst. Serv.* **2020**, *43*, 101120. [CrossRef]
45. Saidi, N.; Spray, C. Ecosystem Services Bundles: Challenges and Opportunities for Implementation and Further Research. *Environ. Res. Lett.* **2018**, *13*, 113001. [CrossRef]
46. Renard, D.; Rhemtulla, J.M.; Bennett, E.M. Historical Dynamics in Ecosystem Service Bundles. *Proc. Natl. Acad. Sci. USA* **2015**, *112*, 13411–13416. [CrossRef]
47. Farrell, C.A.; Coleman, L.; Kelly-Quinn, M.; Obst, C.G.; Eigenraam, M.; Norton, D.; O'donoghue, C.; Kinsella, S.; Delargy, O.; Stout, J.C. Applying the System of Environmental Economic Accounting-Ecosystem Accounting (Seea-Ea) Framework at Catchment Scale to Develop Ecosystem Extent and Condition Accounts. *One Ecosyst.* **2021**, *6*, e65582. [CrossRef]
48. Schröter, M.; Albert, C.; Marques, A.; Tobon, W.; Lavorel, S.; Maes, J.; Brown, C.; Klotz, S.; Bonn, A. National Ecosystem Assessments in Europe: A Review. *Bioscience* **2016**, *66*, 813–828. [CrossRef]
49. Alessandra, L.N.; Ioanna, G.; Karsten, G.; David, B.; Beyhan, E. *Ecosystem and Ecosystem Services Accounts: Time for Applications*; Publications Office of the European Union: Luxembourg, 2021.
50. Blasi, C.; Capotorti, G.; Alós Ortí, M.M.; Anzellotti, I.; Attorre, F.; Azzella, M.M.; Carli, E.; Copiz, R.; Garfì, V.; Manes, F.; et al. Ecosystem Mapping for the Implementation of the European Biodiversity Strategy at the National Level: The Case of Italy. *Environ. Sci. Policy* **2017**, *78*, 173–184. [CrossRef]
51. Laporta, L.; Domingos, T.; Marta-Pedroso, C. Mapping and Assessment of Ecosystems Services under the Proposed Maes European Common Framework: Methodological Challenges and Opportunities. *Land* **2021**, *10*, 1040. [CrossRef]

52. Vallet, A.; Locatelli, B.; Levrel, H.; Wunder, S.; Seppelt, R.; Scholes, R.J.; Oszwald, J. Relationships Between Ecosystem Services: Comparing Methods for Assessing Tradeoffs and Synergies. *Ecol. Econ.* **2018**, *150*, 96–106. [CrossRef]
53. Zhang, Z.; Liu, Y.; Wang, Y.; Liu, Y.; Zhang, Y.; Zhang, Y. What Factors Affect the Synergy and Tradeoff between Ecosystem Services, and How, from a Geospatial Perspective? *J. Clean. Prod.* **2020**, *257*, 120454. [CrossRef]
54. Raudsepp-Hearne, C.; Peterson, G.D.; Bennett, E.M. Ecosystem Service Bundles for Analyzing Tradeoffs in Diverse Landscapes. *Proc. Natl. Acad. Sci. USA* **2010**, *107*, 5242–5247. [CrossRef] [PubMed]
55. Grunewald, K.; Schweppe-Kraft, B.; Syrbe, R.U.; Meier, S.; Krüger, T.; Schorcht, M.; Walz, U. Hierarchical Classification System of Germany's Ecosystems as Basis for an Ecosystem Accounting—Methods and First Results. *One Ecosyst.* **2020**, *5*, e50648. [CrossRef]
56. Sieber, I.M.; Hinsch, M.; Vergílio, M.; Gil, A.; Burkhard, B. Assessing the Effects of Different Land-Use/Landcover Input Datasets on Modelling and Mapping Terrestrial Ecosystem Services—Case Study Terceira Island (Azores, Portugal). *One Ecosyst.* **2021**, *6*, e69119. [CrossRef]
57. Černecký, J.; Gajdoš, P.; Špulerová, J.; Halada, Ľ.; Mederly, P.; Ulrych, L.; Ďuricová, V.; Švajda, J.; Černecká, Ľ.; Andráš, P.; et al. Ecosystems in Slovakia. *J. Maps* **2020**, *16*, 28–35. [CrossRef]
58. Vačkář, D.; Grammatikopoulou, I.; Daněk, J.; Krkoška Lorencová, E. Methodological Aspects of Ecosystem Service Valuation at the National Level. *One Ecosyst.* **2018**, *3*, e25508. [CrossRef]

MDPI
St. Alban-Anlage 66
4052 Basel
Switzerland
Tel. +41 61 683 77 34
Fax +41 61 302 89 18
www.mdpi.com

Land Editorial Office
E-mail: land@mdpi.com
www.mdpi.com/journal/land

www.ingramcontent.com/pod-product-compliance
Lightning Source LLC
Chambersburg PA
CBHW061844210926
51850CB00006B/963
* 9 7 8 3 0 3 6 5 7 5 8 4 1 *